PEM Fuel Cells
THEORY AND PRACTICE

PEM Fuel Cells
THEORY AND PRACTICE

FRANO BARBIR

Amsterdam • Boston • Heidelberg • London
New York • Oxford • Paris • San Diego
San Francisco • Singapore • Sydney • Tokyo

Academic Press is an Imprint of Elsevier

ELSEVIER

Academic Press is an imprint of Elsevier
225 Wyman Street, Waltham, MA 02451, USA
525 B Street, Suite 1800, San Diego, California 92101-4495, USA
84 Theobald's Road, London WC1X 8RR, UK

Notices
Knowledge and best practice in this field are constantly changing. As new research and experience broaden our understanding, changes in research methods, professional practices, or medical treatment may become necessary.

Practitioners and researchers must always rely on their own experience and knowledge in evaluating and using any information, methods, compounds, or experiments described herein. In using such information or methods they should be mindful of their own safety and the safety of others, including parties for whom they have a professional responsibility.

To the fullest extent of the law, neither the Publisher nor the authors, contributors, or editors, assume any liability for any injury and/or damage to persons or property as a matter of products liability, negligence or otherwise, or from any use or operation of any methods, products, instructions, or ideas contained in the material herein.

Library of Congress Cataloging-in-Publication Data
Barbir, Frano, 1954-
    PEM fuel cells : theory and practice / Frano Barbir. – 2nd ed.
        p. cm.
    Includes bibliographical references and index.
    ISBN 978-0-12-387710-9 (alk. paper)
    1. Proton exchange membrane fuel cells. 2. Ion-permeable membranes. 3. Fuel cells.
I. Title. II. Title: Proton exchange membrane fuel cells.
    TK2931.B37 2013
    621.31'2429–dc23

British Library Cataloguing-in-Publication Data
A catalogue record for this book is available from the British Library.

For information on all Academic Press publications
visit our website at http://store.elsevier.com

Printed in the United States of America
09 10 11 12   9 8 7 6 5 4 3 2 1

CONTENTS

FOREWORD

There are two key problems with continued use of fossil fuels, which meet about 80% of the world energy demand today. The first problem is that they are limited in amount and sooner or later will be depleted. According to petroleum company estimates, the production of the most conveniently utilizable fossil fuels, petroleum and natural gas, will peak sometime between 2015 and 2020 and then begin to decrease. This means that there will be a gap between demand and production of fluid fuels beginning around 2015.

The second problem is that fossil fuels are causing serious environmental problems, such as global warming, climate changes, melting of ice caps, rising sea levels, acid rain, pollution, ozone layer depletion, oil spills, forest and agricultural land damage caused by surface mining of coal, and so on. It is estimated that this worldwide environmental damage adds up to some $5 *trillion* per year.

Early in the 1970s a hydrogen energy system was proposed as a solution for these two interconnected global problems. Since then, during the last quarter of the last century, through research and development work in universities and research laboratories around the world, foundations of the hydrogen energy system were established. Earlier in this century, conversion to the hydrogen economy began.

Hydrogen is an excellent energy carrier with many unique properties. It is the lightest, most efficient, and cleanest fuel. One of hydrogen's unique properties is that through electrochemical processes, it can be converted to electricity in fuel cells with higher efficiencies than conversion of fossil fuels to mechanical energy in internal combustion engines or to electrical energy in thermal power plants. This unique property of hydrogen has made hydrogen fuel cells the automotive power plant of choice for car companies. It has also made fuel cells the power plant of choice for companies manufacturing power plants for the future. The reason for higher efficiency of hydrogen fuel cells is that they are electrochemical engines, not heat engines, and as such they are not subject to Carnot cycle limitations. Consequently, it is expected that during the 21st century fuel cells will replace heat engines (internal combustion engines, steam turbines, and gas turbines) as hydrogen replaces fossil fuels.

Research has been and is currently being conducted into several types of fuel cells, such as alkaline fuel cells, proton exchange membrane (PEM) fuel cells, phosphoric acid fuel cells, molten carbonate fuel cells, solid oxide fuel cells, and so forth. Some of these technologies are already commercialized, whereas others are close to commercialization. They are expected to find applications in almost every energy-utilizing plant and/or device, from power plants to cars and homes, from laptop computers to mobile phones.

PEM fuel cells in particular have desirable properties. They operate at relatively low temperatures, which makes them easier to contain and reduces thermal losses. They are also smaller in volume and lighter in weight, making them perfect for automotive and portable applications. That is why all the hydrogen-fueled buses and cars on the market today from major companies are powered by PEM fuel cells and why about 90% of fuel cell research and development work involves PEM fuel cells. Consequently, today's and tomorrow's engineers engaged in energy engineering should be thoroughly knowledgeable about PEM fuel cells, just as the energy engineers of yesterday had to have a thorough knowledge of heat engines.

The author of this book, *PEM Fuel Cells: Theory and Practice*, Dr. Frano Barbir, is a well-established hydrogen energy scientist. After receiving his Ph.D. degree in hydrogen energy, he became and has continued to be closely involved in PEM fuel cell research and development work in fuel cell companies. He has also seen the need for educating engineers in fuel cell technologies. As a result, he has developed and taught senior- and graduate-level courses on fuel cells in two universities. Consequently, he is well qualified to write an authoritative textbook on PEM fuel cells. The book starts with the fundamentals of PEM fuel cells, then covers materials, operations, modeling, design, and applications. It is a well-written, comprehensive, and well-organized look into PEM fuel cells.

Consequently, I strongly recommend this textbook on PEM fuel cells to all senior- and graduate-level engineering students, whether they are mechanical, electrical, chemical, industrial, environmental, or energy engineers, who are studying energy conversion and energy applications. The library of any engineer and researcher involved in power generation, vehicle automotive power plants, and power units for portable systems would also be well served by this textbook.

Dr. T. Nejat Veziroğlu

Director, UNIDO-International Centre for Hydrogen Energy Technologies,
Istanbul, Turkey, February 2005

PREFACE AND ACKNOWLEDGMENTS

The idea for this book came about many years ago. Over the years of working as an engineer, researcher, scientist, and company executive, I gave numerous presentations on fuel cell technology, progress, and perspectives. I immensely enjoyed preparing the presentations, organizing my thoughts and views, and supporting them with facts and graphics. In doing so I always had an audience in mind. I tailored my presentations to suit that audience and designed them to be simple and easy to follow. This book is a direct result of those numerous presentations. I take pleasure in sharing my enthusiasm about this new and exciting technology, and I particularly treasure teaching this new technology to the next generation—young engineers I have worked with and the students I have had the opportunity to advise and teach. After all, they will be the ones implementing this technology and benefiting from it.

For nine years I worked for Energy Partners, a small, privately owned company in West Palm Beach, Florida. It was an amazing educational experience: a sandbox for engineers. Instead of learning from books (at that time there were no books on fuel cells), we learned from the fuel cell itself. We started with a working fuel cell, a cubic-foot "magic box" that sometimes had a mind of its own. My job was to make this box do what we wanted it to do. To do that, I had to learn everything about fuel cells: their theory, materials, components, design, operation, diagnostics, supporting system, and so on—everything I did not get a chance to learn while I was in school. Of course, I could not have done it alone. I enjoyed sharing what I had learned with my young colleagues as much as I enjoyed learning together with them. Soon we made the world's first PEM fuel cell–powered passenger automobile, followed by an extended golf-cart/people transporter and several utility vehicles (John Deere Gators). Every new vehicle was leaps and bounds better than the previous one, and through this process I accumulated knowledge. It is unfortunate that at one time we were expected to start making profit, which the fuel cell technology was not yet capable of. Not at that time. Nevertheless, I am grateful to Mr. John H. Perry, Jr., the owner of Energy Partners, for allowing me to have such fun and to learn so much.

In my career I had a chance to work with and learn from many outstanding engineers and scientists, and I would like to take this opportunity to thank them:

- Prof. T. Nejat Veziroğlu, professor at the University of Miami and now the Director of UNIDO-International Centre for Hydrogen Energy Technologies, for inspiring me with the idea of a hydrogen economy in the time when a hydrogen economy was more like science fiction, and for being my mentor all these years
- The late Prof. Harold J. Plass, Jr., then professor at the University of Miami, for teaching me simplicity in engineering analyses
- The late Prof. Howard T. Odum, then professor at the University of Florida, Gainesville, for the overwhelming depth and clarity of his vision
- Floyd Marken, then chief engineer at Energy Partners, for his wisdom and common-sense engineering
- Dr. Mario Nadal, then principal electrical engineer at Energy Partners, for numerous lunch discussions and for taking care of the electrical side of the fuel cell system with such ease that I never felt the urge to get involved in electrical aspects of fuel cell engineering
- Dr. Hongtan Liu, professor at the University of Miami, my colleague and friend from student days, for coming to work with me and our pioneering work in applying CFD techniques in fuel cell modeling
- Vince Petraglia, then vice president of Energy Partners, for sharing his wealth of fuel cell experience and wisdom, summarized in the following statement: "A fuel cell is sensitive to change of every its color"
- Dr. George Joy, then president of Energy Partners, for believing in me and including me in the company management team
- Trent Molter, one of the founders of Proton Energy Systems and now my colleague at the Connecticut Global Fuel Cell Center, for giving me the opportunity to come to Connecticut with my team of engineers, and for a wealth of information about fuel cells in their early days

Very special thanks go to the following people who made direct contributions in writing this book:
- Dr. Wilson Chiu, assistant professor at the University of Connecticut, who contributed a good chunk of Chapter 7, "Fuel Cell Modeling"
- Niloufar Fekrazad, graduate student at the University of Connecticut, who also contributed to the "Fuel Cell Modeling" chapter
- Xinting Wang and Richard Fu, graduate students at the University of Connecticut, who willingly dug for various data whenever I needed them
- Dr. Haluk Görgün, postdoctoral researcher at the Connecticut Global Fuel Cell Center, for enthusiastically running the lab while I was busy finishing this book and for contributing to Section 9.4, "Electrical Subsystem"

- Ana Barbir, my daughter and also an engineer, who proofread most of my manuscript in its various stages of development, corrected my English grammar, and made many useful comments on how to improve it

These acknowledgments would not be complete without thanking my wife Georgia, my life companion of 30-something years, for taking care of all other aspects of our lives while I was involved in writing this book. Without her constant encouragement and support, this book would never have happened. This book is dedicated to her.

Frano Barbir
University of Connecticut, Storrs, CT, February 200

PREFACE TO THE SECOND EDITION

PEM Fuel Cells: Theory and Practice was written seven years ago as a result of my desire to share with younger generations of engineers and scientists knowledge gained from my experience at the forefront of fuel cell research and development. Most of what I have learned about fuel cells I gathered from hands-on experience in designing, building, testing, and evaluating fuel cell stacks and systems and making them work in practical applications. Over the last seven years I have continued to conduct R&D in PEM fuel cells, yet I only reluctantly agreed to prepare this second edition, as I was afraid that the rapidly developing nature of the field would have rendered the book very outdated. However, in going through the original manuscript I realized that although it is indisputable that fuel cell technology has made tremendous progress in the last seven years, at least the basics of PEM fuel cell engineering and operation covered in the book had not changed so much. That said, there were new materials, new designs, new diagnostic methods, and valuable feedback from years of field experience that needed to be added. As a result, we have developed a much more nuanced understanding of the processes impacting choice of materials, construction, performance, and longevity.

In recognition of the multifaceted character of the field's evolution, the chapter on diagnostics has been rewritten by an expert in this particular area, Dr. Haijiang Wang from the Institute for Fuel Cell Innovation, National Research Council Canada, Vancouver, British Columbia.

Realizing that fuel cell durability is of crucial importance, I decided to add a new chapter on PEM fuel cell durability, and I invited another expert practitioner, Dr. Michael Perry, from United Technologies Research Center, East Hartford, Connecticut, to contribute it. I am taking this opportunity to thank Dr. Wang and Dr. Perry for their valuable contributions.

Advances in new materials, particularly membranes and catalysts, have been addressed in the chapter on main cell components, materials properties, and processes.

In the chapter on fuel cell operating conditions, I have added insights from my colleague Torsten Berning from Aalborg University, Denmark, about dew-point temperature of exhaust gases as a criterion for selection of operating conditions. My thanks are due him as well.

During the last seven years, numerous papers on fuel cell modeling have been published, covering various domains and physical phenomena as well as many new modeling methods and techniques. These models, in conjunction with improved diagnostic techniques, have provided better understanding of the intricacies of fuel cell operation, particularly around the role of water in fuel cell operation. The chapter on fuel cell modeling in the first edition of this book was organized by first presenting the basic governing equations describing conservation of mass, momentum, energy, species, and charge, then by giving their application in modeling examples for various domains, that is, across the membrane, above the channel and land, along the channel, and in three dimensions. I decided to keep the same organization and examples in this revised chapter, since they cover the fundamentals effectively. However, I added an example on modeling water transport in the gas diffusion layer through a pore network model as a representative of the new models that elucidate peculiar water behavior in porous fuel cell structures.

Diagrams, photos, and tables have been updated wherever applicable throughout the book but particularly in the chapter on fuel cell applications. Tables from the U.S. DOE Hydrogen and Fuel Cell Program, showing current status and technical targets for various applications, have been added. All other chapters have been edited once again and corrected or updated where necessary.

I would like to thank my students and post-docs at UNIDO-ICHET, the FESB University of Split, and the University of Wyoming who have pointed out some errors, typos, and unclear statements to help me improve on the content of the first edition.

I use this book as a text to teach a full-semester fuel cell engineering course as well as numerous short courses. In doing so, I have prepared thousands of slides organized to follow the book over the years. These slides will be made available on the book's Website to those who want to use the book to teach a fuel cell course. My advice is to use these slides as a backbone for lectures and spice them up with the newest developments in fuel cell technology as they emerge. At the end of each chapter there are numerical problems that may be assigned as homework. Solutions will also be posted in a password-protected instructor section of the book's Website. In addition, each chapter has a quiz at the end. I have found these quizzes very useful, not only to evaluate students' progress but also to evaluate my effectiveness as a teacher. After each quiz I perform a statistical analysis to identify the questions that most of the students have not answered correctly. I then revisit that part of the lecture.

It is my hope that the publication of this second edition of the book will increase opportunities for interactively sharing the experiences of instructors and students as well as the scientists and engineers all over the world who are involved in making better and more affordable fuel cells. This book is dedicated to all of them.

Finally, I would like to thank Jill Leonard and Tiffany Gasbarrini at Elsevier for their guidance, support, encouragement, and especially their patience throughout the process of preparing this second edition.

Frano Barbir
FESB University of Split, Split, Croatia, June 2012

CHAPTER ONE

Introduction

1.1 WHAT IS A FUEL CELL?

A *fuel cell* is an electrochemical energy converter that converts chemical energy of fuel directly into direct current (DC) electricity. Typically, a process of electricity generation from fuels involves several energy conversion steps:

1. Combustion of fuel converts chemical energy of fuel into heat.
2. This heat is then used to boil water and generate steam.
3. Steam is used to run a turbine in a process that converts thermal energy into mechanical energy.
4. Finally, mechanical energy is used to run a generator that generates electricity.

A fuel cell circumvents all these processes and generates electricity in a single step without involving any moving parts (Figure 1-1). It is this simplicity that attracts attention. Such a device must be simpler, thus less expensive and far more efficient, than the four-step process previously depicted. Is it really? Today—not really! Or better, not yet. But fuel cells are still being developed. This book intends to provide a basis for engineering fuel cell devices. It includes state-of-the-art designs and materials (as they exist at the time of this writing), which are likely to change in the future as this technology continues to develop (perhaps even sooner than the students using this textbook get jobs in the fuel cell industry). However, the engineering basis will not change, at least not dramatically and not so quickly. The knowledge of engineering principles will allow future fuel cell engineers to adopt these new designs and new materials and, we hope, come up with even newer designs and materials. This is what the purpose of an engineering education should be. This book will not teach the principles of thermodynamics, catalysis, electrochemistry, heat transfer, fluid mechanics, or electricity conduction, but it will apply those engineering disciplines in the engineering of a fuel cell as an energy conversion device.

The efficiency of an energy conversion process is one of the most important aspects of that conversion. Some typical questions are usually tackled by engineering textbooks: How much energy of one kind is

PEM Fuel Cells
ISBN 978-0-12-387710-9

1

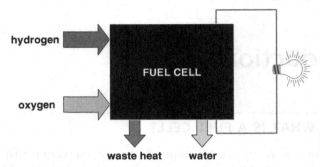

Figure 1-1 A fuel cell generates DC electricity from fuel in one step.

required to generate one unit of energy of another kind? What is the theoretical limit? How close can we come to that limit in practical applications? This last question is where most engineering textbooks fail in providing practical results and real—not theoretical—efficiencies.

As an example, let's examine the *Carnot process*, also called the *Carnot engine*. Every engineering student should know that the Carnot process is the most efficient process to operate between given temperatures. The fact that such an engine cannot be made, and even if it could be made it would have to operate infinitesimally slowly to allow the heat transfer processes to happen with no losses, is perhaps mentioned in some textbooks. But the fact that such an engine would be very efficient at generating no power is never emphasized enough. Yes, the famous Carnot engine would have to operate at efficiencies lower than the famous Carnot efficiency in order to generate useful power. This book emphasizes the efficiency—not only the theoretical efficiency but the efficiency of practical, power-generating devices. That is why there is a subsection on efficiency in almost every chapter. The chapter on fuel cell thermodynamics deals with theoretical fuel cell efficiencies. As important as it is to learn about the Carnot efficiency, it is equally important to learn about theoretical limits in fuel cells. The chapter on fuel cell electrochemistry introduces various losses that are unavoidable because of the physical properties of the materials involved. These losses obviously have an effect on the efficiency of energy conversion. The chapter on fuel cell systems discusses various supporting devices that are needed to get the fuel cell going. Most of those devices need power, which means that some of the power produced by a fuel cell would be used to run those supporting devices, and therefore less net power would actually be delivered by the fuel cell system. This means that the practical efficiency will be somewhat lower than the theoretical one. How much lower? That would depend on the system configuration, design, and selection of auxiliary components. Finally, the efficiency of an energy conversion device in a practical application will probably depend on how that device is used. Does it run all the time at constant power output or does the power output vary? If it varies, how much and how often? These are the reasons that the efficiency is discussed in almost every chapter of this book.

Another important aspect of an energy conversion process is the cost. It is the cost of produced energy that matters in practical applications. Obviously, this cost depends greatly on the efficiency of the energy conversion process and the cost of the consumed (or, thermodynamically, the more correct term is *converted*) energy. The cost of the energy conversion device itself must also be taken into account. The cost of any device depends on the cost of the materials and the efforts (labor) involved to process those materials, make the components, and finally to assemble those components into a working device. Unfortunately, there is not enough information available on fuel cell costs, either materials or labor. One of the reasons the fuel cells are expensive is that they are not being mass produced; one of the reasons they are not being mass produced is that their markets are limited because they are expensive. This "chicken-and-egg" problem is typical for any new technology.

This book interweaves the theory and practice of the fuel cell and fuel cell system design, engineering, and applications. It does not provide a recipe on how to build the best possible fuel cell, but it gives an engineering student an understanding of the basic processes and materials inside a fuel cell. It also supplies enough tools and instructions on how to use them to design a fuel cell or a fuel cell system, or how to select a fuel cell for a particular application. This book does not provide a direct answer to all fuel cell-related questions, but it provides the engineering tools needed to find those answers.

A fuel cell is in some aspects similar to a battery. It has an electrolyte and negative and positive electrodes (Figure 1-2), and it generates DC electricity through electrochemical reactions. However, unlike a battery, a fuel cell requires a constant supply of fuel and oxidant. Also, unlike a battery, the electrodes in a fuel cell do not undergo chemical changes. Batteries generate electricity by the electrochemical reactions that involve the materials that are already in batteries. Because of this, a battery may be *discharged*, which happens when the materials that participate in the electrochemical reactions are depleted. Some batteries are *rechargeable*, which means that the electrochemical reactions may proceed in reverse when external electricity is

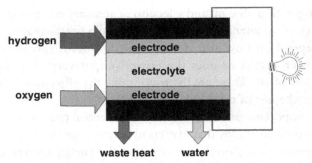

Figure 1-2 A fuel cell is similar to a battery in that it has electrodes and an electrolyte, but it needs a fuel and oxidant supply and it generates waste heat and water.

applied—a process of recharging the battery. A fuel cell cannot be discharged as long as the reactants—fuel and oxidant—are supplied. Typical reactants for fuel cells are hydrogen and oxygen; however, neither has to be in its pure form. Hydrogen may be present either in a mixture with other gases (such as CO_2, N_2, and CO) or in hydrocarbons such as natural gas, CH_4, or even in liquid hydrocarbons such as methanol, CH_3OH. Ambient air contains enough oxygen to be used in fuel cells. Yet another difference between a fuel cell and a battery is that a fuel cell generates by-products—waste heat and water—and the system is required to manage those. (A battery also generates some heat but at a much lower rate that usually does not require any special or additional equipment.)

1.2 A VERY BRIEF HISTORY OF FUEL CELLS

The timeline of fuel cell development history is shown in Figure 1-3. The first observation of a fuel cell effect was made by a German–Swiss scientist, Christian F. Shoenbein, in 1938 [1]. Apparently, based on this work, the first fuel cell was demonstrated by Welsh scientist and barrister Sir William Grove in 1839 [2]. In 1842, Grove developed the first fuel cell, or a *gaseous voltaic battery,* as he called it, which produced electrical energy by combining hydrogen and oxygen [3].

However, in spite of sporadic attempts to make a practical device, the fuel cell remained nothing more than a scientific curiosity for almost a century. In this period, W. F. Ostwald, a Nobel Prize winner in 1909 and founder of the field of physical chemistry, provided much of the theoretical understanding of how fuel cells operate. He realized that energy conversion

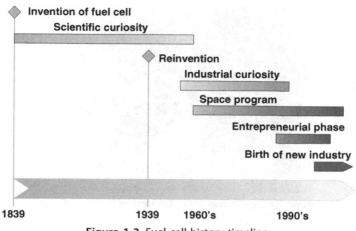

Figure 1-3 Fuel cell history timeline.

in combustion engines is limited by the Carnot efficiency and results in unacceptable levels of atmospheric pollution, whereas the fuel cells, which directly generate electricity, are highly efficient, silent, and generate no pollution. He predicted a technical revolution, although realizing that practical realization of such transition could take a long time [4].

Indeed, it took more than 100 years from Grove's invention of the fuel cell to make a practical device. An English engineer, Francis T. Bacon, started working on practical fuel cells in 1932 and in 1952 completed construction and evaluation of a 5 kW fuel cell stack. However, the first practical fuel cell applications were in the U.S. space program. General Electric developed the first polymer membrane fuel cells, based on the work by Grubb and Niedrach, which were used in the Gemini Program in the early 1960s. This was followed by the Apollo Program, which used the fuel cells to generate electricity for life support, guidance, and communications. These fuel cells were built by Pratt and Whitney based on license taken on Bacon's patents (Figure 1-4). In the mid-1960s General Motors experimented with a fuel cell–powered van (these fuel cells were developed by Union Carbide). Although fuel cells have continued to be successfully used in the U.S. space program until today, they were again "forgotten" for terrestrial applications until the early 1990s. In 1989 Perry Energy Systems, a division of Perry Technologies, working with Ballard, a then emerging Canadian company, successfully demonstrated a polymer electrolyte membrane (PEM) fuel cell–powered submarine (Figure 1-5). In 1993, Ballard Power Systems demonstrated fuel cell–powered buses. Energy

Figure 1-4 Apollo fuel cells. *(Courtesy of UTC Fuel Cells.)*

Figure 1-5 PC1401 by Perry Group, powered by PEM fuel cells (1989). *(Courtesy of Teledyne Energy Systems.)*

Partners, a successor of Perry Energy Systems, demonstrated the first passenger car running on PEM fuel cells in 1993 (Figure 1-6) [5]. The car companies, supported by the U.S. Department of Energy, picked up on this activity and by the end of the century almost every car manufacturer had built and demonstrated a fuel cell–powered vehicle. A new industry was born. The stocks of fuel cell companies, such as Ballard and PlugPower, soared in early 2000 (Figure 1-7) based on a promise of a new energy revolution. (Eventually in 2001 the stocks came down with the rest of the market.) The number of fuel cell–related patents worldwide, but primarily in

Figure 1-6 Energy Partners' GreenCar, the first PEM fuel cell-powered passenger automobile, 1993. *(Courtesy of Teledyne Energy Systems.)*

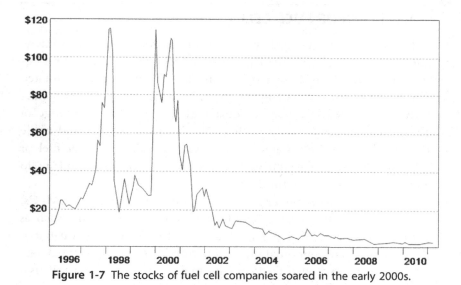

Figure 1-7 The stocks of fuel cell companies soared in the early 2000s.

Japan and the United States, is increasing dramatically (Figure 1-8) [6,7], showing continuous interest and involvement of the scientific and engineering community. This interest continued strongly after the period shown in Figure 1-8 since there were about 5,000 fuel cell-related patent applications in 2008 and about 6,000 in 2009.

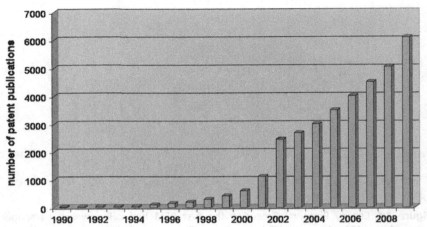

Figure 1-8 Fuel cell patent publications per year worldwide. *(Adapted from [6] and [7].)*

1.3 TYPES OF FUEL CELLS

Fuel cells can be grouped by the type of electrolyte they use, namely:

- *Alkaline fuel cells* (AFCs) use concentrated (85 wt%) KOH as the electrolyte for high-temperature operation (250 °C) and less concentrated (35–50 wt%) for lower-temperature operation (<120 °C). The electrolyte is retained in a matrix (usually asbestos), and a wide range of electrocatalysts can be used (such as Ni, Ag, metal oxides, and noble metals). This fuel cell is intolerant to CO_2 present in either fuel or oxidant. Alkaline fuel cells have been used in the space program (Apollo and Space Shuttle) since the 1960s.

- *Polymer electrolyte membrane* or *proton exchange membrane fuel cells* (PEMFCs) use a thin (<50 µm) proton conductive polymer membrane (such as perfluorosulfonated acid polymer) as the electrolyte. The catalyst is typically platinum supported on carbon with loadings of about 0.3 mg/cm^2, or, if the hydrogen feed contains minute amounts of CO, Pt–Ru alloys are used. Operating temperature is typically between 60 °C and 80 °C. PEM fuel cells are serious candidates for automotive applications as well as small-scale distributed stationary power generation and for portable power applications.

- *Phosphoric acid fuel cells* (PAFCs) use concentrated phosphoric acid (~100%) as the electrolyte. The matrix used to retain the acid is usually SiC, and the electrocatalyst in both the anode and the cathode is platinum. Operating temperature is typically between 150 °C and 220 °C. Phosphoric acid fuel cells are already semicommercially available in container packages (200 kW) for stationary electricity generation (UTC fuel cells). Hundreds of units have been installed all over the world.

- *Molten carbonate fuel cells* (MCFCs) have the electrolyte composed of a combination of alkali (Li, Na, K) carbonates, which is retained in a ceramic matrix of $LiAlO_2$. Operating temperatures are between 600 °C and 700 °C, where the carbonates form a highly conductive molten salt, with carbonate ions providing ionic conduction. At such high operating temperatures, noble metal catalysts are typically not required. These fuel cells are in the precommercial/demonstration stage for stationary power generation.

- *Solid oxide fuel cells* (SOFCs) use a solid, nonporous metal oxide, usually Y_2O_3-stabilized ZrO_2 (YSZ) as the electrolyte. These cells operate at 800 °C to 1,000 °C, where ionic conduction by oxygen ions takes place. Similar to MCFCs, these fuel cells are in the precommercial/demonstration stage for stationary power generation, although smaller units are being developed for portable power and auxiliary power in automobiles.

Figure 1-9 summarizes the basic principles and electrochemical reactions in various fuel cell types.

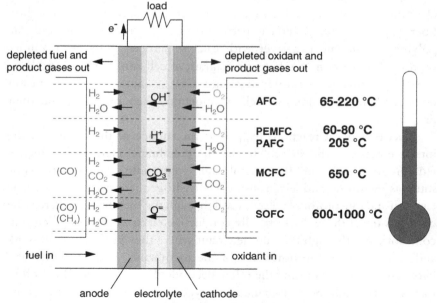

Figure 1-9 Types of fuel cells, their reactions, and operating temperatures.

Sometimes a *direct methanol fuel cell* (DMFC) is categorized as yet another type of fuel cell; however, according to the previous categorization (based on electrolyte), it is essentially a polymer membrane fuel cell that uses methanol instead of hydrogen as a fuel.

1.4 HOW DOES A PEM FUEL CELL WORK?

Although some general engineering principles may be applicable to all fuel cell types, this book is about PEM fuel cells and their operation, design, and applications. PEM stands for *polymer electrolyte membrane* or *proton exchange membrane*. Sometimes they are also called *polymer membrane fuel cells* or simply *membrane fuel cells*. In the early days (the 1960s) they were known as *solid polymer electrolyte* (SPE) fuel cells. This technology has drawn the most attention because of its simplicity, viability, and quick startup as well as the fact that it has been demonstrated in almost any conceivable application, as shown in the following sections.

At the heart of a PEM fuel cell is a polymer membrane that has some unique capabilities. It is impermeable to gases but it conducts protons (hence the name *proton exchange membrane*). The membrane that acts as the electrolyte is squeezed between the two porous, electrically conductive electrodes. These electrodes are typically made of carbon cloth or carbon fiber paper. At the interface between the porous electrode and the polymer membrane is a layer with catalyst particles, typically platinum supported on carbon. A schematic diagram of cell configuration and basic operating principles is shown in Figure 1-10. Chapter 4 deals in greater detail with those major fuel cell components, their materials, and their properties.

Electrochemical reactions happen at the surface of the catalyst at the interface between the electrolyte and the membrane. Hydrogen, which is fed on one side of the membrane, splits into its primary con-stituents—protons and electrons. Each hydrogen atom consists of one electron and one proton. Protons travel through the membrane, whereas the electrons travel through electrically conductive electrodes, through current collectors, and through the outside circuit where they perform useful work and come back to the other side of the membrane. At the catalyst sites between the membrane and the other electrode they meet with the protons that went through the membrane and oxygen that is fed on that side of the membrane. Water is created in the electrochemical reaction and then

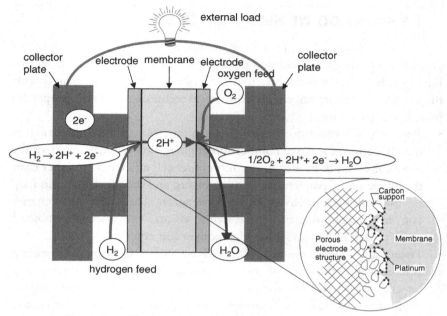

Figure 1-10 The basic principle of operation of a PEM fuel cell.

pushed out of the cell with excess flow of oxygen. The net result of these simultaneous reactions is a current of electrons through an external circuit—direct electrical current.

The hydrogen side is negative and is called the *anode*, whereas the oxygen side of the fuel cell is positive and is called the *cathode*. Chapters 4 and 5 explain in greater detail all the processes involved in making the fuel cell work. Because each cell generates about 1V, as will be shown shortly, more cells are needed in series to generate some practical voltages. Depending on the application, the output voltage may be between 6 V and 200 V or even more. How the cells are stacked up and what the issues are in stack design are discussed in Chapter 6.

A fuel cell stack needs a supporting system (as explained in Chapter 9) to:
- Handle the supply of reactant gases and their exhaust, including the products
- Take care of waste heat and maintain the stack temperature
- Regulate and condition power output
- Monitor the stack vital parameters
- Control the startup, operation, and shutdown of the stack and system components

1.5 WHY DO WE NEED FUEL CELLS?

Fuel cells are a very promising energy technology with a myriad of possible applications, as discussed next and in greater detail in Chapter 10. Fuel cells have many properties that make them attractive compared with the existing, conventional energy conversion technologies. These properties include the following:

- *Promise of high efficiency*. Because fuel cell efficiency is much higher than the efficiency of internal combustion engines, fuel cells are attractive for automobile applications. Furthermore, fuel cell efficiency is higher than the efficiency of conventional power plants, and therefore fuel cells may be used for decentralized power generation. However, new energy conversion technologies, such as hybrid electric vehicles and combined cycle power plants, also have high conversion efficiencies.

- *Promise of low or zero emissions*. Fuel cells operating on hydrogen generate zero emissions; the only exhaust is unused air and water. This quality may be attractive not only for transportation but also for many indoor applications as well as submarines. However, hydrogen is not a readily available fuel, and if a fuel cell is equipped with a fuel processor to generate hydrogen or if methanol is used instead of hydrogen, some emissions are generated, including carbon dioxide. In general, these emissions are lower than those of comparable conventional energy conversion technologies.

- *Issue of national security*. Fuel cells use hydrogen as fuel. Although hydrogen is not a readily available fuel, it may be produced from indigenous sources, either by electrolysis of water or by reforming hydrocarbon fuels. Use of indigenous sources (renewable energy, nuclear, biomass, coal, or natural gas) to generate hydrogen may significantly reduce dependence on foreign oil, which would have an impact on national security. However, widespread use of hydrogen would require establishing a hydrogen infrastructure or the so-called hydrogen economy, which will be discussed in Chapter 11.

- *Simplicity and promise of low cost*. Fuel cells are extremely simple. They are made in layers of repetitive components, and they have no moving parts. For these reasons they have the potential to be mass produced at a cost comparable to that of existing energy conversion technologies, or even lower. To date, fuel cells are still expensive for either automotive or stationary power generation, primarily because of use of expensive materials, such as sulfonated fluoropolymers, used as proton exchanged membranes, and noble metals, such as platinum or ruthenium, used as catalysts. Mass production techniques must still be developed for fabrication of fuel cell components and for the stack and system assembly.

- *No moving parts and promise of long life.* Because a fuel cell does not have any moving parts, it may be expected to exhibit a long life. Current fuel cell technology may reach the lifetime acceptable for automotive applications (3,000–5,000 hours), but cells' durability must be improved by an order of magnitude for use in stationary power generation (where the requirement is >40,000–80,000 hours).
- *Modularity.* Fuel cells are by their nature modular—more power may be generated simply by adding more cells. Mass-produced fuel cells may be significantly less expensive than traditional power plants. Instead of building big power plants, which must be planned well in advance and whose permitting process may be extremely cumbersome, it may be cost-effective to gradually increase generation capacity by adding smaller fuel cells to the grid. Such a concept of distributed generation may not only be cost-effective but also may significantly improve reliability of the power supply.
- *Quiet.* Fuel cells are inherently quiet, which may make them attractive for a variety of applications, such as portable power, backup power, and military applications.
- *Size and weight.* Fuel cells may be made in a variety of sizes—from microwatts to megawatts—which makes them useful in a variety of applications, from powering electronic devices to powering entire buildings. The size and weight of automotive fuel cells approaches those of internal combustion engines, and the size and weight of small fuel cells may offer advantage over the competing technologies, such as batteries for electronic devices.

1.6 FUEL CELL APPLICATIONS

Because of their attractive properties, fuel cells have already been developed and demonstrated in the following applications (some of which are shown in Figure 1-11):

- *Automobiles.* Almost every car manufacturer has already developed and demonstrated at least one prototype vehicle, and many have already gone

Figure 1-11 Collage of fuel cell applications and demonstrations (mid-2004). *(Courtesy of Toyota, Asian Pacific Fuel Cell Technologies, Schatz Energy Center-Humboldt State University, Honda, Aprilia, Teledyne Energy Systems, DaimlerChrysler, Ballard Power Systems, Fuel Cell Propulsion Institute, Plug Power, MTU, Teledyne Energy Systems, Proton Energy Systems, MTI Micro Fuel Cells, and Smart Fuel Cells.)*

through several generations of fuel cell vehicles. Some car manufacturers (General Motors, Toyota, Honda) are working on their own fuel cell technology, and others (DaimlerChrysler, Ford, Nissan, Mazda, Hyundai, Fiat, Volkswagen) acquire fuel cell stacks and systems from fuel cell developers such as Ballard, UTC Fuel Cells, and Nuvera.

- *Scooters and bicycles.* Several companies have demonstrated fuel cell–powered scooters and bicycles using either hydrogen stored in metal hydrides or methanol in direct methanol fuel cells.

- *Golf carts.* Energy Partners demonstrated a fuel cell–powered golf cart in 1994 (it was used in Olympic Village at the 1996 Olympic Games in Atlanta). Schatz Energy Center developed fuel cell–powered golf carts to be used in the city of Palm Desert in California.

- *Utility vehicles.* Energy Partners converted three John Deere Gator utility vehicles to fuel cell power [8] and demonstrated them in service at Palm Springs airport (1996). John Deere continued working with Hydrogenics, Canada, on development of fuel cell–powered electric utility vehicles, including those for lawn maintenance. Forklifts are considered as a viable early market for deployment of fuel cells.

- *Distributed power generation.* Several companies are working on development of small (1–10 kW) fuel cell power systems intended for use in homes, particularly in Japan and Germany. Some of them are combined with boilers to provide both electricity and heat.

- *Backup power.* Ballard made the first attempts to commercialize 1 kW backup power generators in cooperation with Coleman in 2000. Proton Energy Systems demonstrated regenerative fuel cells, combining its own PEM electrolyzer technology with Ballard's Nexa units [9]. A regenerative fuel cell generates its own hydrogen during periods when electricity is available. Backup power, particularly for telecommunications, is considered a viable early market for deployment of fuel cells.

- *Portable power.* Many companies are developing miniature fuel cells as battery replacements for various consumer and military electronic devices. Because of fuel storage issues, most of them use methanol in either direct methanol fuel cells or through microreformers in regular PEM fuel cells.

- *Space.* Fuel cells continue to be used in the U.S. space program, providing power on the space orbiters. Although this proven technology is of the alkaline type, the National Aeronautic and Space Administration (NASA) announced plans to use PEM fuel cells in the future.

- *Airplanes.* In 2008 Boeing announced that it has, for the first time in aviation history, flown a manned airplane powered by hydrogen fuel cells.
- *Boats.* Fuel cells can be used for marine applications and as auxiliary power units (APUs) for pleasure vessels. MTU Friedrichschaffen demonstrated a sailboat on lake Constanze (2004) with four Ballard Nexa fuel cell units. Several other fuel cell-powered boats have been demonstrated in the United States, the United Kingdom, the Netherlands, Germany, and Croatia.
- *Underwater vehicles.* In 1989 Perry Technologies successfully tested the first commercial fuel cell-powered submarine, the two-person observation submersible PC-1401, using Ballard's fuel cell [10]. Siemens has been successfully providing fuel cell engines for large submarines used by the German, Canadian, Italian, and Greek navies.

REFERENCES

[1] Bossel U. The Birth of the Fuel Cell. 1835–1845. Oberrohrdorf, Switzerland: European Fuel Cell Forum; 2000.
[2] Grove WR. On Voltaic Series and the Combination of Gases by Platinum. London and Edinburgh Philosophical Magazine and Journal of Science 1839;14(Series 3): 127–30.
[3] Grove WR. On a Gaseous Voltaic Battery. London and Edinburgh Philosophical Magazine and Journal of Science 1842;21(Series 3):417–20.
[4] Ostwald W. Die wissenschaftliche *Elektrochemie* der Gegenwart und die technische der Zukunft; Z. für Elektrotechnik und Elektrochemie 1894;3. p. 81–84 and 122–125.
[5] Nadal M, Barbir F. Development of a Hybrid Fuel Cell/Battery Powered Electric Vehicle. In: Block DL, Veziroglu TN, editors. Hydrogen Energy Progress X, vol. 3. Coral Gables, FL: International Association for Hydrogen Energy; 1994. p. 1427–40.
[6] Stone C, Morrison AE. From Curiosity to "Power to Change the World". Solid State Ionics 2002;vol. 152–153:1–13.
[7] Fuel Cell Today Patent Review (2008–2009), www.fuelcelltoday.com/analysis/patent (accessed May 2011).
[8] Barbir F, Nadal M, Fuchs M. Fuel Cell Powered Utility Vehicles. In: Buchi F, editor. Proc. Portable Fuel Cell Conference. Switzerland: Lucerne; June 1999. p. 113–26.
[9] Barbir F, Nomikos S, Lillis M. Practical Experiences with Regenerative Fuel Cell Systems. In: Proc. 2003 Fuel Cell Seminar, Miami Beach, FL; 2003.
[10] Blomen LJMJ, Mugerwa MN. Fuel Cell Systems. New York: Plenum Press; 1993.

CHAPTER TWO

Fuel Cell Basic Chemistry and Thermodynamics

A *fuel cell* is an electrochemical energy converter—it converts chemical energy of fuel, typically hydrogen, directly into electrical energy. As such, it must obey the laws of thermodynamics.

2.1 BASIC REACTIONS

The electrochemical reactions in fuel cells happen simultaneously on both sides of the membrane—the anode and the cathode. The basic fuel cell reactions are as follows:

At the anode:

$$H_2 \rightarrow 2H^+ + 2e^- \tag{2-1}$$

At the cathode:

$$1/2 O_2 + 2H^+ + 2e^- \rightarrow H_2O \tag{2-2}$$

Overall:

$$H_2 + 1/2 O_2 \rightarrow H_2O \tag{2-3}$$

These reactions may have several intermediate steps, and there may be some (unwanted) side reactions, but for now these reactions accurately describe the main processes in a fuel cell.

2.2 HEAT OF REACTION

The overall reaction (Equation 2-3) is the same as the reaction of hydrogen combustion. Combustion is an *exothermic process*, which means that there is energy released in the process:

$$H_2 + 1/2 O_2 \rightarrow H_2O + heat \tag{2-4}$$

PEM Fuel Cells
ISBN 978-0-12-387710-9

The heat (or enthalpy) of a chemical reaction is the difference between the heats of formation of products and reactants. For the previous Equation (2-4) this means:

$$\Delta H = (h_f)_{H_2O} - (h_f)_{H_2} - 1/2\,(h_f)_{O_2} \qquad (2\text{-}5)$$

Heat of formation of liquid water is -286 kJ mol^{-1} (at 25 °C) and heat of formation of elements is by definition equal to zero. Therefore:

$$\Delta H = (h_f)_{H_2O} - (h_f)H_2 - 1/2(h_f)_{O_2} = -286 \text{ kJ mol}^{-1} - 0 - 0$$
$$= -286 \text{ kJ mol}^{-1}$$

$$(2\text{-}6)$$

Note that the negative sign for enthalpy of a chemical reaction, by convention, means that heat is being released in the reaction, that is, this is an exothermic reaction. Equation (2-4) may now be rewritten as:

$$H_2 + 1/2O_2 \rightarrow H_2O(1) + 286 \text{ kJ mol}^{-1} \qquad (2\text{-}7)$$

Here a positive sign is used because the enthalpy is placed on the right side of the reaction, clearly meaning heat is a product of the reaction.

This equation is valid at 25 °C only, meaning that both the reactant gases and the product water are at 25 °C. At 25 °C and atmospheric pressure, water is in liquid form.

2.3 HIGHER AND LOWER HEATING VALUE OF HYDROGEN

The enthalpy of a hydrogen combustion reaction (Equation 2-7) (i.e., 286 kJ mol^{-1}) is also called the hydrogen's *heating value*. It is the amount of heat that may be generated by a complete combustion of 1 mol of hydrogen. The measurement of a heating value is conducted in a calorimetric bomb. If 1 mol of hydrogen is enclosed in a calorimetric bomb with $^1/_2$ mol of oxygen, ignited, fully combusted, and allowed to cool down to 25 °C, at atmospheric pressure there will be only liquid water left in the bomb (Figure 2-1). The measurement should show that 286 kJ of heat was released. This is known as hydrogen's *higher heating value*. However, if hydrogen is combusted with sufficient excess of oxygen (or air) and allowed to cool down to 25 °C, the product water will be in the form of vapor mixed with unburned oxygen and/or nitrogen, in case that air was used (Figure 2-2). The measurement should show that less heat was released, exactly 241 kJ. This is known as hydrogen's *lower heating value*.

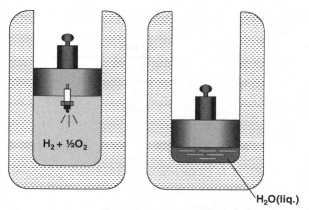

Figure 2-1 Combustion of $H_2 + {}^1/_2O_2$ in a calorimetric bomb—measurement of higher heating value.

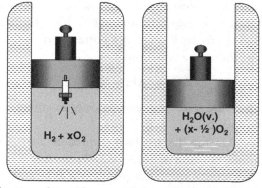

Figure 2-2 Combustion of H_2 with excess O_2 in a calorimetric bomb—measurement of lower heating value.

$$H_2 + 1/2O_2 \rightarrow H_2O(g) + 241 \text{ kJ/mol}^{-1} \qquad (2\text{-}8)$$

The difference between higher and lower heating value is the heat of evaporation of water (at 25 °C):

$$H_{fg} = 286 - 241 = 45 \text{ kJ mol}^{-1} \qquad (2\text{-}9)$$

2.4 THEORETICAL ELECTRICAL WORK

Because there is no combustion in a fuel cell, what is the relevance of hydrogen's heating value (higher or lower) to a fuel cell? Hydrogen heating

Table 2-1 Enthalpies and Entropies of Formation for Fuel Cell Reactants and Products (at 25 °C and 1 atm)

	h_f (kJ mol^{-1})	s_f (kJ mol^{-1}K^{-1})
Hydrogen, H_2	0	0.13066
Oxygen, O_2	0	0.20517
Water (liquid), H_2O (l)	−286.02	0.06996
Water (vapor), H_2O (g)	−241.98	0.18884

value is used as a measure of energy input in a fuel cell. This is the maximum amount of (thermal) energy that may be extracted from hydrogen. However, electricity is produced in a fuel cell. Can all the energy input be converted into electricity? Obviously not! In every chemical reaction some entropy is produced, and because of that, a portion of the hydrogen's higher heating value cannot be converted into useful work—electricity. The portion of the reaction enthalpy (or hydrogen's higher heating value) that can be converted to electricity in a fuel cell corresponds to Gibbs free energy and is given by the following equation:

$$\Delta G = \Delta H - T\Delta S \qquad (2\text{-}10)$$

In other words, there are some irreversible losses in energy conversion due to creation of entropy, ΔS.

Similarly, since ΔH for the reaction (Equation 2-4) is the difference between the heats of formation of products and reactants (Equation 2-5), ΔS is the difference between entropies of products and reactants:

$$\Delta S = (s_f)_{H_2O} - (s_f)_{H_2} - 1/2(s_f)_{O_2} \qquad (2\text{-}11)$$

The values of h_f and s_f for reaction reactants and products at ambient pressure and 25 °C are shown in Table 2-1 [1].

Therefore, at 25 °C, out of 286.02 kJ mol^{-1} of available energy, 237.34 kJ mol^{-1} can be converted into electrical energy and the remaining 48.68 kJ mol^{-1} is converted into heat. At temperatures other than 25 °C, these values are different, as shown in Section 2.6.

2.5 THEORETICAL FUEL CELL POTENTIAL

In general, electrical work is a product of charge and potential:

$$W_{el} = qE \qquad (2\text{-}12)$$

where:

W_{el} = electrical work ($J\ mol^{-1}$)

q = charge (Coulombs mol^{-1})

E = potential (Volts)

The total charge transferred in a fuel cell reaction (Equations 2-1, 2-2, and 2-3) per mol of H_2 consumed is equal to:

$$q = n\ N_{Avg}\ q_{el} \qquad (2\text{-}13)$$

where:

n = number of electrons per molecule of H_2 = 2 electrons per molecule

N_{Avg} = number of molecules per mole (Avogadro's number) = 6.022×10^{23} molecules/mol

q_{el} = charge of 1 electron = 1.602×10^{-19} coulombs/electron

The product of Avogadro's number and charge of 1 electron is known as *Faraday's constant:*

F = 96,485 coulombs/electron-mol

Electrical work is therefore:

$$W_{el} = nFE \qquad (2\text{-}14)$$

As mentioned previously, the maximum amount of electrical energy generated in a fuel cell corresponds to Gibbs free energy, ΔG:

$$W_{el} = -\Delta G \qquad (2\text{-}15)$$

The theoretical potential of a fuel cell is then:

$$E = \frac{-\Delta G}{nF} \qquad (2\text{-}16)$$

Because ΔG, n, and F are all known, the theoretical fuel cell potential of hydrogen/oxygen can also be calculated:

$$E = \frac{-\Delta G}{nF} = \frac{237,340\ \ J\ mol^{-1}}{2 \cdot 96,485\ \ As\ mol^{-1}} = 1.23\ \text{Volts} \qquad (2\text{-}17)$$

At 25 °C, the theoretical hydrogen/oxygen fuel cell potential is 1.23 volts.

2.6 EFFECT OF TEMPERATURE

The theoretical cell potential changes with temperature. Substituting Equation (2-10) into (2-16) yields:

$$E = -\left(\frac{\Delta H}{nF} - \frac{T\Delta S}{nF}\right) \qquad (2\text{-}18)$$

Table 2-2 Enthalpies, Entropies, and Gibbs Free Energy for Hydrogen Oxidation Processes

	ΔH (kJ mol^{-1})	ΔS (kJ mol^{-1}K^{-1})	ΔG (kJ mol^{-1})
$H_2 + {}^1/_2O_2 \rightarrow H_2O(l)$	-286.02	-0.1633	-237.34
$H_2 + {}^1/_2O_2 \rightarrow H_2O(g)$	-241.98	-0.0444	-228.74

Obviously, an increase in the cell temperature results in a lower theoretical cell potential. Note that both ΔH and ΔS are negative (see Table 2-2). In addition, both ΔH and ΔS are functions of temperature:

$$h_T = h_{298.15} + \int_{298.15}^{T} c_p dT \qquad (2\text{-}19)$$

$$S_T = S_{298.15} + \int_{298.15}^{T} \frac{1}{T}c_p dT \qquad (2\text{-}20)$$

Specific heat of any gas is also a function of temperature (Figure 2-3). An empirical relationship may be used [2]:

$$c_p = a + bT + cT^2 \qquad (2\text{-}21)$$

where a, b, and c are the empirical coefficients, different for each gas, as shown in Table 2-3.

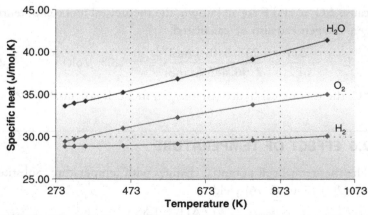

Figure 2-3 Specific heat, C_p, as a function of temperature for hydrogen, oxygen, and water vapor.

Table 2-3 The Coefficients for Temperature Dependency of C_p, in J mol^{-1}K^{-1}, from [2]

	a	b	c
H_2	28.91404	−0.00084	2.01×10^{-6}
O_2	25.84512	0.012987	-3.9×10^{-6}
H_2O (g)	30.62644	0.009621	1.18×10^{-6}

Substituting Equation (2-21) into Equations (2-19) and (2-20) and integrating yields:

$$\Delta H_T = \Delta H_{298.15} + \Delta a(T - 298.15) + \Delta b \frac{\left(T^2 - 298.15^2\right)}{2}$$

$$+ \Delta c \frac{\left(T^3 - 298.15^3\right)}{3} \tag{2-22}$$

$$\Delta S_T = \Delta S_{298.15} + \Delta a \, ln\left(\frac{T}{298.15}\right) + \Delta b(T - 298.15)$$

$$+ \Delta c \frac{\left(T^2 - 298.15^2\right)}{2} \tag{2-23}$$

where Δa, Δb, and Δc are the differences between the coefficients a, b, and c, respectively, for products and reactants, that is:

$$\Delta a = a_{H_2O} - aH_2 - 1/2 a_{O_2}$$

$$\Delta b = b_{H_2O} - bH_2 - 1/2 b_{O_2} \tag{2-24}$$

$$\Delta c = c_{H_2O} - cH_2 - 1/2 c_{O_2}$$

At temperatures below 100 °C, changes of C_p, ΔH, and ΔS are very small (Table 2-4), but at higher temperatures, such as those experienced in solid oxide fuel cells, they must not be neglected. As shown in Table 2-4 and Figure 2-4, the theoretical cell potential decreases with temperature.

Table 2-4 Change of Enthalpy, Gibbs Free Energy, and Entropy of Hydrogen/Oxygen Fuel Cell Reaction with Temperature and Resulting Theoretical Cell Potential

T(K)	ΔH (kJ mol^{-1})	ΔG (kJ mol^{-1})	ΔS (kJ mol^{-1}K^{-1})	E$_{th}$ (V)
298.15	−286.02	−237.34	−0.16328	1.230
333.15	−284.85	−231.63	−0.15975	1.200
353.15	−284.18	−228.42	−0.15791	1.184
373.15	−283.52	−225.24	−0.15617	1.167

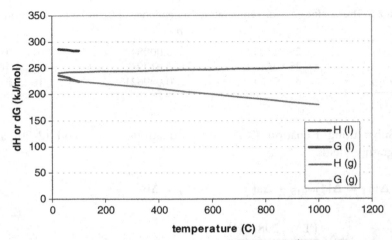

Figure 2-4 Enthalpy and Gibbs free energy of hydrogen/oxygen fuel cell as a function of temperature.

However, in operating fuel cells, in general a higher cell temperature results in a higher cell potential. This is because the voltage losses in operating fuel cells decrease with temperature, and this more than compensates for the loss of theoretical cell potential.

2.7 THEORETICAL FUEL CELL EFFICIENCY

The efficiency of any energy conversion device is defined as the ratio between useful energy output and energy input (Figure 2-5).

In the case of a fuel cell, the useful energy output is the electrical energy produced, and energy input is the enthalpy of hydrogen, that is, hydrogen's higher heating value (Figure 2-6). Assuming that all of the Gibbs free energy can be converted into electrical energy, the maximum possible (theoretical) efficiency of a fuel cell is:

$$\eta = \Delta G / \Delta H = 237.34/286.02 = 83\% \qquad (2\text{-}25)$$

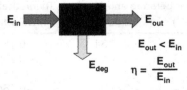

Figure 2-5 Efficiency of any energy conversion process.

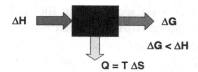

Figure 2-6 Energy inputs and outputs for a fuel cell as an energy conversion device.

Very often, hydrogen's lower heating value is used to express the fuel cell efficiency, not only because it results in a higher number but also to compare it with the fuel cell's competitor—the internal combustion engine, whose efficiency has traditionally been expressed with a lower heating value of fuel. In that case the maximum theoretical fuel cell efficiency would be:

$$\eta = \Delta G / \Delta H_{LHV} = 228.74/241.98 = 94.5\% \qquad (2\text{-}26)$$

The use of lower heating value in both the fuel cell and especially in the internal combustion engine is justified by water vapor being produced in the process, so the difference between higher and lower heating value (heat of evaporation) cannot be used anyway. Although the use of both lower and higher heating values in expressing the efficiency of an energy conversion device is appropriate (as long as it is specified which heating value has been used), the use of lower heating value may become confusing. In the late 1990s the manufacturers of condensing boilers in Germany were claiming that their boilers were more than 100% efficient because they were using the lower heating value of fuel as a measure of energy input. Lower heating value does not account for the heat of condensation of product water, but in this case the heat of condensation was indeed utilized because they were condensing boilers. It is therefore thermodynamically more correct to use the higher heating value because it accounts for all the energy available and it is consistent with the definition of the efficiency, as shown in Figure 2-5.

If both ΔG and ΔH in Equation (2-25) are divided by nF, the fuel cell efficiency may be expressed as a ratio of two potentials:

$$\eta = \frac{-\Delta G}{-\Delta H} = \frac{\dfrac{-\Delta G}{nF}}{\dfrac{-\Delta H}{nF}} = \frac{1.23}{1.482} = 0.83 \qquad (2\text{-}27)$$

where: $\dfrac{-\Delta G}{nF} = 1.23$ V is the theoretical cell potential, and

$\dfrac{-\Delta H}{nF} = 1.482$ V is the potential corresponding to hydrogen's higher heating value, or the thermoneutral potential

As shown in the following sections, the fuel cell efficiency is always proportional to the cell potential and may be calculated as a ratio of the cell potential and the potential corresponding to hydrogen's higher heating value, that is, 1.482 V. The potential corresponding to the lower heating value is 1.254 V.

2.8 CARNOT EFFICIENCY MYTH

Carnot efficiency is the maximum efficiency that a heat engine may have operating between the two temperatures (Figure 2-7).

The Carnot efficiency has little practical value. It is a maximum theoretical efficiency of a hypothetical engine. Even if such an engine could be constructed, it would have to be operated at infinitesimally low velocities to allow the heat transfer to occur. It would be very efficient, but it would generate no power (Figure 2-8), and thus it would be useless. The same applies to the theoretical fuel cell efficiency: The fuel cell operating at theoretical efficiency would generate no current and therefore it would be of no practical value.

It may be shown [3] that the efficiency at maximum power of a Carnot engine is:

$$\eta = 1 - \sqrt{\frac{T_C}{T_H}} \qquad (2\text{-}28)$$

The Carnot efficiency does not apply to fuel cells because a fuel cell is not a heat engine; rather, it is an electrochemical energy converter. For this reason a fuel cell operating at low temperature, such as 60 °C, and discarding heat into the environment at 25 °C may have an efficiency significantly higher than any heat engine operating between the same two temperatures (Figure 2-9). The theoretical efficiency of high–temperature fuel cells may

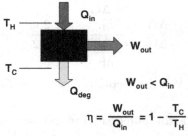

Figure 2-7 Carnot process efficiency.

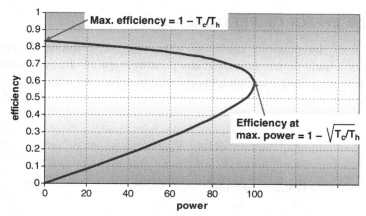

Figure 2-8 Efficiency vs. power curve for a hypothetical Carnot engine.

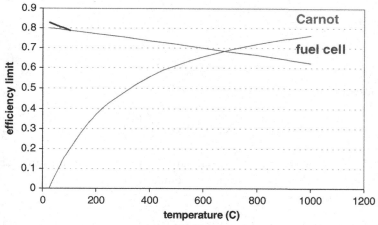

Figure 2-9 Theoretical efficiencies of a Carnot engine and a fuel cell as a function of temperature.

be lower than the theoretical (Carnot) efficiency of a heat engine operating between the same temperatures (Figure 2-9). Some may argue that for any hydrogen/oxygen or hydrogen/air system, the high-temperature source is the temperature of hydrogen/oxygen flame and that in this case no fuel cell efficiency can exceed the Carnot efficiency of an engine using this flame as the heat source. Although this may be correct, it has no relevance to the fuel cell, where there is no flame and the theoretical efficiency is determined by the ratio between Gibbs free energy and enthalpy of the hydrogen/oxygen reaction, regardless of the hydrogen/oxygen flame temperature.

2.9 EFFECT OF PRESSURE

All of the previous equations were valid at atmospheric pressure. However, a fuel cell may operate at any pressure, typically from atmospheric all the way up to 6–7 bar. For an isothermal process, and with a little bit of basic thermodynamics [4], the change in Gibbs free energy may be shown to be:

$$dG = V_m dP \tag{2-29}$$

where:

V_m = molar volume, $m^3\ mol^{-1}$

P = pressure, Pa

For an ideal gas:

$$PV_m = RT \tag{2-30}$$

Therefore:

$$dG = RT\frac{dP}{P} \tag{2-31}$$

After integration:

$$G = G_0 + RT\ ln\left(\frac{P}{P_0}\right) \tag{2-32}$$

where G_0 is Gibbs free energy at standard temperature and pressure (25 °C and 1 atm), and P_0 is the reference or standard pressure (1 atm).

For any chemical reaction:

$$jA + kB \rightarrow mC + nD \tag{2-33}$$

The change in Gibbs free energy is the change between products and reactants:

$$\Delta G = mG_C + nG_D - jG_A - kG_B \tag{2-34}$$

After substituting into Equation (2-32):

$$\Delta G = \Delta G_0 + RT\ ln\left[\frac{\left(\dfrac{P_C}{P_0}\right)^m \left(\dfrac{P_D}{P_0}\right)^n}{\left(\dfrac{P_A}{P_0}\right)^j \left(\dfrac{P_B}{P_0}\right)^k}\right] \tag{2-35}$$

This is known as the *Nernst equation*, where P is the partial pressure of the reactant or product species and P_0 is the reference pressure (i.e., 1 atm or 101.25 kPa).

For the hydrogen/oxygen fuel cell reaction, the Nernst equation becomes:

$$\Delta G = \Delta G_0 + RT \, ln\left(\frac{P_{H_2O}}{P_{H_2}P_{O_2}^{0.5}}\right) \tag{2-36}$$

By introducing Equation (2-18) into Equation (2-36):

$$E = E_0 + \frac{RT}{nF} \, ln\left(\frac{P_{H_2}P_{O_2}^{0.5}}{P_{H_2O}}\right) \tag{2-37}$$

Note that the previous equations are only valid for gaseous products and reactants. When liquid water is produced in a fuel cell, $P_{H_2O} = 1$. From Equation (2-37) it follows that at higher reactant pressures the cell potential is higher. Also, if the reactants are diluted—for example, if air is used instead of pure oxygen—their partial pressure is proportional to their concentration and consequently the cell potential is lower. In case of air versus oxygen, the theoretical voltage loss/gain is:

$$\Delta E = E_{O_2} - E_{Air} = \frac{RT}{nF} ln\left(\frac{P_{O_2}}{P_{Air}}\right)^{0.5} = \frac{RT}{nF} \, ln\left(\frac{1}{0.21}\right)^{0.5} \tag{2-38}$$

At 80 °C this voltage gain/loss becomes 0.012 V. In practice it is much higher, as will be discussed in Chapter 3.

2.10 SUMMARY

The ideal cell potential, if all Gibbs free energy is utilized, is:

$$E_{25C,1atm} = \frac{-\Delta G}{nF} = \frac{237,340 \text{ J mol}^{-1}}{2 \cdot 96,485 \text{ As mol}^{-1}} = 1.23 \text{ Volts} \tag{2-39}$$

Cell potential is a function of temperature and pressure:

$$E_{T,P} = -\left(\frac{\Delta H}{nF} - \frac{T\Delta S}{nF}\right) + \frac{RT}{nF} \, ln\left[\frac{P_{H_2}P_{O_2}^{0.5}}{P_{H_2O}}\right] \tag{2-40}$$

Ignoring the changes of dH and dS with temperature (which has a very small error for temperatures below 100 °C), this equation becomes:

$$E_{T,P} = 1.482 - 0.000845 \ T + 0.0000431 \ T \ ln \left(P_{H_2} P_{O_2}^{0.5} \right) \quad (2\text{-}41)$$

For example, a hydrogen/air fuel cell, operating at 60 °C with reactant gases at atmospheric pressure and with liquid water as a product, is expected to have a potential of:

$$E_{T,P} = 1.482 - 0.000845 \times 333.15 + 0.0000431 \times 333.15 \ ln \left(1 \times 0.21^{0.5} \right)$$

$$= 1.482 - 0.282 - 0.011 = 1.189 \ V$$

$$(2\text{-}42)$$

Note that oxygen concentration (by volume) in air is 21%, and therefore the oxygen partial pressure in this case is 21% of the atmospheric pressure.

The ideal fuel cell efficiency is:

$$\eta = \Delta G / \Delta H = 237.34/286.02 = 83\% \quad (2\text{-}43)$$

or:

$$\eta = E_0/1.482 \ V = 1.23/1.482 = 83\%$$

The ideal efficiency decreases with the temperature. For example, at 60 °C the ideal efficiency of a hydrogen/air fuel cell is:

$$\eta = E_0/1.482 \ V = 1.189/1.482 = 80\%$$

 PROBLEMS

1. Derive Equation (2-29) starting from $G = H - TS$. Explain your assumptions.
2. For a hydrogen/air fuel cell operating at 60 °C with reactant gases at atmospheric pressure and with liquid water as a product, calculate the theoretical cell potential, taking into account the changes of reaction enthalpy and entropy with temperature (Equations 2-22 and 2-23).
3. For a hydrogen/air fuel cell operating at 600 °C with reactant gases at atmospheric pressure and with water vapor as a product, calculate the theoretical cell potential, taking into account the changes of reaction enthalpy and entropy with temperature (Equations 2-22 and 2-23).

Repeat the same process while ignoring those changes and compare the results.

4. Calculate the expected difference in theoretical cell potential between a hydrogen/oxygen fuel cell operating at 80 °C and 5 bar (both reactant gases) and the same fuel cell operating at atmospheric pressure. What if the pressure is increased to 25 bar?

QUIZ

1. A fuel cell is:
   a. A device that stores fuel
   b. A device that converts fuel into electricity
   c. A device that generates fuel
2. A fuel cell catalyst serves to:
   a. Make good electric contacts
   b. Make the electrochemical reactions happen
   c. Generate internal electrical current
3. A fuel cell catalyst must be:
   a. Electrically conductive
   b. An electrical insulator
   c. Electrically conductive or an insulator—it does not matter
4. A porous fuel cell electrode is porous:
   a. So that the fuel cell may be made lightweight
   b. So that it can allow transfer of electrons (through solids) and gases (through voids)
   c. It does not have to be porous
5. Fuel cells must be stacked in order to:
   a. Generate more current than a single cell of the same active area
   b. Generate the same current as a single cell of the same active area but at a higher voltage
   c. Generate both more current and higher voltage than a single cell of the same active area
6. The heating value of fuel is:
   a. A measure of how much heat can be generated in a fuel cell
   b. A measure of how much electricity can be generated in a fuel cell
   c. A measure of how much heat can be generated if the fuel is combusted (not necessarily in a fuel cell)

7. A fuel cell:
   a. Is always more efficient than a Carnot cycle
   b. Has approximately the same efficiency as a Carnot cycle
   c. Ideal efficiency may be compared with a Carnot cycle and it may be higher or lower than the Carnot efficiency, depending on the temperature
8. Theoretical fuel cell potential:
   a. Is higher at higher hydrogen or oxygen pressure
   b. Is lower at higher hydrogen pressure
   c. Does not depend on pressure
9. Theoretical fuel cell potential:
   a. Is higher at higher temperatures
   b. Is lower at higher temperatures
   c. Does not depend on temperature
10. Fuel cell efficiency is:
   a. Proportional to its voltage
   b. Inversely proportional to its voltage
   c. Proportional to the square of voltage divided by the heat-generating rate

REFERENCES

[1] Weast RC, editor. CRC Handbook of Chemistry and Physics. Boca Raton, FL: CRC Press; 1988.
[2] Hirschenhofer JH, Stauffer DB, Engleman RR. Fuel Cells: A Handbook. U.S. Department of Energy, Morgantown Energy Technology Center; January 1994. Revision 3DOE/METC-94/1006.
[3] Kurzon FL, Ahlborn B. Efficiency of a Carnot Engine at Maximum Power Output. American Journal of Physics 1975;43:22–4.
[4] Chen E. Thermodynamics and Electrochemical Kinetics. In: Hoogers G, editor. Fuel Cell Technology Handbook. Boca Raton, FL: CRC Press; 2003.

CHAPTER THREE

Fuel Cell Electrochemistry

3.1 ELECTRODE KINETICS

A fuel cell is an electrochemical energy converter. Its operation is based on the following electrochemical reactions happening simultaneously on the anode and the cathode.

At the anode:

$$H_2 \rightarrow 2H^+ + 2e^- \tag{3-1}$$

At the cathode:

$$1/_2 O_2 + 2H^+ + 2e^- \rightarrow H_2O \tag{3-2}$$

More precisely, the reactions happen on an interface between the ionically conductive electrolyte and the electrically conductive electrode. Because there are gases involved in fuel cell electrochemical reactions, the electrodes must be porous, allowing the gases to arrive, as well as product water to leave, the reaction sites. Note that these are the overall reactions and that in both cases there are several intermediary sequential and parallel steps involved.

3.1.1 Reaction Rate

Electrochemical reactions involve both a transfer of electrical charge and a change in Gibbs energy [1]. The rate of an electrochemical reaction is determined by an activation energy barrier that the charge must overcome in moving from electrolyte to a solid electrode, or vice versa. The speed at which an electrochemical reaction proceeds on the electrode surface is the rate at which the electrons are released or "consumed," which is the electrical current. *Current density* is the current (of electrons or ions) per unit area of the surface. From Faraday's law it follows that current density is proportional to the charge transferred and the consumption of reactant per unit area:

$$i = nFj \tag{3-3}$$

PEM Fuel Cells
ISBN 978-0-12-387710-9

where nF is the charge transferred (Coulombs mol^{-1}) and j is the flux of reactant per unit area (mol s^{-1} cm^{-2}).

Therefore, the reaction rate may be easily measured by a current-measuring device placed externally to the cell. However, the measured current or current density is actually the net current, that is, the difference between forward and reverse current on the electrode. In general, an electrochemical reaction involves either oxidation or reduction of the species:

$$Red \rightarrow Ox + ne^- \qquad (3\text{-}4)$$

$$Ox + ne^- \rightarrow Red \qquad (3\text{-}5)$$

In a hydrogen/oxygen fuel cell the anode reaction is oxidation of hydrogen, Equation (3-1), in which hydrogen is stripped of its electrons, and the products of this reaction are protons and electrons. The cathode reaction is oxygen reduction, Equation (3-2), and water is generated as a product.

On an electrode at equilibrium conditions, that is, when no external current is being generated, both processes, oxidation and reduction, occur at equal rates:

$$Ox + ne^- \leftrightarrow Red \qquad (3\text{-}6)$$

The consumption of the reactant species is proportional to their surface concentration. For the forward reaction of Equation (3-6), which is the reaction described by Equation (3-5), the flux is:

$$j_f = k_f C_{Ox} \qquad (3\text{-}7)$$

where:

k_f = forward reaction (reduction) rate coefficient, (s^{-1}), and

C_{Ox} = surface concentration of the reacting species (mol cm^{-2})

Similarly, for the backward reaction of Equation (3-6), which is the reaction described by Equation (3-4), the flux is:

$$j_b = k_b C_{Rd} \qquad (3\text{-}8)$$

where:

k_b = backward reaction (oxidation) rate coefficient, (s^{-1}), and

C_{Rd} = surface concentration of reacting species (mol cm^{-2})

Each of these two reactions either releases or consumes electrons. The net current generated is the difference between the electrons released and consumed:

$$i = nF(k_f C_{Ox} - k_b C_{Rd}) \qquad (3\text{-}9)$$

At equilibrium, the net current is equal to zero, although the reaction proceeds in both directions simultaneously. The rate at which these reactions proceed at equilibrium is called the *exchange current density*.

3.1.2 Reaction Constants; Transfer Coefficient

From the Transition State Theory [2], it may be shown that the reaction rate coefficient for an electrochemical reaction is a function of the Gibbs free energy [1]:

$$k = \frac{k_B T}{h} \, exp\left(\frac{-\Delta G}{RT}\right) \tag{3-10}$$

where:

k_B = Boltzmann's constant, 1.38×10^{-23} J K^{-1}
h = Planck's constant, 6.626×10^{-34} J s

The Gibbs free energy for electrochemical reactions may be considered to consist of both chemical and electrical terms [1]. In that case, for a reduction reaction:

$$\Delta G = \Delta G_{ch} + \alpha_{Rd} FE \tag{3-11}$$

and for an oxidation reaction:

$$\Delta G = \Delta G_{ch} - \alpha_{Ox} FE \tag{3-12}$$

The subscript *ch* denotes the chemical component of the Gibbs free energy, α is a transfer coefficient, F is the Faraday's constant, and E is the potential. There is a fair amount of confusion in the literature concerning the transfer coefficient, α, and the symmetry factor, β, that is sometimes used. The symmetry factor, β, may be used strictly for a single–step reaction involving a single electron ($n = 1$). Its value is theoretically between 0 and 1, but most typically for the reactions on a metallic surface it is around 0.5. The way in which β is defined requires that the sum of the symmetry factors in the anodic and cathodic direction be unity; if it is β for the reduction reaction, it must be $(1 - \beta)$ for the reverse, the oxidation reaction.

However, both electrochemical reactions in a fuel cell, namely oxygen reduction and hydrogen oxidation, involve more than one step and more than one electron. In that case, at steady state the rate of all steps must be equal, and it is determined by the slowest step in the sequence, which is referred to as the *rate-determining step*. To describe a multistep process, instead of the symmetry factor, β, a rather experimental parameter is used, which is called the *transfer coefficient*, α. Note that in this case $\alpha_{Rd} + \alpha_{Ox}$ does not necessarily have to be equal to unity. Actually, in general $(\alpha_{Rd} + \alpha_{Ox}) = n/\nu$, where n is the number of electrons transferred in the overall reaction and ν is the stoichiometric number defined as the number of times the rate-determining step must occur for the overall reaction to occur once [3].

The forward (reduction) and backward (oxidation) reaction rate coefficients in Equation (3-9) are then, respectively:

$$k_f = k_{0,f} \; exp\left[\frac{-\alpha_{Rd} \; FE}{RT}\right] \tag{3-13}$$

$$k_b = k_{0,b} \; exp\left[\frac{\alpha_{Ox} \; FE}{RT}\right] \tag{3-14}$$

3.1.3 Current Potential Relationship: Butler–Volmer Equation

By introducing into Equation (3-9) the net current, density is obtained:

$$i = nF\left\{k_{0,f}C_{Ox} \; exp\left[\frac{-\alpha_{Rd} \; FE}{RT}\right] - k_{0,b}C_{Rd} \; exp\left[\frac{\alpha_{Ox} \; FE}{RT}\right]\right\} \tag{3-15}$$

At equilibrium, the potential is Er and the net current is equal to zero, although the reaction proceeds in both directions simultaneously. The rate at which these reactions proceed at equilibrium is called the *exchange current density* [1,4]:

$$i_0 = nFk_{0,f}C_{Ox} \; exp\left[\frac{-\alpha_{Rd} \; FE_r}{RT}\right] = nFk_{0,b}C_{Rd} \; exp\left[\frac{\alpha_{Ox} \; FE_r}{RT}\right] \tag{3-16}$$

By combining Equations (3-15) and (3-16), we obtain a relationship between the current density and potential:

$$i - i_0\left\{exp\left[\frac{-\alpha_{Rd} \; F(E-E_r)}{RT}\right] - exp\left[\frac{\alpha_{Ox} \; F(E-E_r)}{RT}\right]\right\} \tag{3-17}$$

This is known as the Butler–Volmer equation, where E_r is the reversible or equilibrium potential. Note that the reversible or equilibrium potential at the fuel cell anode is 0 V by definition [5], and the reversible potential at the fuel cell cathode is 1.229 V (at 25 °C and atmospheric pressure) and it does vary with temperature and pressure, as shown in Chapter 2. The difference between the electrode potential and the reversible potential is called *over-potential*. It is the potential difference required to generate current.

The Butler–Volmer equation (3-17) is valid for both anode and cathode reaction in a fuel cell:

$$i_a = i_{0,a} \left\{ exp\left[\frac{-\alpha_{Rd,a}F(E_a - E_{r,a})}{RT} \right] - exp\left[\frac{\alpha_{Ox,a}F(E_a - E_{r,a})}{RT} \right] \right\} \quad (3\text{-}18)$$

and

$$i_c = i_{0,c} \left\{ exp\left[\frac{-\alpha_{Rd,c}F(E_c - E_{r,c})}{RT} \right] - exp\left[\frac{\alpha_{Ox,c}F(E_c - E_{r,c})}{RT} \right] \right\} \quad (3\text{-}19)$$

The overpotential on the anode is positive $(E_a > E_{r,a})$, which makes the first term of Equation (3-18) negligible in comparison with the second term, that is, the oxidation current is predominant and the equation may be reduced to:

$$i_a = -i_{0,a}\, exp\left[\frac{\alpha_{Ox,a}F(E_a - E_{r,a})}{RT} \right] \quad (3\text{-}20)$$

Note that the resulting current has a negative sign, which denotes that the electrons are leaving the electrode (net oxidation reaction).

Similarly, the overpotential on the cathode is negative $(E_c < E_{r,c})$, which makes the first term of Equation (3-19) much larger than the second term, that is, the reduction current is predominant and the equation may be reduced to:

$$i_c = i_{0,c}\, exp\left[\frac{-\alpha_{Rd,c}F(E_c - E_{r,c})}{RT} \right] \quad (3\text{-}21)$$

Note that Equations (3-20) and (3-21) are not valid for very small values of i.

The transfer coefficients in the previous equations for hydrogen/oxygen fuel cells using Pt catalyst seem to have a value around 1. Note that in some literature there is an n parameter in the previous equations [1,5] denoting the number of electrons involved. In that case, clearly, on the fuel cell anode side, $n = 2$, and on the cathode side, $n = 4$, and it is the product of $n\alpha$ that has a value around 1. Larminie and Dicks [6] list a value of $\alpha = 0.5$ for the hydrogen fuel cell anode (with two electrons involved) and $\alpha = 0.1$ to 0.5 for the cathode (with four electrons involved). Newman [7] specifies α in the range between 0.2 and 2.

3.1.4 Exchange Current Density

Exchange current density, i_0, in electrochemical reactions is analogous to the rate constant in chemical reactions. Unlike the rate constants, exchange current density is concentration dependent (as can be seen directly from Equation 3-16). It is also a function of temperature (from Equation 3-10). The effective exchange current density (per unit of electrode geometrical area) is also a function of electrode catalyst loading and catalyst specific surface area. If the reference exchange current density (at reference temperature and pressure) is given per actual catalyst surface area, the effective exchange current density at any temperature and pressure is given by the following equation [8]:

$$
i_0 = i_0^{ref} a_c L_c \left(\frac{P_r}{P_r^{ref}} \right)^\gamma exp \left[-\frac{E_C}{RT} \left(1 - \frac{T}{T_{ref}} \right) \right] \qquad (3\text{-}22)
$$

where:

i_0^{ref} = reference exchange current density (at reference temperature and pressure, typically 25 °C and 101.25 kPa) per unit catalyst surface area, $A\ cm^{-2} Pt$

a_c = catalyst-specific area (theoretical limit for Pt catalyst is 2400 $cm^2\ mg^{-1}$, but state-of-the-art catalyst has about 600–1000 $cm^2\ mg^{-1}$, which is further reduced by incorporation of catalyst in the electrode structures by up to 30%)

L_c = catalyst loading (state-of-the-art electrodes have 0.3–0.5 mgPt cm^{-2}; electrodes with catalyst loading lower than 0.1 mgPt cm^{-2} have been demonstrated)

P_r = reactant partial pressure, kPa

P_r^{ref} = reference pressure, kPa

γ = pressure dependency coefficient (0.5 to 1.0)

E_C = activation energy, 66 kJ mol^{-1} for oxygen reduction on Pt [8]

R = gas constant, 8.314 J $mol^{-1} K^{-1}$

T = temperature, K

T_{ref} = reference temperature, 298.15 K

The product $a_c L_c$ is also called *electrode roughness*, meaning the catalyst surface area, cm^2, per electrode geometric area, cm^2. Instead of the ratio of partial pressures, a ratio of concentrations at the catalyst surface may be used as well.

Exchange current density is a measure of an electrode's readiness to proceed with the electrochemical reaction. If the exchange current density is high, the surface of the electrode is more active. In a hydrogen/oxygen fuel cell, the exchange current density at the anode is much larger (several orders of magnitude) than at the cathode. The higher the exchange current density, the lower the energy barrier that the charge must overcome in moving from electrolyte to the catalyst surface, and vice versa. In other words, the higher the exchange current density, the more current is generated at any overpotential.

Because the anode exchange current density in hydrogen/oxygen fuel cells is several orders of magnitude larger than the cathode current density ($\sim 10^{-3}$ versus $\sim 10^{-9}$ A cm^{-2}Pt, at 25 °C and 1 atm, for acid electrolyte), the overpotential on the cathode is much larger than the anode over-potential. For that reason, very often the cell potential/current relationship is approximated solely by Equation (3–21).

3.2 VOLTAGE LOSSES

If a fuel cell is supplied with reactant gases but the electrical circuit is not closed (Figure 3-1a), it will not generate any current, and one would expect the cell potential to be at, or at least close to, the theoretical cell potential for given conditions (temperature, pressure, and concentration of reactants). However, in practice this potential, called the *open circuit potential*, is significantly lower than the theoretical potential, usually less than 1 V. This suggests that there are some losses in the fuel cell, even when no external current is generated. When the electrical circuit is closed with a load (such as a resistor) in it, as shown in Figure 3-1b, the potential is expected to drop even further as a function of current being generated, due to unavoidable

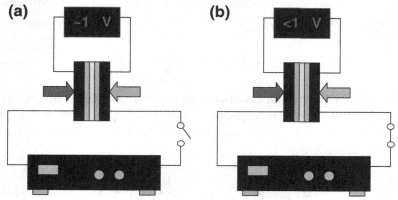

Figure 3-1 Fuel cell with a load: (a) in open circuit; (b) load connected.

losses. There are different kinds of voltage losses in a fuel cell caused by the following factors:

- Kinetics of the electrochemical reactions
- Internal electrical and ionic resistance
- Difficulties in getting the reactants to reaction sites
- Internal (stray) currents
- Crossover of reactants

Although mechanical and electrical engineers prefer to use voltage losses, (electro)chemical engineers use terms such as *polarization* or *overpotential*. They all have the same physical meaning: the difference between the electrode potential and the equilibrium potential. From the electrochemical engineer's point of view, this difference is the driver for the reaction, and from a mechanical or electrical engineer's point of view, this represents the loss of voltage and power.

3.2.1 Activation Polarization

Some voltage difference from equilibrium is needed to get the electro-chemical reaction going, as shown previously (Equation 3-17). This is called *activation polarization*, and it is associated with sluggish electrode kinetics. The higher the exchange current density, the lower the activation polarization losses. These losses happen at both anode and cathode; however, oxygen reduction requires much higher overpotentials, that is, it is a much slower reaction than hydrogen oxidation.

As discussed earlier, at relatively high negative overpotentials (i.e., potentials lower than the equilibrium potential) such as those at the fuel cell cathode, the first term in the Butler–Volmer equation becomes predominant, which allows for expression of potential as a function of current density (from Equation 3-21):

$$\Delta V_{act,c} = E_{r,c} - E_c = \frac{RT}{\alpha_c F} ln \left(\frac{i}{i_{0,c}} \right) \tag{3-23}$$

Figure 3-2 shows typical activation polarization for oxygen reduction on Pt.

Similarly, at the anode at positive overpotentials (i.e., higher than the equilibrium potential), the second term in the Butler–Volmer equation becomes predominant:

$$\Delta V_{act,a} = E_a - E_{r,a} = \frac{RT}{\alpha_a F} ln \left(\frac{i}{i_{0,a}} \right) \tag{3-24}$$

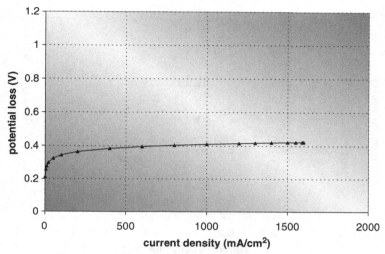

Figure 3-2 Voltage loss due to polarization activation.

Note that by definition, in electrochemistry the reversible potential of the hydrogen oxidation reaction is zero at all temperatures [5]. That is why the standard hydrogen electrode is used as a reference electrode. Therefore, for hydrogen anodes $E_{r,a} = 0$ V. Activation polarization of the hydrogen oxidation reaction is much smaller than activation polarization of the oxygen reduction reaction.

A simplified way to show the activation losses is to use the so-called *Tafel equation*:

$$\Delta V_{act} = a + b\,\log(i) \qquad (3\text{-}25)$$

where term *b* is called the *Tafel slope*. The Tafel equation (Equation 3-25) is purely empirical; however, one may notice that it has the same form as Equations (3-23) and (3-24). By comparing Equations (3-23) and (3-24) with Equation (3-25), it is apparent that the coefficients *a* and *b* in the Tafel equation become:

$$a = -2.3\,\frac{RT}{\alpha F}\,log(i_o), \quad \text{and} \quad b = 2.3\,\frac{RT}{\alpha F}.$$

Note that at any given temperature the Tafel slope depends solely on a transfer coefficient, α. For $\alpha = 1$, the Tafel slope at 60 °C is ~60 mV per decade, what is typically found for oxygen reduction on Pt.

If voltage–current relationship is plotted in a logarithmic scale, the main parameters, *a*, *b*, and i_0, are easily detectable (as shown in Figure 3-3).

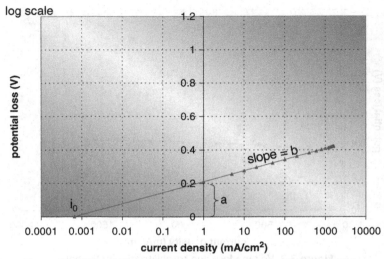

Figure 3-3 Potential loss due to activation polarization in log scale.

If these activation polarizations were the only losses in a fuel cell, the cell potential would be:

$$E_{cell} = E_c - E_a = E_r - \Delta V_{act,c} - \Delta V_{act,a} \qquad (3\text{-}26)$$

$$E_{cell} = E_r - \frac{RT}{\alpha_c F} ln\left(\frac{i}{i_{0,c}}\right) - \frac{RT}{\alpha_a F} ln\left(\frac{i}{i_{0,a}}\right) \qquad (3\text{-}27)$$

If anode polarization is neglected, the previous equation becomes:

$$E_{cell} = E_r - \frac{RT}{\alpha F} ln\left(\frac{i}{i_0}\right) \qquad (3\text{-}28)$$

which has the same form as the Tafel equation (3-25).

3.2.2 Internal Currents and Crossover Losses

Although the electrolyte, a polymer membrane, is not electrically conductive and is practically impermeable to reactant gases, some small amount of hydrogen will diffuse from anode to cathode, and some electrons may also find a "shortcut" through the membranes. Because each hydrogen molecule contains two electrons, this fuel crossover and the so-called internal currents are essentially equivalent. Each hydrogen molecule that diffuses through the polymer electrolyte membrane and reacts with oxygen on the cathode side of the fuel cell results in two fewer electrons in the generated current of electrons that travels through an external circuit. These losses may appear insignificant in fuel cell operation because the rate of hydrogen permeation or electron crossover is several orders of magnitude lower than the hydrogen consumption rate or total electrical current generated. However, when the fuel cell is at open circuit potential or when it operates at very low current densities, these losses may have a dramatic effect on cell potential, as shown in Figure 3-4.

The total electrical current is the sum of external (useful) current and current losses due to fuel crossover and internal currents:

$$I = I_{ext} + I_{loss} \qquad (3\text{-}29)$$

Current divided by the electrode active area, A, is current density, $A\ cm^{-2}$:

$$i = \frac{I}{A} \qquad (3\text{-}30)$$

Therefore:

$$i = i_{ext} + i_{loss} \qquad (3\text{-}31)$$

If this total current density is used in the equation that approximates the cell potential (Equation 3-28), the following equation results:

$$E_{cell} = E_r - \frac{RT}{\alpha F} ln \left(\frac{i_{ext} + i_{loss}}{i_0} \right) \qquad (3\text{-}32)$$

Therefore, even if the external current is equal to zero, such as at open circuit, the cell voltage may be significantly lower than the reversible cell

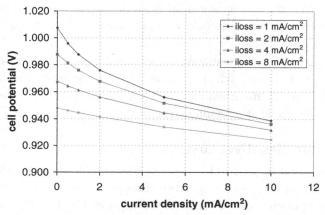

Figure 3-4 Effect of internal currents and/or hydrogen crossover loss on open circuit potential.

potential for given conditions. Indeed, open circuit potential of hydrogen/ air fuel cells is typically below 1 V, most likely about 0.94 to 0.97 V (depending on operating pressure and the state of membrane hydration):

$$E_{cell,OCV} = E_r - \frac{RT}{\alpha F} ln\left(\frac{i_{loss}}{i_0}\right) \qquad (3\text{-}33)$$

Although hydrogen crossover and internal currents are equivalent, they physically have different effects in a fuel cell. The loss of electrons occurs after the electrochemical reaction has taken place and therefore the effect on both anode and cathode activation polarization would have the effect as depicted by Equation (3-32). Hydrogen that permeates the membrane does not participate in the electrochemical reaction on the anode side, and in that case the total current resulting from the electrochemical reaction would be the same as the external current. However, hydrogen that permeates the membrane to the cathode side may react with oxygen on the surface of the catalyst in reaction $H_2 + \frac{1}{2}O_2 \rightarrow H_2O$ and as a result would "depolarize" the cathode, that is, reduce the cathode (and cell) potential. Equations (3-32) and (3-33) are therefore only an approximation.

In addition, oxygen may permeate the membrane as well, although the oxygen permeation rate is much lower than the hydrogen permeation rate. The effect on fuel cell performance would be similar to that of hydrogen crossover loss, but in this case the anode would be "depolarized."

Hydrogen crossover is a function of membrane permeability, membrane thickness, and hydrogen partial pressure (i.e., hydrogen concentration) difference across the membrane as the main driving force. A very low open circuit potential (significantly below 0.9 V) may indicate either a hydrogen leak or an electrical short.

As the fuel cell starts generating current, hydrogen concentration in the catalyst layer decreases, which reduces the driving force for hydrogen permeation through the membrane. That is one of the reasons these losses are mainly negligible at operating currents.

3.2.3 Ohmic (Resistive) Losses

Ohmic losses occur because of resistance to the flow of ions in the electrolyte and resistance to the flow of electrons through the electrically conductive fuel cell components. These losses can be expressed by Ohm's law:

$$\Delta V_{ohm} = iR_i \qquad (3\text{-}34)$$

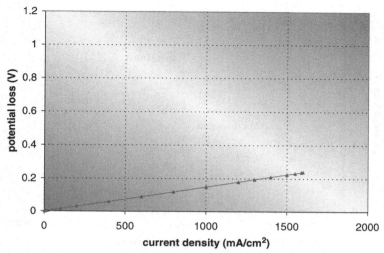

Figure 3-5 Resistive (ohmic) losses in the fuel cell ($R_i = 0.15$ Ohm-cm^2).

where:

i = current density, $A\ cm^{-2}$, and

R_i = total cell internal resistance (which includes ionic, electronic, and contact resistance, Ω cm^2):

$$R_i = R_{i,i} + R_{i,e} + R_{i,c} \qquad (3\text{-}35)$$

Electronic resistance is almost negligible, even when graphite or graphite/polymer composites are used as current collectors. Ionic and contact resistances are approximately of the same order of magnitude [8,9]. Both will be discussed later in Chapter 4. Typical values for *Ri* are between 0.1 and 0.2 Ω cm^2. Figure 3-5 shows typical resistive losses in the fuel cell (for $R_i = 0.15\ \Omega$ cm^2). Note that Figure 3-5 is in the same scale as Figure 3-6 for an easy comparison of the magnitude of these losses.

3.2.4 Concentration Polarization

Concentration polarization occurs when a reactant is rapidly consumed at the electrode by the electrochemical reaction so that concentration gradients are established. We learned before that the electrochemical reaction potential changes with partial pressure of the reactants, and this relationship is given by the Nernst equation:

$$\Delta V = \frac{RT}{nF} ln\left(\frac{C_B}{C_S}\right) \qquad (3\text{-}36)$$

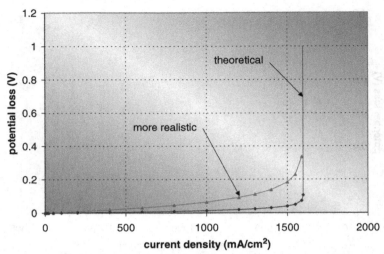

Figure 3-6 Concentration polarization losses in the fuel cell.

where:

C_B = bulk concentration of reactant, mol cm^{-3}

C_S = concentration of reactant at the surface of the catalyst, mol cm^{-3}

According to Fick's law of one dimensional diffusion, the flux of reactant is proportional to concentration gradient:

$$N = \frac{D \cdot (C_B - C_S)}{\delta} A \qquad (3\text{-}37)$$

where:

N = flux of reactants, mol s^{-1}

D = diffusion coefficient of the reacting species, cm^{2}s^{-1}

A = electrode active area, cm^2

δ = diffusion distance, cm

In steady state, the rate at which the reactant is consumed in the electrochemical reaction is equal to the diffusion flux (Equation 3-3):

$$N = \frac{I}{nF} \qquad (3\text{-}38)$$

By combining Equations (3-37) and (3-38), the following relationship is obtained:

$$i = \frac{nf \cdot D \cdot (C_B - C_S)}{\delta} \qquad (3\text{-}39)$$

The reactant concentration at the catalyst surface thus depends on current density—the higher the current density, the lower the surface concentration. The surface concentration reaches zero when the rate of consumption becomes equal to the diffusion rate; in other words, the reactant is consumed at the same rate as it is reaching the surface, and as a result the concentration of reactant and the catalyst surface is equal to zero. The current density at which this happens is called the *limiting current density*. A fuel cell cannot produce more than the limiting current because there are no more reactants at the catalyst surface. Therefore, for $C_S = 0$, $i = i_L$, and the limiting current density is then:

$$i_L = \frac{nFDC_B}{\delta} \tag{3-40}$$

By combining Equations (3-36), (3-39), and (3-40), a relationship for voltage loss due to concentration polarization is obtained:

$$\Delta V_{conc} = \frac{RT}{nF} ln\left(\frac{i_L}{i_L - i}\right) \tag{3-41}$$

The previous equation would result in a sharp drop of cell potential as the limiting current is approached (as shown in Figure 3-6). However, because of nonuniform conditions over the porous electrode area, the limiting current is almost never experienced in practical fuel cells. To experience a sharp drop of cell potential when the limiting current density is reached, the current density would have to be uniform over the entire electrode surface, which is almost never the case, since the electrode surface consists of discrete particles. Some particles may reach the limiting current density while the remaining particles may operate normally. Limiting current density may be experienced at either cathode or anode.

Another reason that a sharp distinguished drop in cell voltage at the limiting current does not show in practical fuel cells is that the exchange current density is a function of concentration of reactant at the catalyst surface, C_S. As the current density approaches limiting current density, the surface concentration, and consequently the exchange current density, approaches zero (as follows from Equation 3-22), which causes additional voltage loss, as follows from Equation (3-27) or Equation (3-28).

An empirical equation better describes the polarization losses, as suggested by Kim et al. [10]:

$$\Delta V_{conc} = c \cdot exp\left(\frac{i}{d}\right) \tag{3-42}$$

where c and d are empirical coefficients (values of $c = 3 \times 10^{-5}$ V and $n = 0.125$ A cm^{-2} have been suggested [6], but clearly, these coefficients are dependent on the conditions inside the fuel cell and therefore would have to be determined experimentally for every fuel cell).

3.3 CELL POTENTIAL: POLARIZATION CURVE

Figure 3-7 shows the proportions between the three types of losses in the fuel cell. Activation losses are by far the largest losses at any current density.

Activation and concentration polarization can occur at both anode and cathode. The cell voltage is therefore:

$$V_{cell} = E_r - (\Delta V_{act} + \Delta V_{conc})_a - (\Delta V_{act} + \Delta V_{conc})_c - \Delta V_{ohm} \quad (3\text{-}43)$$

By introducing Equations (3-23), (3-24), (3-34), and (3-41) into Equation (3-43), a relationship between fuel cell potential and current density, the so-called *fuel cell polarization curve*, is obtained:

$$E_{cell} = E_{r,T,P} - \frac{RT}{\alpha_c F} ln\left(\frac{i}{i_{O,c}}\right) - \frac{RT}{\alpha_a F} ln\left(\frac{i}{i_{0,a}}\right)$$

$$\quad (3\text{-}44)$$

$$- \frac{RT}{nF} ln\left(\frac{i_{L,c}}{i_{L,c} - i}\right) - \frac{RT}{nF} ln\left(\frac{i_{L,a}}{i_{L,a} - i}\right) - iR_i$$

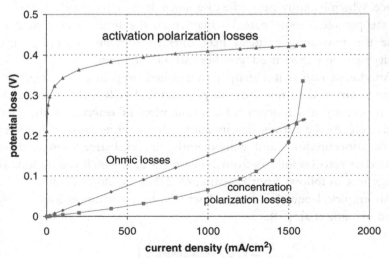

Figure 3-7 Voltage losses in the fuel cell.

In addition, Equation (3-32) may be taken into account, which would account for hydrogen crossover and internal current losses:

$$E_{cell} = E_{r,T,P} - \frac{RT}{\alpha_c F} \ln \left(\frac{i_{ext} + i_{loss}}{i_{0,c}} \right) - \frac{RT}{\alpha_a F} \ln \left(\frac{i_{ext} + i_{loss}}{i_{0,a}} \right)$$

$$- \frac{RT}{nF} \ln \left(\frac{i_{L,c}}{i_{L,c} - i_{ext} - i_{loss}} \right) - \frac{RT}{nF} \ln \left(\frac{i_{L,a}}{i_{L,a} - i_{ext} - i_{loss}} \right)$$

$$- (i_{ext} + i_{loss}) R_{i,j} - i_{ext} (R_{i,e} + R_{i,c})$$

$$(3\text{-}45)$$

An additional complication is that the exchange current densities, as well as hydrogen crossover losses, are proportional to local reactant concentration in the catalyst layer, which decreases with increased reaction rate, that is, increased current density. If a similar argument is used as in Equations (3-35) through (3-40), which assumes a linear decrease of reactant concentration with increased current density, the following equations may be derived:

$$i_{loss}(i_{ext}) = i_{loss}^{ref} \frac{i_L - i_{ext}}{i_L + i_{loss}^{ref}} \qquad (3\text{-}46)$$

and

$$i_0(i_{ext}) = \frac{i_0^{ref}}{i_L} (i_L - i_{ext} - i_{loss}) \qquad (3\text{-}47)$$

where i^{ref}_{loss} is the current loss when the surface concentration is equal to bulk concentration. Similarly, i^{ref}_0 is defined as the exchange current density when the surface concentration is equal to bulk concentration. Note that in case of hydrogen crossover or internal current loss, the surface concentration may not be equal to bulk concentration when there is no external current ($i_{ext} = 0$).

A sufficiently accurate approximation of the fuel cell polarization curve may be obtained by the following equation:

$$E_{cell} = E_{r,T,P} - \frac{RT}{\alpha F} \ln \left(\frac{i}{i_0} \right) - \frac{RT}{nF} \ln \left(\frac{i_L}{i_L - i} \right) - i R_i \qquad (3\text{-}48)$$

that has the same form and same parameters as Equation (3-44) but assumes that the anode losses are negligible compared with the cathode losses. If the anode activation losses cannot be ignored, the fuel cell polarization

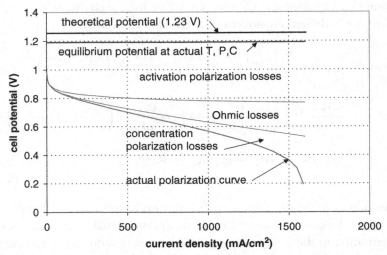

Figure 3-8 Voltage losses in the fuel cell and the resulting polarization curve.

curve may still be expressed with the previous Equation (3-48), but in that case:

$$\frac{1}{\alpha} = \frac{1}{\alpha_a} + \frac{1}{\alpha_c} \quad \text{and} \quad i_0 = i_{0,a}{}^{\alpha/\alpha_a} \cdot i_{0,c}{}^{\alpha/\alpha_c}$$

Figure 3-8 shows how the cell polarization curve is formed by subtracting the activation polarization losses, ohmic losses, and concentration polarization losses from the equilibrium potential. Anode and cathode activation losses are lumped together, but as mentioned before, a majority of the losses occur on the cathode because of sluggishness of the oxygen reduction reaction.

3.4 DISTRIBUTION OF POTENTIAL ACROSS A FUEL CELL

Figure 3-9 illustrates potential distribution in a hydrogen/air fuel cell over the cell cross-section [8]. At open circuit, when there is no current being generated, the anode is at reference or zero potential and the cathode is at the potential corresponding to the reversible potential at given temperature, pressure, and oxygen concentration. As soon as the current is being generated, the cell potential, measured as the difference between cathode and anode solid phase potentials (*solid phase* means electrically conductive parts), drops because of various losses, as discussed earlier.

Note that the cell potential is equal to the reversible cell potential (or the equilibrium potential, E_{eq}) reduced by the potential losses:

$$E_{cell} = E_r - E_{loss} \qquad (3\text{-}49)$$

where the losses are composed of activation and concentration polarization on both anode and cathode and of ohmic losses, as discussed earlier:

$$E_{loss} = (\Delta V_{act} + \Delta V_{conc})_a + (\Delta V_{act} + \Delta V_{conc})_c + \Delta V_{ohm} \qquad (3\text{-}50)$$

The cell potential is equal to the difference between the cathode and the anode solid-state potentials:

$$E_{cell} = E_c - E_a \qquad (3\text{-}51)$$

where the cathode potential is:

$$E_c = E_{r,c} - (\Delta V_{act} + \Delta V_{conc})_c \qquad (3\text{-}52)$$

and the anode potential is:

$$E_a = E_{r,a} + (\Delta V_{act} + \Delta V_{conc})_a$$
$$E_{r,a} = 0 \text{ (by definition)} \qquad (3\text{-}53)$$

All these potentials may be tracked down in Figure 3-9.

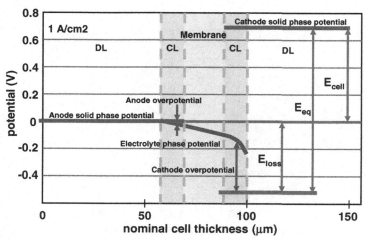

Figure 3-9 Potential distribution in fuel cell cross-section.

3.5 SENSITIVITY OF PARAMETERS IN POLARIZATION CURVE

A *polarization curve* is the most important characteristic of a fuel cell and its performance. It has numerous parameters, even in its shorter version (Equation 3-48):

$$E_{cell} = E_{r,T,P} - \frac{RT}{\alpha F} \ln \left(\frac{i + i_{loss}}{i_0} \right) - \frac{RT}{nF} \ln \left(\frac{i_L}{i_L - i} \right) - iR_i \quad (3-54)$$

It would be useful to see what effect each of the parameters has on the polarization curve shape. As a baseline, the following parameters were selected that resulted in a realistic fuel cell polarization curve (Figure 3-10):

Fuel: Hydrogen
Oxidant: Air
Temperature: 333 K
Pressure: 101.3 kPa (atmospheric)
Gas constant, R: 8.314 J mol^{-1} K^{-1}
Transfer coefficient, α: 1
Number of electrons involved, n: 2
Faraday's constant, F: 96,485 C mol^{-1}
Current loss, i_{loss}: 0.002 A cm^{-2}
Reference exchange current density, i_0: 3 × 10^{-6} A cm^{-2}
Limiting current density, i_L: 1.6 A cm^{-2}
Internal resistance, R_i: 0.15 Ohm-cm^2

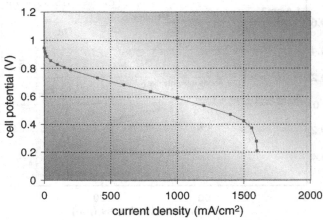

Figure 3-10 A typical fuel cell polarization curve.

When the previous numerical values are plugged into Equation (3-54), the baseline polarization curve shown in Figure 3-10 results.

3.5.1 Effect of Transfer Coefficient/Tafel Slope

The transfer coefficient, α, has a strong effect on fuel cell performance. Although its typical value is about 1 (note that some books use αn instead of just α; in that case $\alpha n \approx 1$), Figure 3-11 shows fuel cell performance with $\alpha = 0.5$ and $\alpha = 1.5$.

Transfer coefficient is the determining factor for the Tafel slope. The Tafel slope is a parameter in Equation (3-25) defined as:

$$b = 2.3 \, \frac{RT}{\alpha F}$$

With the selected numerical values, the Tafel slope is 0.066 V/decade, which is a typical value for hydrogen/oxygen fuel cells. For $\alpha = 0.5$ and $\alpha = 1.5$, the Tafel slope would be 0.132 and 0.044 V/decade, respectively. Figure 3-12 shows the polarization curves for three different values of the Tafel slope. For convenience, the curves are plotted in the log scale with the cell potential corrected for resistive losses, and concentration polarization losses have been neglected. In that case the polarization curve becomes a straight line. Higher Tafel slopes result in lower performance.

3.5.2 Effect of Exchange Current Density

Figure 3-13 shows polarization curves for three different values of exchange current density. For each order of magnitude higher exchange current

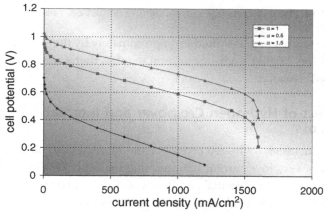

Figure 3-11 Effect of transfer coefficient on fuel cell performance.

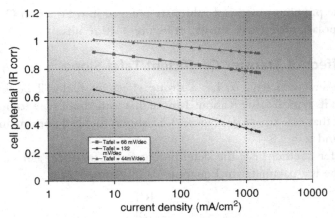

Figure 3-12 Effect of Tafel slope on fuel cell polarization curve.

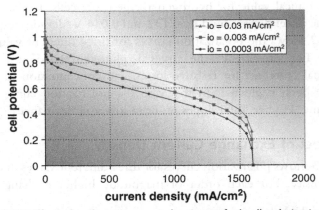

Figure 3-13 Effect of exchange current density on fuel cell polarization curve.

density, the entire curve is shifted up approximately by the value of b, that is, the Tafel slope. Higher exchange current density, therefore, results in better fuel cell performance.

3.5.3 Effect of Hydrogen Crossover and Internal Current Loss

As already discussed, hydrogen crossover and internal current losses have an effect at only very low current densities (Figure 3-14). These losses reduce the cell's open circuit potential and the potential at current densities below 100 mA cm^{-2}. Typical values for hydrogen crossover and internal current losses are several mA cm^{-2}. Even an order of magnitude higher losses do not

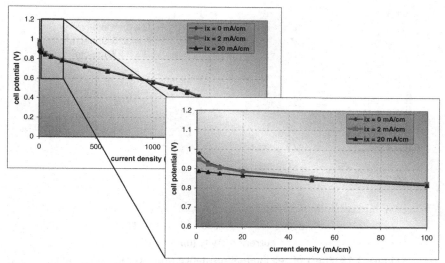

Figure 3-14 Effect of internal currents and/or hydrogen crossover on the fuel cell polarization curve.

have a significantly larger effect on the fuel cell polarization curve at higher current densities.

3.5.4 Effect of Internal Resistance

Resistive or *ohmic losses* are directly proportional to current density. Typical values for internal resistance are between 0.1 and 0.2 Ohm-cm^2, as shown in Figure 3-15. The values above 0.2 Ohm-cm^2 would indicate inadequate selection of cell materials, insufficient contact pressure, or severe drying of the membrane.

3.5.5 Effect of Limiting Current Density

Limiting current density only has an effect at very high current densities approaching the limiting current density (Figure 3-16). At low current densities there is almost no effect, that is, the three polarization curves for three different limiting currents fall on top of each other.

3.5.6 Effect of Operating Pressure

An increase in cell operating pressure results in higher cell potential due to:
1. The Nernst equation:

$$E = E_0 + \frac{RT}{nF} ln\left(\frac{P_{H_2}\ P_{O_2}^{0.5}}{P_{H_2O}}\right) \qquad (3\text{-}55)$$

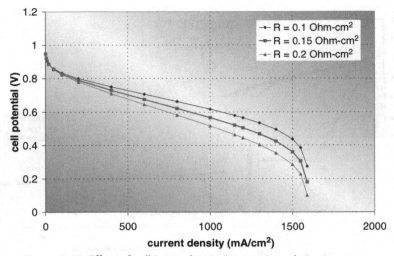

Figure 3-15 Effect of cell internal resistance on its polarization curve.

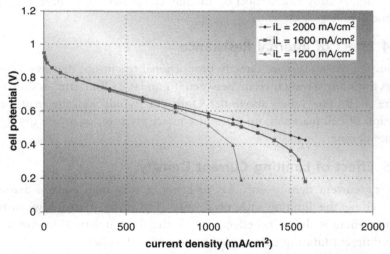

Figure 3-16 Effect of limiting current density on fuel cell polarization curve.

2. An increase in exchange current density due to increased concentration of reactant gases in the electrode(s). Remember, exchange current density is proportional to surface concentration (Equation 3-16), which in turn is directly proportional to pressure. Exchange current density at

pressure different from reference/ambient pressure, as shown in Equation (3-22), is:

$$i_0 = i_{0,P_0} \left(\frac{P}{P_0}\right)^\gamma \qquad (3\text{-}56)$$

An expected cell potential gain at elevated pressure is:

$$\Delta V = \frac{RT}{nF} \ln \left[\left(\frac{P_{H2}}{P_0}\right) \left(\frac{P_{O2}}{P_0}\right)^{0.5} \right] + \frac{RT}{\alpha F} \ln \left(\frac{P}{P_0}\right)^\gamma \qquad (3\text{-}57)$$

For given conditions, an expected gain resulting from a pressure increase from atmospheric to 200 kPa is 34 mV, and from atmospheric to 300 kPa it is 55 mV(assuming $\gamma = 1$). This gain applies to any current density, which results in an elevated polarization curve at elevated pressures, as shown in Figure 3-17. In addition, elevated pressure may have an effect on limiting current density by improving mass transfer of gaseous species. However, operation at elevated pressure with hydrogen/air fuel cells results in additional energy requirements for operation of an air compressor, which may offset the voltage gain (this topic is discussed in Chapter 9, "Fuel Cell System Design"). When both hydrogen and oxygen are supplied from pressurized tanks, there is no additional energy requirement and operation at elevated pressure may be advantageous; the only limitation may be the construction of the fuel cell.

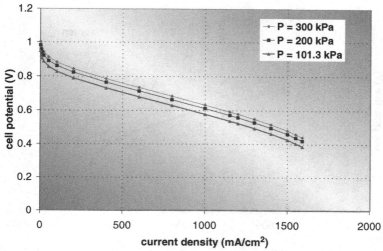

Figure 3-17 Effect of operating pressure on fuel cell polarization curve.

3.5.7 Air vs. Oxygen

A similar effect may be expected if pure oxygen is used instead of air. Because oxygen concentration in air is only 21%, operation with pure oxygen results in a gain similar to elevating the air pressure by a factor of 1/0.21.

An expected gain is:

$$\Delta V = \frac{RT}{nF} ln \left[\left(\frac{1}{0.21} \right)^{0.5} \right] + \frac{RT}{\alpha F} ln \left(\frac{1}{0.21} \right) \qquad (3\text{-}58)$$

At given conditions, the calculated gain is 56 mV. In addition, operation with pure oxygen usually does not result in noticeable concentration polarization; therefore the gain at higher current density is even larger than calculated, 56 mV, as shown in Figure 3-18.

3.5.8 Effect of Operating Temperature

The effect of operating temperature on fuel cell performance cannot be predicted simply by the equations describing the polarization curve derived previously. The temperature appears explicitly and implicitly in every term of the polarization curve, and in some cases the increased temperature may result in voltage gain and in some cases in voltage loss. Increased temperature results in theoretical potential loss due to $T\Delta S/nF$

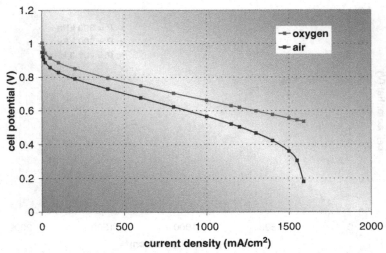

Figure 3-18 Effect of oxygen concentration on fuel cell polarization curve.

(Equation 2.18 and Table 2.4). It also results in a higher Tafel slope, which in turn results in potential loss (Figure 3-12). On the other side, an increased temperature results in higher exchange current density (from Equation 3-22) and better ionic conductivity in the membrane (as will be discussed in Chapter 4) and significantly improves mass transport properties. In addition, at higher temperatures gases may contain larger amounts of water vapor, thus reducing chances of flooding with liquid water. Fuel cell performance usually improves with elevated temperatures, but only up to a certain temperature, which may differ from cell to cell depending on its construction and operation conditions. Figure 3-19 shows the results of an experiment in which the cell temperature gradually increased from −10 °C to 60 °C, and the resulting polarization curves clearly indicate voltage gain with increased temperature [11].

3.6 FUEL CELL EFFICIENCY

The *fuel cell efficiency* is defined as a ratio between the electricity produced and the hydrogen consumed. Of course, the two must be in the same units, such as watts or kilowatts.

$$\eta = \frac{W_{el}}{W_{H2}} \qquad (3\text{-}59)$$

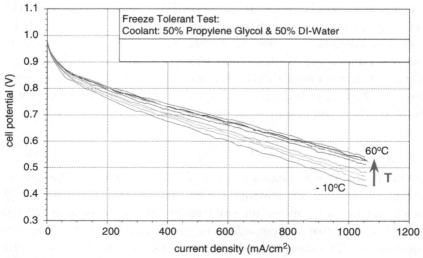

Figure 3-19 Effect of operating temperature on fuel cell polarization curve [11].

Electricity produced is simply a product between voltage and current:

$$W_{el} = I \cdot V \tag{3-60}$$

where I is the current in amperes and V is the cell potential in volts. Hydrogen consumed is (according to Faraday's law) directly proportional to current:

$$N_{H2} = \frac{I}{nF} \tag{3-61}$$

where N_{H2} is in mol s^{-1}, and

$$W_{H2} = \Delta H \frac{I}{nF} \tag{3-62}$$

where:

W_{H2} = energy value of hydrogen consumed in Joules per second (watts)

ΔH = hydrogen's higher heating value (286 k Jmol^{-1})

It should be noted that $\Delta H/nF$ has a dimension of volts, and for $\Delta H = 286$ kJ mol^{-1} it has a value of 1.482 V, which is the so-called *thermoneutral potential*.

By combining Equations (3-59) through (3-62), the fuel cell efficiency is simply directly proportional to cell potential:

$$\eta = \frac{V}{1.482} \tag{3-63}$$

Sometimes, instead of hydrogen's higher heating value (HHV), $\Delta H = 286$ kJ/mol, the lower heating value (LHV) is used ($\Delta H_{LHV} = 241$ kJ/mol). The difference between the higher and lower heating values is the heat of product water condensation. Because the product water may leave the fuel cell in either form, that is, as liquid or as vapor, both values are correct; however, the type of heating value used to calculate the efficiency must be specified. The lower heating value efficiency is:

$$\eta_{LHV} = \frac{V}{1.254} \tag{3-64}$$

If some hydrogen is lost either because of hydrogen diffusion through the membrane, because of combining with oxygen that diffused through the membrane, or because of internal currents, hydrogen consumption will be higher than that corresponding to generated current (as shown in Equation 3-61). Consequently, the fuel cell efficiency would be somewhat lower than given by Equation (3-63). Typically, this loss is very low, on the order of magnitude of a few mA cm^{-2}, and therefore it affects the fuel cell efficiency only at very low current densities (as shown in Figure 3-14). The fuel cell efficiency is then a product of voltage efficiency and current efficiency:

$$\eta = \frac{V}{1.482} \frac{i}{(i + i_{loss})} \tag{3-65}$$

If hydrogen is supplied to the cell in excess of that required for the reaction stoichiometry, this excess will leave the fuel cell unused. In case of pure hydrogen, this excess may be recirculated back into the stack so it does not change the fuel cell efficiency (not accounting for the power needed for the hydrogen recirculation pump), but if hydrogen is not pure (such as in reformate gas feed), unused hydrogen leaves the fuel cell and does not participate in the electrochemical reaction. The fuel cell efficiency is then:

$$\eta = \frac{V}{1.482} \eta_{fu} \tag{3-66}$$

where η_{fu} is fuel utilization, which is equal to $1/S_{H2}$, where S_{H2} is the hydrogen stoichiometric ratio, that is, the ratio between the amount of hydrogen actually supplied to the fuel cell and that consumed in the electrochemical reaction:

$$S_{H2} = \frac{N_{H2,act}}{N_{H2,theor}} = \frac{nF}{I} N_{H2,act} \tag{3-67}$$

Well-designed fuel cells may operate with 83% to 85% fuel utilization when operated with reformate and above 90% when operated with pure hydrogen. Note that the current efficiency term in Equation (3-65) is included in fuel utilization, η_{fu} in Equation (3-66).

3.7 IMPLICATIONS AND USE OF FUEL CELL POLARIZATION CURVE

The polarization curve is the most important characteristic of a fuel cell. It may be used for diagnostic purposes as well as for sizing and controlling a fuel cell. In addition to a potential–current relationship, other information about the fuel cell may also become available just by rearranging the potential–current data.

3.7.1 Other Curves Resulting from Polarization Curve

For example, *power* is a product of potential and current (Equation 3-60). Similarly, power density (in W/cm^2) is a product of potential and current density:

$$w = V \cdot i \qquad\qquad (3\text{-}68)$$

Power density vs. current density may be plotted together with the polarization curve on the same diagram (Figure 3-20). Such a plot shows that there is a maximum power density a fuel cell may reach. It does not make sense to operate a fuel cell at a point beyond this maximum power point because the same power output may be obtained at a lower current and higher potential. Although the graph in Figure 3-20 shows maximum power density of about $0.6\ W/cm^2$, power densities in excess of $1\ W/cm^2$ have been reported with PEM fuel cells.

If cell potential is plotted vs. power density (Figure 3-21), the same information is available—there is a maximum power that the cell can reach. Because the fuel cell efficiency is directly proportional to the cell potential (Equations 3-63 and 3-65), the same graph shows a very useful piece of information: a relationship between the cell efficiency and power density. For a fuel cell with a polarization curve like the one shown in Figure 3-20, the maximum power is reached at efficiency of 33%. This is significantly

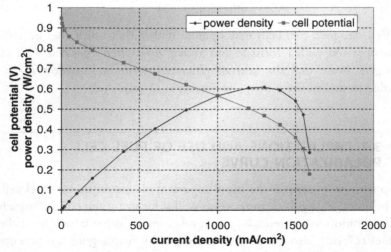

Figure 3-20 Typical fuel cell polarization curve and resulting power curve.

lower than the maximum theoretical fuel cell efficiency of 83%. A higher efficiency may be reached with the same fuel cell but at significantly lower power densities. This means that for a required power output a fuel cell may be made larger (with a larger active area) and more efficient, or more compact but less efficient, by selecting the operating point anywhere on the polarization curve or on the efficiency–power density plot. Typically, a fuel cell is rarely sized anywhere close to the maximum power density. More commonly, the operating point is selected at cell potential around 0.7 V. For the graph in Figure 3-21, this would result in power density of 0.36 W/cm^2 and efficiency of 47%. For applications where a higher efficiency is required, a higher nominal cell potential may be selected (0.8 V or higher), which would result in a 55% to 60% efficient fuel cell, but the power density would be $<0.1 \text{ W/cm}^2$. Similarly, for applications in which fuel cell size is important, a lower nominal cell potential may be selected (around 0.6 V), which would result in a higher power density, that is, a smaller fuel cell.

Although Figure 3-21 shows that fuel cell efficiency above 60% may be possible, albeit at very low current and power densities, in practice that is rarely the case. At very low current densities, hydrogen crossover and internal current losses, although very small, become important and the efficiency–power curve flattens. For the particular case, a maximum efficiency of ~55% is reached (Figure 3-22).

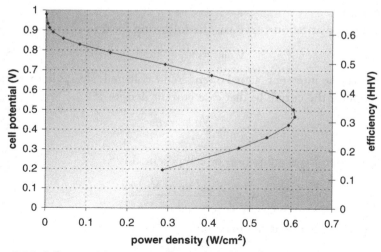

Figure 3-21 Cell potential vs. power density for a fuel cell with the polarization curve from Figure 3-20.

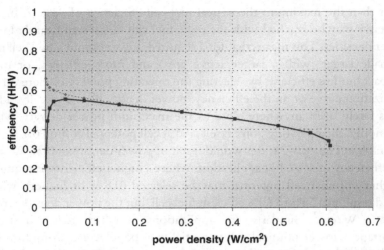

Figure 3-22 Fuel cell efficiency vs. power density curve; solid line with and dashed line without internal current and/or hydrogen crossover losses.

3.7.2 Linear Approximation of Polarization Curve

Sometimes quick calculations regarding fuel cell efficiency–power–size relationships need to be made. Unlike an equation that describes the fuel cell polarization curve (such as Equation 3-54), a linear approximation may be easily manipulated. For most fuel cells and their practical operating range, a linear approximation is actually a very good fit, as shown in Figure 3-23.

A linear polarization curve has the following form:

$$V_{cell} = V_0 - k \cdot i \qquad (3\text{-}69)$$

where V_0 is the intercept (actual open circuit voltage is always higher) and k is the slope of the curve. In that case current density is:

$$i = \frac{V_0 - V_{cell}}{k} \qquad (3\text{-}70)$$

and power density as a function of cell potential is:

$$w = \frac{V_{cell}(V_0 - V_{cell})}{k} \qquad (3\text{-}71)$$

It can be shown that the maximum power density of:

$$w_{max} = \frac{V_0^2}{4k} \qquad (3\text{-}72)$$

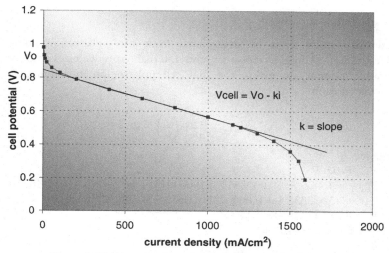

Figure 3-23 Linearization of a fuel cell polarization curve.

is reached at cell potential:

$$V_{cell}\big|_{w_{max}} = \frac{V_0}{2} \qquad (3\text{-}73)$$

3.7.3 Use of Polarization Curve for Fuel Cell Sizing

Example

The H_2/air fuel cell polarization curve is given with the following parameters:

$\alpha = 1$

$i_0 = 0.001$ mA cm^{-2}

$R_i = 0.2$ Ω cm^2

Operating conditions:

T = 60 °C

P = 101.3 kPa

Operating point is selected at 0.6 V.

Active area is 100 cm^2.

a. Calculate nominal power output.

Solution

Power output is:

$W_{el} = V_{cell} \times i \times A$

$V_{cell} = 0.6$ V

$A = 100$ cm^2

$i = ?$

Current density must be determined from the polarization curve. Because no hydrogen crossover and internal current losses and no limiting current are given, the fuel cell polarization curve may be calculated from:

$$V_{cell} = E_r - \frac{RT}{\alpha F} ln\left(\frac{i}{i_0}\right) - iR_i$$

where:

$V_{cell} = 0.6$ V

$E_{r,P,T} = 1.482 - 0.000845T + 0.0000431T \, ln(P_{H2}P_{O2^{0.5}}) = 1.482 - 0.000845 \times 333.15 + 0.0000431 \times 333.15 \times ln(0.21)^{0.5} = 1.189$ V

$R = 8.314$ J mol^{-1}K^{-1}

$T = 60\,°C = 333.15$ K

$\alpha = 1$

$n = 2$

$F = 96,485$ C mol^{-1}

$i_0 = 0.001$ mA cm^{-2}

$R_i = 0.2\ \Omega$ cm^2

Current density cannot be explicitly calculated from the previous equation, but it can be calculated:

- Graphically by plotting the polarization curve
- By iteration
- By linear approximation

This fuel cell's polarization curve with graphical solution for current density at 0.6 V is shown in Figure 3-24.

The resulting current density at 0.6 V is:

$$i = 970 \text{ mA cm}^{-2}$$

Fuel cell power output is then:

$$W_{el} = V_{cell} \times i \times A = 0.6 \times 0.970 \times 100 = 58.2 \text{ W}$$

b. The engineers improved fuel cell performance by improving internal resistance to $R_i = 0.15$ Ohm-cm^2. Calculate power gain at 0.6 V.

Solution

A new polarization curve is shown in Figure 3-25. From there, the new current density is:

$$i = 1.25 \text{ A cm}^{-2}$$

New fuel cell power output is:

$$W_{el} = V_{cell} \times i \times A = 0.6 \times 1.25 \times 100 = 75.0 \text{ W}$$

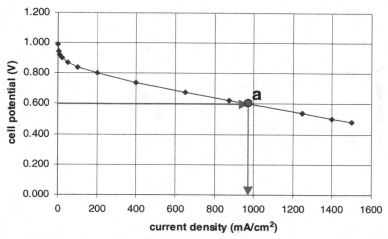

Figure 3-24 Resulting polarization curve and operating point *a* for the example.

Power gain is:

$$\Delta W = 75.0\ W - 58.2\ W = 16.8\ W \text{ or } 28.9\%$$

c. The engineers realized that there was not enough air flow to operate this fuel cell at a higher current density. Calculate the power and efficiency gain if the improved fuel cell is to be operated at the original current density.

Solution

$$V_{cell} = E_r - \frac{RT}{\alpha F}\ ln\left(\frac{i}{i_0}\right) - iR_i$$

$$V_{cell} = 1.189 - \frac{8.314 \times 333.15}{1 \times 96,485}\ ln\left(\frac{970}{0.001}\right) - 0.97 \times 0.15 = 0.648\ V$$

New fuel cell power output is:

$$W_{el} = V_{cell} \times i \times A = 0.648 \times 0.97 \times 100 = 62.9\ W$$

Power gain in respect to the original fuel cell is:

$$\Delta W = 62.9\ W - 58.2\ W = 4.7\ W \quad \text{or} \quad 8\%$$

The efficiency before improvement was:

$$\eta = V_{cell}/1.482 = 0.6/1.482 = 0.405$$

The efficiency after improvement is:

$$\eta = V_{cell}/1.482 = 0.648/1.482 = 0.437$$

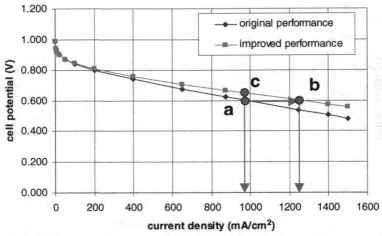

Figure 3-25 Improved polarization curve and operating points *b* and *c* for the example.

d. The engineers realized that there is no need for additional power. Calculate the efficiency gain if the improved fuel cell is to be operated at the original power output.

Fuel cell power output is:

$$58.2 = V_{cell} \times i \times 100$$

Another V_{cell}–i relationship is obtained from the polarization curve:

$$V_{cell} = 1.189 - \frac{8.314 \times 333.15}{1 \times 96,485} \, ln\left(\frac{i}{0.001}\right) - i \times 0.15$$

These two equations with two unknown, namely V_{cell} and i, may be solved by iteration or graphically, and the solution is:

$V_{cell} = 0.666$ V

$i = 875$ mA cm^{-2}

Note that points *a* and *d* (the original and new operating points) in Figure 3-26 lie on two different polarization curves but on the same constant power line.

The new efficiency is:

$$\eta = V_{cell}/1.482 = 0.666/1.482 = 0.449$$

Therefore, an improvement of internal cell resistance from $0.2\,\Omega$ cm^2 to $0.1\,\Omega$ cm^2 for this fuel cell resulted in an efficiency increase from 0.405 to 0.449 (or about 10%) while keeping the same power output of 58.2 W.

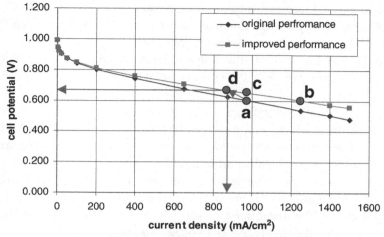

Figure 3-26 Operating point *d* for the example.

PROBLEMS

1. A fuel cell operates at $60\,°C$ and ambient pressure. Exchange current density, i_0, is $0.005\ mA/cm^2$. Assume charge transfer coefficient to be 1.0.
   a. Calculate the activation voltage losses at $1.5\ A/cm^2$.
   b. If the resistive losses at $1.5\ A/cm^2$ are ½ of the activation losses, calculate the cell resistance, R_i.
   c. Assume no concentration polarization losses and calculate cell voltage at $1.5\ A/cm^2$, at $1\ A/cm^2$, and at $0.5\ A/cm^2$.
2. A fuel cell voltage of 0.8 V was measured at $0.2\ A/cm^2$ and 0.6 V at $0.8\ A/cm^2$. The size of the fuel cell is $100\ cm^2$. Approximate the fuel cell with a linear polarization curve ($V = V_0 - k \cdot i$). Calculate:
   a. Maximum power output
   b. Cell voltage and current density at ½ maximum power
   c. Efficiency at maximum power
   d. Efficiency at ½ maximum power
3. Cathode exchange current density is $1 \times 10^{-10}\ A/cm^2$ of Pt surface (measured with oxygen at $25\,°C$ and atmospheric pressure). Calculate the expected current density at 0.9 V (iR corrected) if an MEA is prepared with catalyst specific surface area of $640\ cm^2/mg$ and with Pt

loading of 0.4 mg/cm^2, and the cell operates with H$_2$/Air at 60 °C and 300 kPa. What potential gain may be expected at same current density if the Pt loading on the cathode is increased to 2 mg/cm^2?

4. An H$_2$/O$_2$ fuel cell has the following polarization curve parameters: $i_0 = 0.003$, $\alpha = 0.5$, $R_i = 0.15$ Ohm-cm^2. The fuel cell operates at 65 °C and 1 bar.
   a. Calculate the cell voltage at 1 A/cm^2.
   b. Calculate voltage gain if the cell is to be operated at 6 bar (take into account both the ideal voltage gain and increase in exchange current density proportional to pressure increase).

5. An H$_2$/air fuel cell operates at 80 °C and 1 atm. Exchange current density at these conditions is 0.0012 mA/cm^2 of the electrode area. Pt loading is 0.4 mg/cm^2. The charge transfer coefficient is 1.0. Electrode area is 100 cm^2.
   a. Calculate the theoretical cell voltage for these conditions.
   b. When the voltmeter is connected to this fuel cell, it shows an open circuit voltage of 0.975 V. Calculate current density loss due to hydrogen crossover or due to internal currents.

6. Hydrogen flow rate into a 100-cm^2 fuel cell is 0.0018 g/s. What would be the maximum current that could be produced by this fuel cell?.

7. A 100-cm^2 hydrogen/oxygen fuel cell operates at 0.5 A/cm^2 and 0.7 V. Hydrogen flow rate is kept proportional to current generated at stoichiometric ratio of 1.5. Hydrogen crossover and internal current loss account for 2 mA/cm^2.
   a. Calculate the fuel cell efficiency at these conditions.
   b. Calculate the hydrogen flow rate at the entrance.
   c. Calculate the hydrogen flow rate at the fuel cell exit.

QUIZ

1. The largest voltage losses in a fuel cell in normal operation are due to:
   a. Activation
   b. Concentration/mass transport difficulties
   c. Resistance

2. Higher exchange current density:
   a. Means more voltage losses
   b. Means less voltage losses
   c. Has nothing to do with voltage losses

3. Concentration polarization means:
   a. Concentration of reactants at the catalyst site is too high
   b. Reactants reach the catalyst site at an insufficient rate
   c. Reactant flow rate is higher than it should be
4. Resistance in a fuel cell is:
   a. Ionic resistance through the membrane
   b. Electric resistance through electrically conductive parts
   c. Both ionic and electric
5. Leak of hydrogen through a membrane would:
   a. Reduce the fuel cell voltage, particularly at low or no external currents
   b. Increase the cell voltage
   c. Increase the cell current
6. Pressure increase in a fuel cell typically results in:
   a. Voltage gain
   b. Voltage loss
   c. Lower efficiency
7. Higher Pt loading typically:
   a. Does not change the cell voltage
   b. Results in voltage loss
   c. Results in voltage gain
8. iR corrected cell voltage is:
   a. Cell voltage plus voltage loss attributable to resistance
   b. Cell voltage minus voltage loss attributable to resistance
   c. Resistance loss plus activation voltage losses
9. Selecting an operating point on the polarization curve of a fuel cell at a higher voltage means:
   a. More power output than the same size fuel cell with a lower selected nominal cell voltage
   b. Larger fuel cell for the same power output
   c. Higher current density
10. A fuel cell runs on H_2 and air and generates 10 W. If pure O_2 is used to replace the air, and current and voltage are adjusted so that the fuel cell continues to deliver 10 W, the fuel cell would:
    a. Operate at higher current than before O_2 introduction
    b. Operate at lower current than before O_2 introduction
    c. Initially show an increase in cell potential, but then it would come back to the original current and cell voltage so that the power output does not change

REFERENCES

[1] Chen E. Thermodynamics and Electrochemical Kinetics. In: Hoogers G, editor. Fuel Cell Technology Handbook. Boca Raton, FL: CRC Press; 2003.

[2] Atkins PW. Physical Chemistry. Sixth ed. Oxford University Press, Oxford; 1998.

[3] Gileadi E. Electrode Kinetics for Chemists, Chemical Engineers and Material Scientists. New York: VCH Publishers; 1993.

[4] Bockris J O'M, Srinivasan S. Fuel Cells: Their Electrochemistry. New York: McGraw-Hill; 1969.

[5] Bard AJ, Faulkner LR. Electrochemical Methods. New York: John Wiley & Sons; 1980.

[6] Larminie J, Dicks A. Fuel Cell Systems Explained. Second ed. Chichester, England: John Wiley & Sons; 2003.

[7] Newman JS. Electrochemical Systems. Second ed. NJ: Prentice Hall, Englewood Cliffs; 1991.

[8] Gasteiger HA, Gu W, Makharia R, Matthias MF. Catalyst Utilization and Mass Transfer Limitations in the Polymer Electrolyte Fuel Cells, Tutorial. Orlando, FL: Electrochemical Society Meeting; 2003.

[9] Barbir F, Braun J, Neutzler J. Properties of Molded Graphite Bi-Polar Plates for PEM Fuel Cells. International Journal on New Materials for Electrochemical Systems 1999:197–200. No. 2.

[10] Kim J, Lee S-M, Srinivasan S, Chamberlain CE. Modeling of Proton Exchange Fuel Cell Performance with an Empirical Equation. Journal of Electrochemical Society 1995; Vol. 142:2670–4. No. 8.

[11] Fuchs M, Barbir F. Development of Advanced, Low-Cost PEM Fuel Cell Stack and System Design for Operation on Reformate Used in Vehicle Power Systems, Transportation Fuel Cell Power Systems, 2000 Annual Progress Report. Washington, DC: U.S. Department of Energy, Office of Advanced Automotive Technologies; October 2000. 79–84.

CHAPTER FOUR

Main Cell Components, Material Properties, and Processes

4.1 CELL DESCRIPTION

The heart of a fuel cell is a *polymer*, a proton-conductive membrane. On each side of the membrane is a porous electrode. The electrodes must be porous because the reactant gases are fed from the back and must reach the interface between the electrodes and the membrane, where the electrochemical reactions take place in the so-called catalyst layers or, more precisely, on the catalyst surface. Technically, the catalyst layer may be a part of the porous electrode or part of the membrane, depending on the manufacturing process. The multilayer assembly of the membrane sandwiched between the two electrodes is commonly called the *membrane electrode assembly,* or MEA. The MEA is then sandwiched between the collector/separator plates—*collector* because they collect and conduct electrical current, and *separator* because in multicell configuration they separate the gases in the adjacent cells. At the same time, in multicell configuration they physically/electrically connect the cathode of one cell to the anode of the adjacent cell, and that is why they are also called the *bipolar plates.* They provide the pathways for flow of reactant gases (so-called *flow fields*), and they also provide the cell's structural rigidity.

The following processes take place inside the fuel cell (the numbers correspond to those in Figure 4-1):

1. Gas flow through the channels; some convective flows may be induced in the porous layers
2. Gas diffusion through porous media
3. Electrochemical reactions, including all the intermediary steps
4. Proton transport through proton-conductive polymer membrane
5. Electron conduction through electrically conductive cell components
6. Water transport through polymer membrane including both electrochemical drag and back diffusion
7. Water transport (both vapor and liquid) through porous catalyst layer and gas diffusion layers

PEM Fuel Cells
ISBN 978-0-12-387710-9

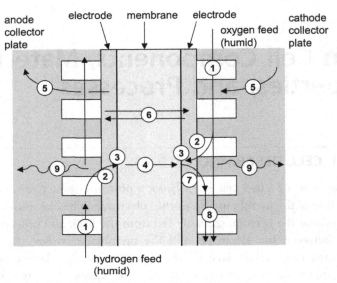

anode   electrode   membrane   electrode   cathode
collector   oxygen feed   collector
plate   (humid)   plate

hydrogen feed
(humid)

Figure 4-1 Main cell components (not in scale) and processes.

8. Two-phase flow of unused gas carrying water droplets
9. Heat transfer, including both conduction through solid components of the cell and convection to reactant gases and cooling medium

Obviously, the design of the components and properties of materials must accommodate these processes with minimum obstruction and losses. Because in some of the components more than one process takes place, very often with conflicting requirements, the properties and the design must be optimized. For example, the gas diffusion layer must be optimized so that the reactant gas may easily diffuse, yet at the same time that water, which travels in the opposite direction, does not accumulate in the pores. On top of that, the diffusion layer (or current collector layer, as it is sometimes called) must be both electrically and thermally conductive. Similar requirements may be established for almost every fuel cell component. Although a fuel cell seems to be a very simple device, numerous processes take place simultaneously. It is therefore important to understand those processes, their mutual interdependence, and their dependence on component design and material properties.

First fuel cell law: One cannot change only one parameter in a fuel cell; change of one parameter causes a change in at least two other parameters, and at least one of them has an opposite effect of the one expected to be seen.

4.2 MEMBRANE

A fuel cell membrane must exhibit relatively high proton conductivity, must present an adequate barrier to mixing of fuel and reactant gases, and must be chemically and mechanically stable in the fuel cell environment [1]. Typically, the membranes for PEM fuel cells are made of *perfluorocarbonsulfonic acid* (PSA) ionomer. This is essentially a copolymer of tetrafluorethylene (TFE) and various perfluorosulfonate monomers. The best-known membrane material is Nafion,™ made by Dupont, which uses perfluorosulfonylfluoride ethyl-propyl-vinyl ether (PSEPVE). Figure 4-2 shows the chemical structure of a perfluorosulfonate ionomer such as Nafion. Similar materials have been developed and sold as either a commercial or development product by other manufacturers such as Fumatech (Fumion), Asahi Glass (Flemion), Asahi Chemical (Aciplex), Chlorine Engineers ("C" membrane), and Dow Chemical. W. L. Gore and Associates have developed a composite/reinforced membrane made up of a Teflon-like component that provides mechanical strength and dimensional stability and a perfluorosulfonic acid component that provides protonic conductivity.

The SO_3H group is ionically bonded, so the end of the side chain is actually an SO^-_3 ion with an H^+ ion. This is why such a structure is called an *ionomer*. Because of their ionic nature, the ends of the side chains tend to cluster within the overall structure of the membrane. Although the Teflon-like backbone is highly hydrophobic, the sulphonic acid at the end of the side chain is highly hydrophyllic. The hydrophyllic regions are created around the clusters of sulphonated side chains. This is why this kind of material absorbs relatively large amounts of water (in some cases up to 50% by weight). H^+ ions' movement within well-hydrated regions makes these materials proton conductive.

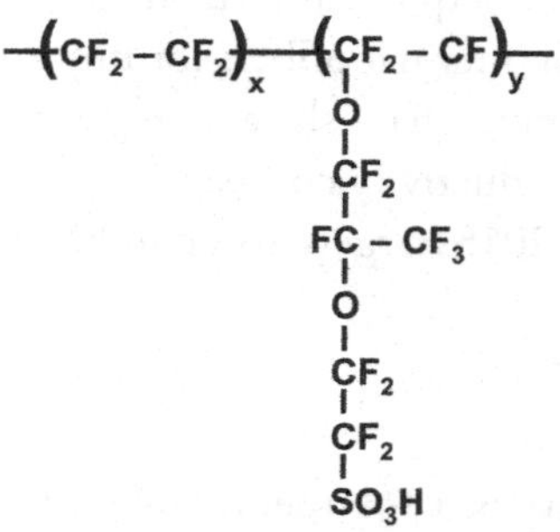

$$-\!\!\left(CF_2-CF_2\right)_x\!\!-\!\!\left(CF_2-CF\right)_y\!\!-$$
$$|$$
$$O$$
$$|$$
$$CF_2$$
$$|$$
$$FC-CF_3$$
$$|$$
$$O$$
$$|$$
$$CF_2$$
$$|$$
$$CF_2$$
$$|$$
$$SO_3H$$

Figure 4-2 Structure of PFSA polymer (Nafion™).

Nafion membranes come extruded in different sizes and thicknesses. They are marked with a letter N, followed by a three- or four-digit number. The first two digits represent equivalent weight divided by 100, and the last digit or two is the membrane thickness in mills (1 mill $= 1/1000$ inch $= 0.0254$ mm). Nafion is available in several thicknesses, namely 2, 3.5, 5, 7, and 10 mills (50, 89, 127, 178, 254 μm, respectively). For example, Nafion N117 has equivalent weight of 1100 and is 7 mills (0.178 mm) thick. The equivalent weight (EW) in geq^{-1} of a polymer membrane can be expressed by the following equation:

$$EW = 100\,n + 446 \tag{4-1}$$

where n is the number of TFE groups on average per PSEPVE monomer [2].

EW is practically a measure of ionic concentration within the ionomer. Typical EW for Nafion membranes is 1100, although materials with EW as low as 700 have been synthesized and studied. Copolymers with EW greater than approximately 1500 geq^{-1} do not have ionic conductivity sufficient for practical fuel cell applications, whereas those with EW lower than 700 usually have poor mechanical integrity.

Since 2004, chemically stabilized versions of the ionomer have been obtained by fluorination of end groups according to proprietary procedures [3]. The membranes made with this improved ionomer exhibit substantially lower fluoride ion release compared to the unstabilized polymer—a sign of improved chemical durability. They are produced by dispersion casting and are being sold as Nafion NR211 and NR212, thus 1 and 2 mill thick, respectively (25 and 50 μm, respectively). The equivalent weight is between 990 and 1050 [3]. Properties of Nafion N115, N117, N1110, NR211, and NR212 are shown in Table 1 [4,5].

3M has developed a new ionomer, perfluoroimide acid (PFIA), which has very low EW (625 geq-1) and stabilizing additives for improved chemical stability and polymer nanofibers for improved mechanical stability [6]. This new membrane has shown superior mechanical stability, chemical stability, and conductivity compared to other available membranes. It has met U.S. DOE 2015 targets for conductivity and other physical properties.

4.2.1 Water Uptake

The protonic conductivity of a polymer membrane is strongly dependent on membrane structure and its water content. The water content in a membrane is usually expressed as grams of water per gram of polymer dry weight or as number of water molecules per sulfonic acid groups present in the polymer, $\lambda = N(H_2O)/N(SO_3H)$. The maximum amount of water in the membrane strongly depends on the state of water used to equilibrate the membrane. It has been noticed that a Nafion membrane equilibrated with liquid water (i.e., boiled in water) takes roughly up to 22 water molecules per sulfonate group, whereas the maximum water uptake from the vapor phase is only about 14 water molecules per sulfonate group. In addition, water uptake from the liquid phase is dependent on the membrane pretreatment. Zawodzinski et al. [7,8] have shown that water uptake after the membrane had been completely dried out at 105 °C is significantly smaller than if the membrane had been dried out at room temperature; $\lambda = 12$ to 16, depending on the temperature of rehydration for the membrane previously dried out at 105 °C vs. $\lambda = 22$ and independent of the temperature of rehydration for the membrane previously dried at room temperature. This may be explained by the polymer morphological changes at elevated temperatures. Indeed, for an experimental Dow membrane, which has a somewhat higher glass transition temperature than Nafion, the effect of drying the membrane at 105 °C is less pronounced than for a Nafion membrane, that is, after rehydration at 80 °C the membrane exhibited the same water uptake ($\lambda = 25$) as the membrane previously dried out at room temperature.

Water uptake from the vapor phase may be more relevant for fuel cell operation, where the reactant gases are humidified and water is present in the vapor phase. The shape of a generic isotherm for ion-exchange polymers is shown in Figure 4-3. It may be noticed that there are two distinct steps in water sorption from the gas phase, namely:
1. At the low vapor activity region, $a_{H2O} = 0.15–0.75$, water uptake increases to about $\lambda = 5$, and
2. At the high vapor activity region, $a_{H2O} = 0.75–1.0$, water uptake increases sharply to about 14.4.

The first step corresponds to uptake of water by solvation via the ions in the membrane, whereas the second step corresponds to water that fills the pores and swells the polymer. It is important to notice that the resulting water uptake from the fully saturated vapor phase (with $a_{H2O} = 1$) is significantly lower than that from the liquid phase (also $a_{H2O} = 1$), that is, $\lambda = 14$ vs. $\lambda = 22$, respectively. This phenomenon was first reported in 1903 by Schroeder and is therefore called *Schroeder's paradox* [1]. A possible explanation of this difference in uptake from vapor and liquid phases is that

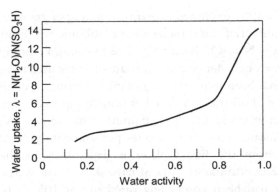

Figure 4-3 Water uptake in proton conductive membranes at 30 °C [9].

sorption from the vapor phase involves condensation of water inside the polymer, most probably on the strongly hydrophobic polymer backbone, and the resulting uptake is lower than if sorption and imbibition occurred directly from the liquid phase [1].

Based on experimental results, Zawodzinski et al. [8] fit a polynomial equation to obtain a relationship between water activity on the faces of the membrane and water content:

$$\lambda = 0.043 + 17.18\,a - 39.85\,a^2 + 36\,a^3 \qquad (4\text{-}2)$$

where a is water vapor activity. Assuming that the gas mixture behaves as an ideal gas, the water vapor activity can be replaced by relative humidity, thus p/p_{sat}, where p is water partial pressure, and p_{sat} is the saturation pressure at a given temperature.

4.2.2 Physical Properties

Water uptake results in the membrane swelling and changes its dimensions, which is a very significant factor for fuel cell design and assembly. From Table 4-1, which shows some critical Nafion membrane properties at various water contents [4,5], we can see that the dimensional changes are in the order of magnitude of 10%, which must be taken into account in cell design and during the installation of the membrane in the cell.

In 1995, W. L. Gore and Associates, Inc., introduced the Gore–Select membrane, a new microreinforced *expanded polytetrafluorethylene* (ePTFE) polymer electrolyte membrane targeted specifically at PEMFC applications [9]. The microreinforcements allow Gore–Select membranes to utilize ionomers that do not have sufficient mechanical properties

Table 4-1 Properties of Nafion™ Membranes[4,5]

Property	Type of Membrane		
	N115, N117, N1110		NR211, NR212
Density, g cm^{-3}	1,98		1,97
Tensile modulus, MPa			
50% RH, 23 °C	249		
Water soaked, 23 °C	114		
Water soaked, 100 °C	64		
Tensile strength,[1] MPa			
50% RH, 23 °C	43 (N115) MD	32 (N115) TD	23 (NR211) MD 28 (NR211) TD
			32 (NR212) MD 32 (NR212) TD
Water soaked, 23 °C	34 (N115) MD	26 (N115) TD	
Water soaked, 100 °C	25 (N115) MD	24 (N115) TD	
Conductivity	0.10 [2]		0.105 @ 25 °C − 0.116 @ 100 °C [3]
Ionic exchange capacity,[4] meq g^{-1}	0.91		0.95 − 1.01
Equivalent weight,[3] g eq^{-1}	1100		990 − 1050

(Continued)

Table 4-1 Properties of Nafion™ Membranes[4,5]—cont'd

Property	Type of Membrane	
	N115, N117, N1110	NR211, NR212
Hydrogen crossover,[5] ml·min⁻¹·cm⁻²	5	(NR211) < 0.020 (NR212) < 0.010
Water content,[6] % water		5 ± 3%
Water uptake,[7] % water	38	50 ± 3%
Thickness change, % increase		
From 50%RH, 23°C to Water soaked 23°C	10	
From 50%RH, 23°C to Water soaked 100°C	14	
Linear expansion,[8] % increase		
From 50%RH, 23°C to Water soaked 23°C	10	10
From 50%RH, 23°C to Water soaked 100°C	15	15

Footnotes:
[1] MD Machine direction, TD Transversal direction; measurements taken at 23°C and 50%RH.
[2] Conductivity measurements as described by Zawodzinski et al. [10] Membrane conditioned in 100°C water for 1 hour. Measurement cell immersed in 25°C deionized water during experiment. Membrane impedance (real) taken at zero imaginary impedance.
[3] Not reported by Dupont. Data taken from [3].
[4] A base titration procedure measures the equivalence of sulfonic acid in the polymer and used the measurements to calculate acid capacity or equivalent weight of the membrane.
[5] Hydrogen crossover measured at 22°C 100%RH and 344.7 kPa (50 psi) pressure difference.
[6] Water content in the membrane conditioned at 23°C and 50%RH compared to dry weight basis.
[7] Water uptake from dry membrane to water soaked at 100°C for 1 hour (dry weight basis).
[8] Average of MD and TD values; MD expansion is slightly less than TD for N-type membranes and similar to TD for NR-type membranes.

(i.e., ionomers with EW < 1,000). These membranes exhibit much higher strength, better dimensional stability, lower gas permeability, and higher conductivity than comparable nonreinforced Nafion membranes [9].

4.2.3 Protonic Conductivity

Protonic conductivity is the most important function of the polymer membranes used in fuel cells. The charge carrier density in an ionomeric proton conductive membrane of EW 1100 is similar to that in 1 M aqueous sulfuric acid solution. Remarkably, proton mobility in a fully hydrated membrane is only one order of magnitude lower than the proton mobility in aqueous sulfuric acid solution. As a result, the protonic conductivity of a fully hydrated membrane is about 0.1 S cm^{-1} at room temperature. The conductivity of PFSA membranes is a strong function of water content and temperature (Figures 4-4 and 4-5). Above $\lambda = 5$, the relationship between water content and protonic conductivity is almost linear. Below $\lambda = 5$, there is very little water uptake (Figure 4-3), which may suggest that there is not enough water in the clusters around the ends of the sulphonated side chains and because of that, protons are

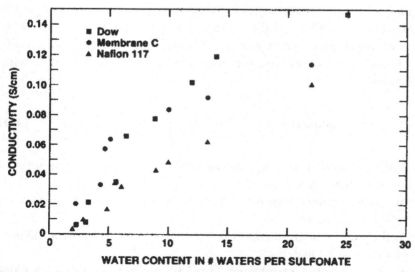

Figure 4-4 Conductivity of various proton conductive membranes at 30 °C as a function of state of membrane hydration [8]. *(Reprinted by permission of the Electrochemical Society.)*

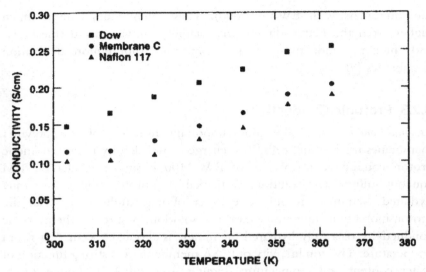

Figure 4-5 Conductivity of various proton conductive membranes immersed in water as a function of temperature [8]. *(Reprinted by permission of the Electrochemical Society.)*

sequestered by the sulphonate groups. Note that conductivity at $\lambda = 14$ (membrane equilibrated with water vapor) is about 0.06 S cm^{-1}. Protonic conductivity dramatically increases with temperature (Figure 4-5) and at 80 °C reaches 0.18 S cm^{-1} for a membrane immersed in water. Based on these measurements, Springer et al. [11] correlated the ionic conductivity (in S cm^{-1}) to water content and temperature with the following expression:

$$\kappa = (0.005139\,\lambda - 0.00326)exp\left[1268\left(\frac{1}{303} - \frac{1}{T}\right)\right] \qquad (4\text{-}3)$$

Zawodzinski [12] has suggested that there are several possible ways of ionic conductivity of Nafion-like materials (Figure 4-6):
- At very low water contents ($\lambda \sim 2$–4), hydronium ions (H_3O^+) move via vehicle mechanism.
- As the water content increases ($\lambda \sim 5$–14), easier movement of hydronium ions is facilitated.
- At fully hydrated membranes ($\lambda > 14$), water in interfacial regions screens weakly bound water from ion–dipole interactions, and both water and ions move freely.

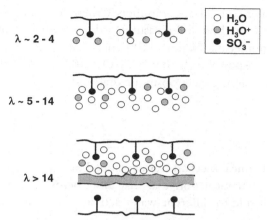

Figure 4-6 Mechanisms of water and hydronium ion movement through a PFSA ionomer at various hydration levels, as suggested by Zawodzinski [12].

4.2.4 Water Transport

There are several mechanisms of water transport across a polymer membrane. Water is produced on the cathode side as the result of the electrochemical reaction (Equation 2.2). The rate of water generation (in $mols^{-1}cm^{-2}$) is:

$$N_{H2O,gen} = \frac{i}{2F} \qquad (4\text{-}4)$$

where i is current density (A/cm^2), and F is Faraday's constant.

Water is dragged from the anode to the cathode by protons moving through the electrolyte, as described previously. This is called *electroosmotic drag*. The flux of water due to electroosmotic drag (in $mols^{-1}cm^{-2}$) is:

$$N_{H2O,drag} = \xi(\lambda)\frac{i}{F} \qquad (4\text{-}5)$$

where ξ is the electroosmotic drag coefficient defined as number of water molecules per proton. In general, this coefficient is a function of membrane hydration (λ). This number has been reported over the years with considerable variation [1], depending on the method of measurement and data fitting. One way to measure the electroosmotic drag is to pass current through the membrane and monitor the level of a water column [13]. Using this method, Le Conti et al. [13] reported drag coefficients in the range of two to three water molecules per proton for membrane water content in the range of $15 \leq \lambda \leq 25$. They concluded that the drag coefficient decreases linearly with water content of the immersed membranes. Zawodzinski et al. [9] used the same method and measured a drag coefficient of 2.5 for fully hydrated and immersed Nafion 1100 membranes. At $\lambda = 11$ the drag coefficient was measured to be 0.9 $N(H_2O)/N(H^+)$. The following linear relationship between the electroosmotic drag and membrane water content is proposed [8]:

$$\zeta(\lambda) = \frac{2.5\,\lambda}{22} \qquad (4\text{-}6)$$

An electrochemical method was developed by Fuller and Newman [14] based on electrochemical potential that arises across a membrane sample exposed at each side to different water activities. In that case the electrochemical drag is:

$$\xi(\lambda) = \frac{F\Delta\Phi}{RT\,log\dfrac{a_{H2O,r}}{a_{H2O,1}}} \qquad (4\text{-}7)$$

where $\Delta\Phi$ is the measured potential, a_{H2O} is water activity, and subscripts r and l refer to the right and left sides of the membrane. This method is suitable for vapor-equilibrated membranes. Fuller and Newman [14] reported an essentially constant drag coefficient of 1.4 $N(H_2O)/N(H^+)$ in the range of $5 \le \lambda \le 14$. In the range $0 \le \lambda \le 5$, the coefficient gradually drops to zero. Zawodzinski et al. [15] used a wider range of water activities and found a drag coefficient of 1.0 $N(H_2O)/N(H^+)$ in the range of $1.4 \le \lambda \le 14$.

Water generation and electroosmotic drag would create a large concentration gradient across the membrane. Because of this gradient, some water diffuses back from the cathode to the anode. The rate of water diffusion (in $mol\,s^{-1}cm^{-2}$) is:

$$N_{H2O,diff} = D(\lambda)\frac{\Delta c}{\Delta z} \qquad (4\text{-}8)$$

where D is the water diffusion coefficient in ionomer of water content λ, and $\Delta c/\Delta z$ is a water concentration gradient along the z-direction (through the membrane). Measurement of diffusion coefficient of water through a polymer membrane is not simple. Several methods have been applied, such as:

• Water uptake dynamics, used by Yeo and Eisenberg [16], which resulted in diffusion coefficient in the range of 1×10^{-6} to 10×10^{-6} $cm^2\,s^{-1}$, increasing with temperature in the range of 0–99 °C, with activation energy of 18.8 kJ mol^{-1}. Similar results were also reported by Eisman [17].

- Radiotracer and electrochemical techniques, developed by Verbrugge and coworkers [18], which resulted in self-diffusion coefficient of water in the range of 6–10 × 10^{-6} cm^2 s^{-1} in fully hydrated Nafion membranes at room temperature.

- Pulsed field gradient neutron magnetic resonance (NMR) used by Slade et al. [19] and Zawodzinski et al. [10] resulted in self-diffusion coefficients of water close to 10 × 10^{-6} cm^2 s^{-1} for fully hydrated Nafion samples. Zawodzinski et al. [10] also measured the self-diffusion coefficients in Nafion membranes equilibrated with water vapor and found that the diffusion coefficient decreases from 6 × 10^{-6} cm^2 s^{-1} to 0.6 × 10^{-6} cm^2 s^{-1} as water content in the membrane decreased from $\lambda = 14$ to $\lambda = 2$. (These measurements were conducted at 30 °C.)

It should be noted that the radiotracer and pulsed field gradient NMR techniques measure the self-diffusion coefficient of water, D_S, rather than the Fickian or interdiffusion coefficient of water through the polymer membrane, D, and some correction is required because it is the Fickian water diffusion coefficient that is the proper transport property to use in macroscopic studies of water diffusion [20]. The relationship is:

$$D = \frac{\partial(ln\ a)}{\partial(ln\ C_w)}\ D_S \qquad (4\text{-}9)$$

where a is the thermodynamic activity of water and C_w, $mol\ cm^{-3}$ is water concentration in the membrane:

$$C_w = \frac{\rho_m}{EW}\ \lambda \qquad (4\text{-}10)$$

where ρ_m is membrane density, $g\ cm^{-3}$, and EW, $g\ eq^{-1}$, is polymer-equivalent weight.

Motupally et al. [21] studied water transport in Nafion membranes and, by comparing the variation in literature values of water diffusion coefficients with their own experiments and the results of Zawodzinski et al. [10], suggested the following relationships:

$$D(\lambda) = 3.1 \times 10^{-3}\ \lambda(e^{0.28\ \lambda} - 1)exp\left(\frac{-2436}{T}\right) \quad \text{for}\ \ 0 < \lambda < 3$$

$$(4\text{-}11)$$

$$D(\lambda) = 4.17 \times 10^{-4} \, \lambda (161 \, e^{-\lambda} + 1) exp\left(\frac{-2436}{T}\right) \quad \text{for} \quad 3 < \lambda < 17$$

$$(4\text{-}12)$$

Based on observation that Equation (4-10) overestimates back diffusion relative to electroosmotic drag, Nguyen and White [22] suggested another relationship:

$$D(\lambda) = \left(0.0049 + 2.02 \, a - 4.53 \, a^2 + 4.09 \, a^3\right) D^0 exp\left(\frac{2416}{303} - \frac{2416}{T}\right)$$

for a $\leq$ 1

$$(4\text{-}13)$$

$$D(\lambda) = [1.59 + 0.159(a - 1)] D^0 \, exp\left(\frac{2416}{303} - \frac{2416}{T}\right) \quad \text{for} \quad a > 1$$

$$(4\text{-}14)$$

with the value for D^0 suggested to be 5.5×10^{-7} cm^2 s^{-1}.

Husar et al. [23] experimentally found that the actual diffusion coefficient in a PEM fuel cell is closer to the Nguyen and White [22] relationship (Equation 4-13).

In addition to diffusion due to concentration gradient, water may be hydraulically pushed from one side of the membrane to the other if there is a pressure difference between the cathode and the anode. The rate of hydraulic permeation (in mols^{-1}cm^{-2}) is [1]:

$$N_{H2O,hyd} = k_{hyd}(\lambda)\frac{\Delta P}{\Delta z} \qquad (4\text{-}15)$$

where k_{hyd} is the hydraulic permeability coefficient of the membrane of water content λ, and $\Delta P/\Delta z$ is a pressure gradient along the z-direction (through the membrane).

For a thin membrane, water back diffusion may be sufficient to counteract the anode-drying effect due to the electroosmotic drag. However, for a thicker membrane, drying may occur on the anode side. This was very vividly demonstrated by Büchi and Scherer [24], who created thick membranes by combining several layers of Nafion membranes. They showed that the membrane resistance is independent of current density for the membranes up to 120 μm, but it does increase for thicker membranes (Figure 4-7).

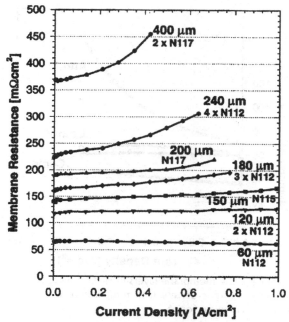

Figure 4-7 *In situ* resistance of Nafion membranes with different thicknesses as function of current density (cell temperature 60 °C) [24]. *(Reprinted by permission of the Electrochemical Society.)*

Because the thicker membranes consisted of several layers, it was possible to measure the resistance of individual layers. The only layer that exhibited resistance increase with current density was the layer next to the anode (Figure 4-8). This clearly indicates that drying as a result of electroosmotic drag occurred close to the anode, because back diffusion was not sufficient to counteract the electroosmotic drag.

Janssen and Overvelde [25] studied the net water transport in an operating fuel cell with Nafion 105 and 112 membranes and found the effective drag (net water transport across the membrane) to be much smaller than reported previously. The values of − 0.3 to + 0.1 were reported (negative values refer to the back diffusion being higher than the electroosmotic drag) and were found to greatly depend on anode humidification. No significant dependence was found on reactants' stoichiometry, and pressure differential, indicating that the hydraulic permeation may be negligible for these membranes. As expected, Nafion 112 membrane exhibited somewhat lower net drag than thicker Nafion 105 membrane.

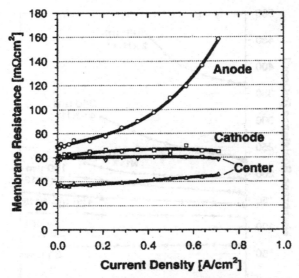

Figure 4-8 Resistance of the membrane layers in H_2/O_2 fuel cell with four Nafion 112 layers as a function of current density (cell temperature 60 °C) [24]. *(Reprinted by permission of the Electrochemical Society.)*

4.2.5 Gas Permeation

In principle, the membrane should be impermeable to reactant species to prevent their mixing before they have had a chance to participate in the electrochemical reaction. However, because of the membrane's essentially porous structure, its water content, and solubility of hydrogen and oxygen in water, some gas does permeate the membrane.

Permeability is a product of diffusivity and solubility:

$$P_m = D \times S \qquad\qquad (4\text{-}16)$$

Because diffusivity is expressed in $cm^2\ s^{-1}$ and solubility in $mol\ cm^{-3}\ Pa^{-1}$, permeability would have unit of $mol\ cm\ s^{-1} cm^{-2}\ Pa^{-1}$. A common unit for permeability is Barrer.

$$1\ Barrer\ =\ 10^{-10} cm^3\ cm\ s^{-1}\ cm^{-2} cm\ Hg^{-1}$$

Solubility of hydrogen in Nafion [26] is $S_{H2} = 2.2 \times 10^{-10}$ $mol\ cm^{-3}\ Pa^{-1}$ and is fairly independent of temperature, and diffusivity is a function of temperature [26,27]:

$$D_{H2} = 0.0041\ exp\left(-\frac{2602}{T}\right) \qquad\qquad (4\text{-}17)$$

The oxygen solubility (in mol cm^{-3} Pa^{-1}) is a function of temperature, and it is given by the following equation [27,28]:

$$S_{O2} = 7.43 \times 10^{-12} \, exp\left(\frac{666}{T}\right) \qquad (4\text{-}18)$$

Oxygen diffusivity (in cm^2 s^{-1}) is [26,27]:

$$D_{O2} = 0.0031 \, exp\left(-\frac{2768}{T}\right) \qquad (4\text{-}19)$$

Permeabilities of various gases through dry Nafion is shown in Figure 4-9 [29]. As expected, hydrogen has one order of magnitude higher permeability than oxygen. Permeability through wet Nafion is another order of magnitude higher, as shown in Figure 4-10. Expectedly, permeation through wet Nafion is somewhat lower than permeation through

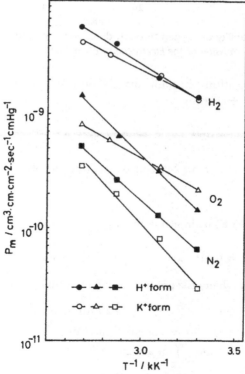

Figure 4-9 Permeability of hydrogen, oxygen, and nitrogen for a Nafion 125 membrane [29]. *(Reprinted by permission of the Electrochemical Society.)*

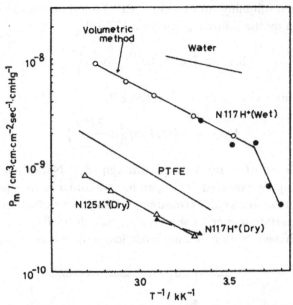

Figure 4-10 Permeability of oxygen through dried and hydrated Nafion membranes [29]. *(Reprinted by permission of the Electrochemical Society.)*

water, and permeation through dry Nafion is somewhat lower than permeation through Teflon.

Example
Calculate hydrogen permeability in Barrers through Nafion at 25 °C and 101.3 kPa (1 atm).

Solution
At 25 °C (298.15 K) hydrogen diffusivity is:

$$D_{H2} = 0.0041 \, exp\left(-\frac{2602}{298.15}\right) = 6.65 \times 10^{-7} \text{ cm}^2 \text{ s}^{-1}$$

Hydrogen permeability is a product of diffusivity and permeability:

$$P_m = 6.65 \times 10^{-7} \text{ cm}^2 \text{ s}^{-1} \times 2.2 \times 10^{-10} \text{ mol cm}^{-3} \text{ Pa}^{-1}$$

$$= 1.44 \times 10^{-16} \text{ mol cm s}^{-1} \text{ cm}^{-2} \text{ Pa}^{-1}$$

Molar volume (cm^3 mol^{-1}) for any gas is:

$$V_m = \frac{RT}{P}$$

where:

R = universal gas constant = 8.314 J mol^{-1} K^{-1}

P = pressure = 101,300 Pa

T = temperature = 298.15 K

$$V_m = \frac{8.314 \times 298.15}{101,300} = 0.02447 \text{ m}^3 \text{ mol}^{-1} = 24,470 \text{ cm}^3 \text{ mol}^{-1}$$

Permeability of hydrogen through Nafion at 25 °C and 101.3 kPa (1 atm) is therefore:

P_m = 1.44 × 10^{-16} mol cm s^{-1} cm^{-2} Pa^{-1} × 24, 470 cm^3 mol^{-1} × 1350 Pa/cm Hg

= 47.7 × 10^{-10} cm^3 cm s^{-1} cm^{-2} cm Hg^{-1} = 47.7 Barrers

The permeation rate, in addition to permeability, obviously should be proportional to pressure and the area of the membrane and inversely proportional to membrane thickness. The permeation rate is then:

$$N_{gas} = P_m \frac{AP}{d} \qquad (4\text{-}20)$$

For example, the permeation rate of hydrogen through 100 cm^2 of Nafion 112 at 25 °C and 300 kPa is:

$$N_{gas} = 1.44 \times 10^{-16} \times \frac{100 \times 300 \times 10^3}{50.8 \times 10^{-4}} = 8.5 \times 10^{-7} \text{ mol s}^{-1}$$

Hydrogen permeation may also be expressed in A cm^{-2}:

$$N_{H2} = \frac{I}{2F} \Rightarrow i = \frac{2FN_{H2}}{A} = \frac{2 \times 96,485 \times 8.5 \times 10^{-7}}{100} = 0.0016 \text{ A cm}^{-2}$$

4.2.6 High-Temperature Membranes

There is a need to operate the fuel cells at temperatures above 100 °C in order to improve tolerance to carbon monoxide that might be present in the fuel stream (see Chapter 9) but also to reduce the size of the heat rejection equipment in automobiles. Operation above 100 °C would also eliminate the water management problem because all water would be in vapor phase. However, PFSA membranes cannot be operated at higher temperatures above 100 °C without affecting their durability.

An approach to increase the water retention of the polymer at high temperature and low relative humidity is to include hydrophilic additives into the ionomer [30], such as phosphotungstic acid, Zr phosph(on)ates, SiO_2 or TiO_2 [31,32,33], which may lead to better cell performance if the composition and morphology of the additive and composite are well designed.

An entirely different approach is that of "water-free" acid-doped polymers for operating at temperatures far beyond 100 °C. Water as the proton solvent is thereby replaced by the acid. In this realm, phosphoric acid (H_3PO_4) doped polybenzimidazole (PBI) membranes constitute the most advanced technology [34]. Membranes with extremely high H_3PO_4 content of 85% by weight are obtained through a sol–gel process [35]. The heterocycle of the PBI is involved in the proton transport process because it provides free electron pairs for proton binding. The operating temperature of H_3PO_4/PBI-based PEFCs is between 160 °C and 200 °C. Although the material is essentially an intrinsic proton conductor, the presence of water greatly improves conductivity. One difficulty faced in the operation of such membranes is the transition through the liquid water regime below 100 °C during startup and shutdown, which leads to a leaching-out of the acid if water condensation is allowed to happen [30].

4.3 ELECTRODES

A *fuel cell electrode* is essentially a thin catalyst layer pressed between the ionomer membrane and porous, electrically conductive substrate. It is the layer where the electrochemical reactions take place. More precisely, the electrochemical reactions take place on the catalyst surface. Because there are three kinds of species that participate in the electrochemical reactions—namely, gases, electrons, and protons—the reactions can take place on a portion of the catalyst surface where all three species have access. Electrons travel through electrically conductive solids, including the catalyst itself, but it is important that the catalyst particles are somehow electrically connected to the substrate. Protons travel through the ionomer; therefore the catalyst must be in intimate contact with the ionomer. Finally, the reactant gases travel only through voids; therefore the electrode must be porous to allow gases to travel to the reaction sites. At the same time, product water must be effectively removed; otherwise the electrode would flood and prevent oxygen access.

As shown graphically in Figure 4-11a, the reactions take place at the three-phase boundary, namely ionomer, solid, and void phases. However,

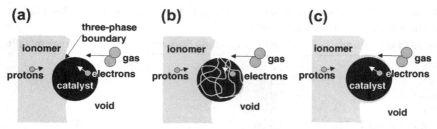

Figure 4-11 Graphical representation of the reaction sites.

this boundary has an infinitesimally small area (essentially it is a line, not an area) that would result in infinitely large current densities. In practice, because some gas may permeate the polymer, the reaction zone is larger than a single three-phase boundary line. The reaction zone may be enlarged by either "roughening" the surface of the membrane or by incorporating ionomer in the catalyst layer (as shown in Figure 4-11b). In an extreme case, the entire catalyst surface may be covered by a thin ionomer layer (Figure 4-11c), except for some allowance for electrical contacts. Obviously, the ratios between the catalyst area covered by the ionomer to the catalyst area opened to void to the catalyst area contacting other catalyst particles or electrically conductive support must be optimized.

The most common catalyst in PEM fuel cells for both oxygen reduction and hydrogen oxidation reactions is platinum. In the early days of PEMFC development large amounts of Pt catalyst were used (up to 28 mg cm^{-2}). In the late 1990s, with the use of supported catalyst structure, this was reduced to 0.3–0.4 mg cm^{-2}. It is the catalyst surface area that matters, not the weight, so it is important to have small platinum particles (4 nm or smaller) with large surface area finely dispersed on the surface of catalyst support, typically carbon powders (cca 40 nm) with high mesoporous area (> 75 m^2 g^{-1}). Typical support material is Vulcan XC72R by Cabot, but other carbons such as Black Pearls BP 2000, Ketjen Black International, or Chevron Shawinigan have been used.

To minimize the cell potential losses due to the rate of proton transport and reactant gas permeation in the depth of the electrocatalyst layer, this layer should be made reasonably thin. At the same time, the metal active surface area should be maximized, for which the Pt particles should be as small as possible. For the first reason, higher Pt/C ratios should be selected ($> 40\%$ by wt.); however, smaller Pt particles and consequently larger metal areas are achieved with lower loading (as shown in Table 4-2). Paganin et al. [36] found that the cell's performance remained virtually unchanged when

Table 4-2 Achievable Pt Active Area for Various Pt/Carbon Compositions Using Ketjen Carbon Black-Supported Catalyst [37]

wt.% Pt on Carbon	XRD Pt Crystallite Size, nm	Active Area* m²/gPt
40	2.2	120
50	2.5	105
60	3.2	88
70	4.5	62
Unsupported Pt black	5.5—6	20—25

*CO chemisorption.

the Pt/C ratio was varied from 10% to 40% with a Pt loading of 0.4mg/cm^2. However, the performance deteriorated as the Pt/C ratio was increased beyond 40%. This indicates a negligible change in the active catalyst area between Pt/C ratios of 10% and 40% and significant decrease in catalyst active area beyond 40% Pt/C (also seen in Table 4-2). Table 4-2 shows the achievable catalyst active area for various Pt/carbon compositions [37] (using a Ketjen Carbon Black-supported electrocatalyst).

In general, higher Pt loading results in voltage gain (Figure 4-12) [38], assuming equal utilization and reasonable thickness of the catalyst layer. However, when current density is calculated per area of Pt surface, there is almost no difference in performance, that is, all the polarization curves fall on top of each other (Figure 4-13) [38]. Note that the Tafel slope is about 70 mV/decade.

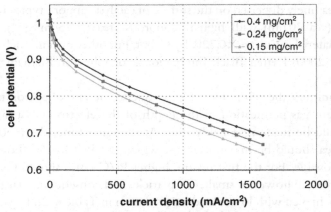

Figure 4-12 Effect of Pt loading on fuel cell polarization curve (H_2/O_2 fuel cell).

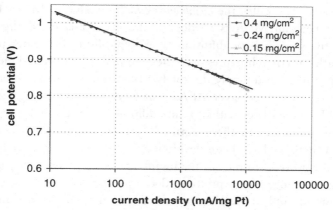

Figure 4-13 Cell performance per unit of Pt electrocatalyst.

The key to improving the PEM fuel cell performance is not in increasing the Pt loading but rather in increasing Pt utilization in the catalyst layer.

The catalyst surface active area may be greatly increased if the ionomer is included in the catalyst layer, either by painting it with solubilized PFSA in a mixture of alcohols and water or preferably by premixing catalyst and ionomer in a process of forming the catalyst layer. Zawodzinski et al. [39] have shown that there is an optimum amount of ionomer in the catalyst layer—around 28% by weight (Figure 4–14). Similar findings were reported by Qi and Kaufman [40] and Sasikumar et al. [41]

In principle, there are two ways of preparing the catalyst layer and its attachment to the ionomer membrane. Such a combination of membrane

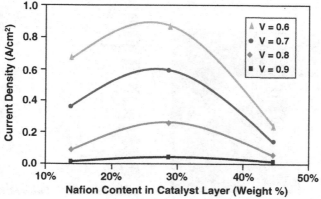

Figure 4-14 Effect of Nafion content in catalyst layer on fuel cell performance [39].

and catalyst layers is called the *membrane electrode assembly*, or MEA. The first way to prepare an MEA is to deposit the catalyst layer to the porous substrate, the so-called gas diffusion layer, typically carbon fiber paper or carbon cloth, and then hot-press it to the membrane. The second method of preparing an MEA is to apply the catalyst layers directly to the membrane, forming a so-called three-layer MEA or catalyzed membrane. A gas diffusion layer may be added later, either as an additional step in MEA preparation (in that case a five-layer MEA is formed) or in a process of stack assembly.

Several methods have been developed for deposition of a catalyst layer on either the porous substrate or the membrane, such as spreading, spraying, sputtering, painting, screen printing, decaling, electrodeposition, evaporative deposition, and impregnation reduction. There are several manufacturers of MEAs, including Dupont, 3M, Johnson Matthey, W. L. Gore & Associates, and BASF. Their manufacturing processes are typically trade secrets.

Recently there has been some progress reported on development of new catalysts and catalyst layer structures. 3M has developed a nanostructured thin film catalyst (Figure 4-15) that has a completely different structure from the conventional carbon supported catalyst. The NSTF $Pt_{68}Co_{29}Mn_3$ catalyst fundamentally has higher specific activity for oxygen reduction and

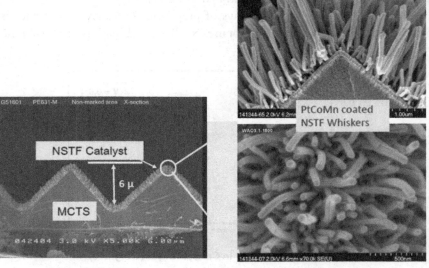

Figure 4-15 3M nanostructured thin film catalyst. *(Courtesy of 3M.)*

removes all durability issues with carbon supports, demonstrates much lower losses due to Pt dissolution and membrane chemical attack, and has significant high-volume all-dry roll-good manufacturing advantages [42]. It exceeds the previous DOE 2015 target of 0.2 g-Pt/kW in a full-size short stack with 0.05 mg cm^{-2} of PGM on the anode and 0.1 mg cm^{-2} on the cathode [42].

A new promising catalyst $Pt_{1-x}Ni_x$ was found to be unique in showing an extraordinarily sharp peak in ORR activity (60% higher than for the NSTF standard $Pt_{68}Co_{29}Mn_3$ alloy) as a function of the as-made composition around $x = 0.69 \pm 0.02$, determined gravimetrically [43].

High-performance fuel cell electrocatalysts for the oxygen reduction reaction (ORR) comprising contiguous Pt monolayer (ML) on stable, inexpensive metal or alloy supported on solid tetrahedral or hollow nanoparticles, nanowires, nanorods, and carbon nanotubes are being developed. PtML/Pd9Au/C and PtML/Pd/C seem to be practical electrocatalysts, needing only 10 grams of Pt and about 15–20 grams of Pd for a 100 kW fuel cell [44].

The researchers at Los Alamos National Laboratory developed a family of nonprecious metal catalysts that approach the performance of platinum-based systems at a cost sustainable for high-power fuel cell applications, possibly including automotive power. They used polyaniline as a precursor to a carbon-nitrogen template for high-temperature synthesis of catalysts incorporating iron and cobalt. The most active materials in the group catalyze the oxygen reduction reaction at potentials within ~60 millivolts of that delivered by state-of-the-art carbon-supported platinum, combining their high activity with remarkable performance stability for non–precious metal catalysts (700 hours at a fuel cell voltage of 0.4 volts) as well as excellent four-electron selectivity (which means hydrogen peroxide yield < 1.0%) [45].

4.4 GAS DIFFUSION LAYER

A layer between the catalyst layer and bipolar plates is called a *gas diffusion layer, electrode substrate*, or *diffusor/current collector*. Although it does not directly participate in the electrochemical reactions, a gas diffusion layer in PEM fuel cells has several important functions:

- It provides a pathway for reactant gases from the flow field channels to the catalyst layer, allowing them access to the entire active area (not just to those adjacent to the channels).

- It provides a pathway for product water from the catalyst layer to the flow field channels.
- It electrically connects the catalyst layer to the bipolar plate, allowing the electrons to complete the electrical circuit.
- It also serves to conduct heat generated in the electrochemical reactions in the catalyst layer to the bipolar plate, which has means for heat removal (as discussed in Chapter 6).
- It provides mechanical support to the MEA, preventing it from sagging into the flow field channels.

The required properties of the gas diffusion layer follow from its functions:
- It must be sufficiently porous to allow flow of both reactant gases and product water (note that these fluxes are in opposite directions). Depending on the design of the flow field, diffusion in both through-plane and in-plane is important.
- It must be both electrically and thermally conductive, again in both through-plane and in-plane. Interfacial or contact resistance is typically more important than bulk conductivity.
- Because the catalyst layer is made of discrete small particles, the pores of the gas diffusion layer facing the catalyst layer must not be too big.
- It must be sufficiently rigid to support the "flimsy" MEA. However, it must have some flexibility to maintain good electrical contacts.

These somewhat conflicting requirements are best met by carbon fiber-based materials such as carbon fiber papers and woven carbon fabrics or cloths. Figure 4-16 shows two typical kinds of gas diffusion media, namely fiber paper and carbon cloth [46]. Table 4-3 shows the properties of gas diffusion layers made of carbon paper and carbon cloth, as reported by various manufacturers. From Table 4-3 it follows that the thickness of

Figure 4-16 SEM micrographs of the E-TEK EC-CC1-060 carbon cloth (left) and of Toray TGP-H-090 5% wet-proof PTL carbon paper (right).

Table 4-3 Properties of Typical Fuel Cell Gas Diffusion Layers

Company	Material	Thickness cm	Density g/cm³	Weight g/m²	Porosity %	Electrical Resistivity Through-Plane	Electrical Resistivity In-Plane Ohmcm
Toray	Carbon Fiber Papers						
	TGP-H-060	0.019	0.44	84	78	0.080	0.0058
	TGP-H-090	0.028	0.44	123	78	0.080	0.0056
	TGP-H-120	0.037	0.45	167	78	0.080	0.0047
Spectracorp	2050-A	0.026	0.48	125		2.692	0.012
	2050-L	0.02	0.46	92		7.500	0.022
	2050-HF	0.026	0.46	120		3.462	0.014
Ballard	AvCarb P50	0.0172	0.28	48		0.564	
	AvCarb P50T	0.0172	0.28	48		0.564	
SGL Carbon	10-BA	0.038	0.22	84	88	0.263	
	10-BB	0.042	0.30	125	84	0.357	
	20-BA	0.022	0.30	65	83	0.455	
	20-BC	0.026	0.42	110	76	0.538	
	21-BA	0.02	0.21	42	88	0.550	
	21-BC	0.026	0.37	95	79	0.577	
	30-BA	0.031	0.31	95	81	0.323	
	30-BC	0.033	0.42	140	77	0.394	
	31-BA	0.03	0.22	65		0.317	
	31-BC	0.034	0.35	120	82	0.441	
E-TEK	LT 1100-N	0.018	0.50	90		0.360	
	LT 1200-W	0.0275	0.73	200		0.410	
	LT 1400-W	0.04	0.53	210		0.500	
	LT 2500-W	0.043	0.56	240		0.550	
Carbon cloth	AvCarb	0.038	0.31	118		0.132	0.009
Ballard	1071 HCB*						

*Measurements performed and reported by General Motors [47].

various gas diffusion materials varies between 0.017 to 0.04 cm, density varies between 0.21 to 0.73 g cm^{-3}, and porosity varies between 70% and 80%.

4.4.1 Treatments and Coatings

Diffusion media are generally made hydrophobic to avoid flooding in their bulk. Typically, both cathode and anode gas diffusion media are PTFE treated. A wide range of PTFE loadings have been used in PEMFC diffusion media (5% to 30%), most typically by dipping the diffusion media into a PTFE solution, followed by drying and sintering.

Hydrophobic properties of gas diffusion media are rarely reported by the manufacturers. These properties are often tailored to a specific cell design and must be measured and correlated to the cell performance. Typically this involves the measurement of contact angle on the surface by either a Sessile drop method or Wilhelmy methods.

Figure 4-17 shows a fuel cell performance with treated and untreated cathode diffusion media [47]. The untreated one was susceptible to flooding, especially at higher current densities.

In addition, the interface with the adjacent catalyst layer may also be fitted with a coating or a microporous layer to ensure better electrical contacts as well as efficient water transport in and out of the diffusion layer. This layer (or layers) consists of carbon or graphite particles mixed with PTFE binder. The resulting pores are between 0.1 and 0.5 μm, thus they are much smaller than the pore size of the carbon fiber papers (20–50 μm).

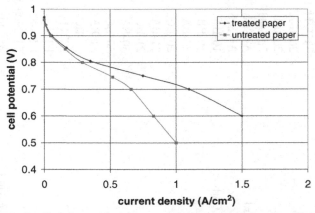

Figure 4-17 Fuel cell performance with treated and untreated carbon fiber paper [47]. 50 cm^2, H$_2$/Air; 80 °C, 270 kPa, 2.0/2.0 stoichiometry, 100%/50% an/ca relative humidity.

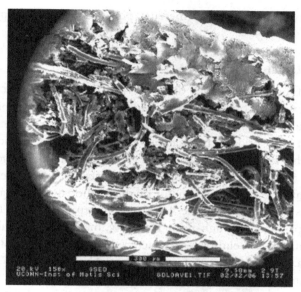

Figure 4-18 A nonwoven gas diffusion media with a microporous layer on top of it.

Figure 4-18 shows nonwoven gas diffusion media with a microporous layer on top of it. The small pore size helps improve the electrical contacts with the adjacent catalyst layer. However, the primary role of this microporous layer is to facilitate effective wicking of liquid water from the cathode catalyst layer into the diffusion media, resulting in much smaller water droplets that are less likely to clog and flood the gas diffusion media bulk.

4.4.2 Porosity

Gas diffusion media are by definition porous. Porosity is typically between 70% and 80%, as shown in Table 4-3. Porosity of a gas diffusion layer may be easily calculated from its areal weight, its thickness, and the density of the solid phase (for carbon-based materials, ρ_{real} varies between 1.6 and 1.95 g cm^{-3}). The porosity, ε, also depends on the compressed thickness:

$$\varepsilon = 1 - \frac{W_A}{\rho_{real}d} \qquad (4\text{-}21)$$

where:

    W_A = areal weight (g cm^{-2})
    ρ_{real} = solid phase density
    d = thickness (either compressed or uncompressed)

Porosity may be measured by mercury porosimetry or by capillary flow porometry [47].

4.4.3 Electrical Conductivity

One of the functions of the gas diffusion layer is to connect electrically the catalyst layer with the bipolar plate. Because only a portion of the bipolar plate makes the contact (the other portion is open for access of reactant gases), the gas diffusion layer bridges the channels and redistributes electrical current. Because of this, both through-plane and in-plane resistivities of gas diffusion material are important. Through-plane resistivity, ρ_z, often includes both bulk and contact resistance, depending on the method used in measurements. It is obvious from data in Table 4-3 that some manufacturers (Toray, for example) report true through-plane resistivity, measured, for example, with mercury contacts to eliminate contact resistance, whereas others apply a method that includes the contact resistance. Mathias et al. [47] measured through-plane resistivity of Toray TGP-H-060 and confirmed the manu-facturer's data of 0.08 Ωcm. They also measured the total through-plane resistance of 0.009 Ωcm^2, which, if divided by the thickness (0.019 cm), would result in 0.473 Ωcm (which is close to data reported by other manu-facturers in Table 4-3). In-plane resistivity of common gas diffusion media, ρ_{xy}, typically measured by a four-point probe method, is typically about an order of magnitude lower than the through-plane value. The relevance of through-plane and in-plane resistivity is discussed in Chapter 6, "Stack Design," and the importance of contact resistance between the gas diffusion layer and bipolar plates is discussed in this chapter in Section 4.5, "Bipolar Plates."

4.4.4 Compressibility

In a fuel cell, a gas diffusion layer is compressed to minimize the contact resistance losses. Both carbon papers and carbon cloths are relatively soft and easily deformable materials. Cloth is more compressible than paper, as shown in Figure 4-19. When exposed to a cyclic compression test they both exhibit a weakening of the material—the first compression stress-strain curve is different from the one resulting from subsequent cycles [47].

4.4.5 Permeability

Effective diffusion coefficients in typical PEMFC diffusion media include the effects of material porosity and tortuosity. In most cases they reflect bulk

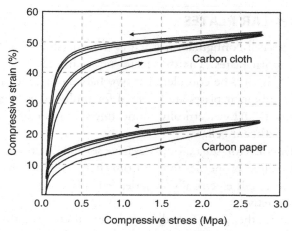

Figure 4-19 Stress-strain curves for carbon paper (Toray TGP-H-060) and carbon cloth (Textron 1071HCB) [47].

as opposed to Knudsen diffusion because the pore diameters are several orders of magnitude higher than the mean free path of gas molecules. However, Knudsen diffusion may be prevalent in microporous layers where the pore size can approach the gas molecule mean free path. Convective flow resistance of the diffusion media is given as either a Gurley number or a Darcy coefficient. The Gurley number is the time required to pass a specified volume of flow through a sample at a given pressure drop. The Darcy coefficient relates to the pressure drop, which, according to Darcy's law, is proportional to the volumetric flow rate:

$$Q = K_D \frac{A}{\mu l} \Delta P \tag{4-22}$$

where:

Q = volumetric flow rate, $m^3 s^{-1}$
K_D = Darcy coefficient, m^2
A = cross-sectional area perpendicular to the flow, m^2
μ = gas viscosity, $kg\ m^{-1}\ s^{-1}$
l = length of the path (thickness of diffusion media), m
ΔP = pressure drop, Pa

For an uncompressed Toray TGP-H-060 carbon fiber paper, the Darcy coefficient of $5-10 \times 10^{-12}\ m^2$ has been reported [47]. Approximately the same value has been reported for in-plane flow through the same material compressed to 75% of its original thickness [47].

4.5 BIPOLAR PLATES

In a single-cell configuration (such as the one shown in Figure 4-1) there are no bipolar plates. The two plates on each side of the membrane electrode assembly may be considered as two halves of a bipolar plate. The fully functioning bipolar plates are essential for multicell configurations (as shown in Figure 4-20) by electrically connecting the anode of one cell to the cathode of the adjacent cell.

The bipolar collector/separator plates have several functions in a fuel cell stack. Their required properties follow from their functions, namely:

- They connect cells electrically in series—therefore, they must be electrically conductive.
- They separate the gases in adjacent cells—therefore, they must be impermeable to gases.
- They provide structural support for the stack—therefore, they must have adequate strength, yet they must be lightweight.
- They conduct heat from active cells to the cooling cells or conduits—therefore, they must be thermally conductive.
- They typically house the flow field channels—therefore, they must be conformable.

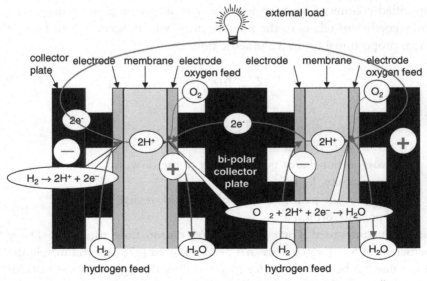

Figure 4-20 Bipolar plate connects and separates two adjacent cells.

In addition, they must be corrosion resistant in the fuel cell environment, yet they must not be made of "exotic" and expensive materials. To keep the cost down, not only must the material be inexpensive, the manufacturing process must be suitable for mass production.

Some of these requirements may contradict each other; therefore, selection of the material involves an optimization process. The resulting material may not be the best in any of the property categories, but it is the one that best satisfies the optimization criteria (typically the lowest cost per kWh of electricity produced). Table 4-4 summarizes bipolar plate requirements.

4.5.1 Materials

One of the first materials used for PEM fuel cell bipolar plates was graphite, primarily because of its demonstrated chemical stability in the fuel cell environment. Graphite is inherently porous, which may be detrimental in fuel cell applications. Those plates therefore must be impregnated to make them impermeable. This material is still used in laboratory fuel cells (primarily in single cells). However, machining of graphite plates is not an easy task and may be prohibitively expensive for most fuel cell applications. It should be mentioned that one fuel cell manufacturer (UTC Fuel Cells) used porous graphite plates for water management inside the fuel cell stack.

In general, two families of materials have been used for fuel cell bipolar plates: graphite-based (including graphite/composite) materials and metallic materials.

Table 4-4 Bipolar Plate Requirements [48]

Property	Requirements	Comment
Electrical conductivity	$> 100 \text{ S cm}^{-1}$	Bulk conductivity
Corrosion rate	$< 16 \text{ }\mu\text{A cm}^{-2}$	
Hydrogen permeability	$< 2 \times 10^{-6} \text{ cm}^3 \text{ cm}^{-2} \text{ s}^{-1}$	@80 °C, 3 atm
Compressive strength	$> 2 \text{ MPa}$	
Thermal conductivity	$> 20 \text{ W/mK}$	Strong function of stack design—some designs may require higher
Tolerance	$< 0.05 \text{ mm}$	
Cost	$< \$10/\text{kW}$	Including both material and fabrication
Weight	$< 1 \text{ kg/kW}$	

Metallic Plates

The bipolar plates are exposed to a very corrosive environment inside a fuel cell (pH 2 to 3 and temperature 60 °C to 80 °C). Typical metals such as aluminum, steel, titanium, or nickel would corrode in a fuel cell environment, and dissolved metal ions would diffuse into the ionomer membrane, resulting in lowering of ionic conductivity and reducing the fuel cell life. In addition, a corrosion layer on the surface of a bipolar plate would increase electrical resistance. Because of these issues, metallic plates must be adequately coated with a noncorrosive yet electrically conductive layer, such as graphite, diamond-like carbon, conductive polymer, organic self-assembled polymers, noble metals, metal nitrides, metal carbides, indium doped tin oxide, and so on. The effectiveness of the coating in protecting the bipolar plate from the corrosive PEM fuel cell environment depends on (i) corrosion resistance of coating, (ii) micropores and microcracks in the coating layer, and (iii) difference between the coefficient of thermal expansion of the base material and coating. Metallic plates are suitable for mass manufacturing (stamping, embossing) and, because they can be made very thin (< 1 mm), result in compact and lightweight stacks. The need for protective coating and the problems associated with those in fuel cell operation are the major drawbacks for the metallic plates in PEM fuel cells.

Graphite-Composite Plates

Carbon composite bipolar plates have been made using thermoplastics (polypropylene, polyethylene, or polyvinylidenefluoride) or thermoset resins (phenolic, epoxies, and vinyl esters) with fillers (such as carbon/graphite powder, carbon black, or coke-graphite) and with or without fiber reinforcements. These materials are typically chemically stable in fuel cell environments, although some thermosets may leach and consequently deteriorate. Depending on the rheological properties of these materials, they are suitable for compression molding, transfer molding, or injection molding. Very often, a careful optimization of the material's composition and properties is required, involving a trade-off between manufacturability (i.e., cost) and functional properties (i.e., electrical conductivity). For example, a replacement of a compression-molded material with bulk resistivity of 2.9 mOhm-cm with an injection-molded material with bulk resistivity of 26 mOhm-cm has been considered [49] because of the improvements in manufacturing speed (20 seconds of cycle time for injection molding vs. 20 minutes for thermoplastic compression molding). Important properties that must be considered in designing and manufacturing graphite/composite bipolar plates are tolerances, warping, and skinning effect (accumulation of polymer at the surface of the plate as a result of the molding process). High-speed molding processes can meet the cost targets, and the materials (graphite and polymer) are not expensive. These plates, especially those with fluoropolymers, have unsurpassed chemical stability in the fuel cell environment. However, they are bulky (minimum thickness ~ 2 mm) and relatively brittle (which may be a problem for high-speed automated stack assembly processes). Although their electrical conductivity is several orders of magnitude lower than the conductivity of the metallic plates, the bulk resistive losses are on an order of a magnitude of several millivolts.

Composite Graphite/Metallic Plates

A sandwich of two embossed graphite foils with a thin metallic sheet in between has been patented by Ballard [50]. This concept combines the advantages of both graphite (corrosion resistance) and metallic plates (impermeability and structural rigidity) and results in a lightweight, durable, and easy-to-manufacture bipolar plate. It should also be mentioned that graphite foil has very low contact resistance due to its conformability.

4.5.2 Properties

The most important properties of various metallic and graphite/composite bipolar plate materials are summarized in Tables 4-5 and 4-6, respectively.

One of the most important properties of the fuel cell bipolar plates is their electrical conductivity. One should distinguish between the bulk and total conductivity or resistivity, the latter including both bulk and interfacial contact components. In an actual fuel cell stack, contact (interfacial) resistance is more important than bulk resistance.

Bulk electrical resistivity of the plates may be measured with a four-point probe method for sheet resistivity, as described by Smits [52]. The

Table 4-5 Selected Properties of Metallic Bipolar Plate Materials

Property	Unit	Materials			
		SS	Al	Ti	Ni
Density	$g\ cm^{-3}$	7.95	2.7	4.55	8.94
Bulk el. conductivity	$S\ cm^{-1}$	14,000	377,000	23,000	146,000
Thermal conductivity	$W\ m^{-1}\ K^{-1}$	15	223	17	60.7
Thermal expansion	$\mu\ mm^{-1}\ K^{-1}$	18.5	24	8.5	13

(Data taken from [51]).

Table 4-6 Selected Properties of Some Graphite/Composite Bipolar Plate Materials

Property	Unit	Materials and Manufacturers			
		Graphite POCO	BBP 4 SGL	PPG 86 SGL	BMC940 BMC
Density	g cm^{-3}	1.78	1.97	1.85	1.82
Bulk el. conductivity	S cm^{-1}	680	200	56	100
Thermal conductivity	W m^{-1} K^{-1}	95	20.5	14	19.2
Thermal expansion	μ mm^{-1} K^{-1}	7.9	3.2	27	30
Tensile strength	MPa	60			30
Flexural strength	MPa	90	50	35	40
Compressive strength	MPa	145	76	50	

experimental setup is shown in Figure 4-21. This method allows measurement of bulk resistivity of thin plates by applying a geometry-dependent correction factor to the measured value of voltage drop and applied current:

$$\rho = k \frac{V}{I} t \qquad (4\text{-}23)$$

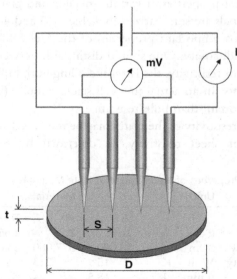

Figure 4-21 Experimental setup for measuring sample resistivity with four-point probe method [49].

Table 4-7 Correction Factor for Bulk Resistivity Measurement of Thin Round Samples with Four-Point Probe [52]

| | | | | | t/S | | | | | |
D/S	<0.4	0.4	0.5	0.6	0.7	0.8	1	1.25	1.666	2
3	2.2662	2.2651	2.2603	2.2476	2.2240	2.1882	2.0881	1.9240	1.6373	1.4359
4	2.9289	2.9274	2.9213	2.9049	2.8744	2.8281	2.6987	2.4866	2.1161	1.8558
5	3.3625	3.3608	3.3538	3.3349	3.3000	3.2468	3.0982	2.8548	2.4294	2.1305
7.5	3.9273	3.9253	3.9171	3.8951	3.8543	3.7922	3.6186	3.3343	2.8375	2.4883
10	4.1716	4.1695	4.1608	4.1374	4.0940	4.0281	3.8437	3.5417	3.0140	2.6431
15	4.3646	4.3624	4.3533	4.3288	4.2834	4.2145	4.0215	3.7055	3.1534	2.7654
20	4.4364	4.4342	4.4249	4.4000	4.3539	4.2838	4.0877	3.7665	3.2053	2.8109
40	4.5076	4.5053	4.4959	4.4706	4.4238	4.3525	4.1533	3.8270	3.2567	2.8560
∞	4.5324	4.5301	4.5206	4.4952	4.4481	4.3765	4.1762	3.8480	3.2747	2.8717

where:

ρ = resistivity (Ωcm)

k = correction factor, function of D/S and t/S, where D = sample diameter, S = spacing of probes, and t = sample thickness; values for correction factor are tabulated in Table 4-7 [52]

V = measured voltage (V)

I = applied current (A)

t = sample thickness (cm)

However, bulk resistivity is not a significant source of voltage loss in fuel cells, even for relatively high-resistivity plates. For example, a 3-mm thick molded graphite/composite plate with bulk resistivity as high as 8 mΩcm would result in about 2.4 mV voltage loss at 1A/cm^2. Much higher resistance results from the interfacial contacts, such as between the bipolar plate and the gas diffusion layer.

The interfacial contact resistance may be determined by sandwiching a bipolar plate between the two gas diffusion layers [53] (or a gas diffusion layer between the two bipolar plates [54]) and then passing electrical current through the sandwich and measuring voltage drop (see Figure 4-22). The total voltage drop (or resistance, $R = V/I$) in this experiment is a strong function of clamping pressure. There are several serial resistances in this experiment, namely, contact resistance between the gold contact plates and the gas diffusion media, $R_{Au\text{-}GDL}$; bulk (through-plane) resistance of the gas diffusion media, R_{GDL}; contact resistance between the gas diffusion media and bipolar plate, $R_{GDL\text{-}BP}$; and bulk resistance of the bipolar plate, R_{BP}, as illustrated in Figure 4-23:

$$R_{mes} = 2R_{Au-GDL} + 2R_{GDL} + 2R_{GDL-Bp} + R_{BP} \qquad (4\text{-}24)$$

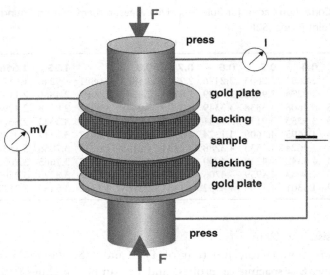

Figure 4-22 Schematic diagram of experimental setup for measuring total (contact and bulk) resistance of graphite and molded graphite composite samples, including electrode-backing layers [49, 53].

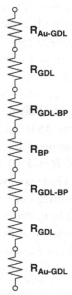

Figure 4-23 Resistances involved in measurement from Figure 4-20.

The bulk resistance of the gas diffusion media, R_{GDL}, and of the bipolar plate, R_{BP}, should be known from independent measurements or from manufacturer's specifications, and the unwanted contact resistance between the gold contact plate and gas diffusion media may be determined by an additional measurement involving just the gas diffusion media between the two gold contact plates:

$$R'_{mes} = 2R_{Au-GDL} + R_{GDL} \tag{4-25}$$

The contact resistance is then:

$$R_{GDL-BP} = (R_{mes} - R'_{mes} - R_{BP} - R_{GBL})/2 \tag{4-26}$$

Bulk resistance of the bipolar plate and the gas diffusion media should be independent of the clamping force, but the contact resistance is obviously a strong function of the clamping force. Figure 4-24 shows the contact resistance for several gas diffusion media [54]. At 2 MPa, contact resistance for carbon fiber paper is about 3 mΩcm. Carbon cloth has lower contact resistance, that is, about 2 mΩcm. Almost identical results were reported by Mathias et al. [47] using a slightly different measuring procedure.

Interfacial contact resistance depends not only on contact (clamping) pressure but also on the surface characteristics and the effective conductivities of the two surfaces in contact. Using the fractal geometric description of surface topographies, Majumdar and Tien [55] arrived at a relationship between the contact resistance and clamping pressure. Mishra et al. [54]

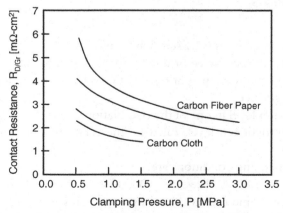

Figure 4-24 Results of the contact resistance measurements between the gas diffusion media and graphite composite bipolar plate [54].

modified the Majumdar–Tien relationship to fit the case of a relatively soft material, such as gas diffusion layer, with hard material, such as bipolar plate:

$$R = \frac{A_a K G^{D-1}}{\kappa L^D} \left[\frac{D}{(2-D)p*} \right]^{\frac{D}{2}} \tag{4-27}$$

where:

R = contact resistance, Ωm^2

A_a = apparent contact area at the interface, m^2

K = geometric constant

G = topothesy of a surface profile, m

D = fractal dimension of a surface profile

κ = effective electrical conductivity of two surfaces, Sm^{-1}

$$\frac{1}{\kappa} = \frac{1}{2} \left(\frac{1}{\kappa_1} + \frac{1}{\kappa_2} \right) \tag{4-28}$$

L = scan length, m

$p*$ = dimensionless clamping pressure (ratio of actual clamping pressure and compressive modulus of gas diffusion layer)

After obtaining the geometric parameters from the surface profilometric scans and plugging them in Equation (4-21), Mishra et al. [54] obtained reasonably good agreements with measurements of contact resistance.

 PROBLEMS

1. Nafion conductivity is 0.1 S/cm when fully hydrated.
   a. Calculate the resistance of the Nafion 117 membrane.
   b. Calculate the resistance of two Nafion 112 layers; assume no interfacial resistance.
   c. If the water level in one of the two Nafion 112 layers drops to 50% of the fully hydrated level, calculate the total resistance through both layers.
2. Fuel cell operating conditions are:
   Temperature = 80 °C, Pressure = 300 kPa, both gases fully saturated
   Calculate cell ionic resistance and hydrogen crossover for two different membranes, namely Nafion 112 and 117.

3. Resistance measurement of a 65-cm^2 fuel cell with Nafion membrane shows 2.61 mOhm. The reactant gases are fully saturated at a cell operating temperature of 70 °C. Estimate the membrane thickness, assuming that $\frac{2}{3}$ of the above resistance is protonic resistance.

4. Calculate:
   a. Permeability of hydrogen at 60 °C.
   b. What would be the permeation rate through 200 cm^2 of Nafion 115 if the pressure of hydrogen is 300 kPa?
   c. What would be the equivalent current loss?

5. A fuel cell stack is put together that has 10 cells, 300-cm^2 active area each, using Nafion 115. Before it can be operated, it must be checked for leaks.
   a. What would be an acceptable hydrogen leak rate (if the test is performed at room temperature and at pressure of 0.5 bar above atmospheric)? Assume that Nafion is fully saturated with water.
   b. If Nafion was dry after assembly, would you expect higher or lower permeation rate than if it was wet?
   c. What would be the permeation rate in operation when the operating temperature is 80 °C and the operating pressure is 200 kPa?
   d. What would be the equivalent current loss?

6. Two fuel cells of equal active area operate at same operating conditions (H$_2$/O$_2$, 80 °C, 1 atm, both gases fully humidified) and generate the same current. However, cell #1 has 30 mV higher potential than cell #2. The only possible difference between the two cells is the catalyst loading. The first cell has Pt loading of 0.4 mg/cm^2 Pt. What is the Pt loading of cell #2?

7. If 1% of the cars produced in the world used PEM fuel cells with platinum loading of 0.1 mg/cm^2, what would be the annual demand for platinum? See if you can compare it with today's world platinum consumption. (Assume 75 kW fuel cell per car with operation at 1.3 A/cm^2 at 0.65 V per cell.)

 QUIZ

1. Nafion works in fuel cells because:
   a. It is so thin
   b. It is full of water
   c. It conducts protons

2. Higher water content in a Nafion membrane results in:
    a. Higher ionic conductivity
    b. Lower ionic conductance because the membrane becomes thicker
    c. Higher ionic resistivity
3. Membrane thickness has an effect on:
    a. Ionic conductance
    b. Gas permeation
    c. Both ionic conductance and gas permeation
4. Smaller catalyst particles result in:
    a. More surface area, thus higher exchange current density
    b. Thicker catalyst layer
    c. Higher catalyst loading
5. In a gas diffusion layer on the cathode (air) side:
    a. Oxygen and water flow in opposite directions
    b. Oxygen and water flow in the same direction
    c. Oxygen and water mix together
6. Bipolar plate bulk resistance is a:
    a. Function of its thickness
    b. Function of material conductivity
    c. Function of both thickness and conductivity
7. The advantage of graphite-based bipolar plates over metallic plates is:
    a. Better conductivity
    b. Better corrosion resistance
    c. Lighter weight
8. Catalyst loading is:
    a. A process of loading the catalyst on the support particles
    b. Quantity of catalyst per unit electrode geometric area
    c. Quantity of catalyst per cell
9. Electroosmotic drag:
    a. Moves water from anode to cathode
    b. Moves water from cathode to anode
    c. Moves water in both directions as a function of pressure difference
10. Contact or clamping pressure has an effect on:
    a. Bulk resistance of the bipolar plate
    b. Resistance in the membrane
    c. Interfacial resistance

REFERENCES

[1] Gottesfeld S, Zawodzinski TA. Polymer Electrolyte Fuel Cells. In: Alkire RC, Gerischer H, Kolb DM, Tobias CW, editors. Advances in Electrochemical Science and Engineering, vol. 5. New York: Wiley-VCH; 1997.

[2] Doyle M, Rajendran G. Perfluorinated Membranes. In: Vielstich W, Lamm A, Gastegier HA, editors. Handbook of Fuel Cells: Fundamentals, Technology, and Applications. Fuel Cell Technology and Applications, vol. 3. New York: John Wiley & Sons; 2003. p. 351–95.

[3] Peron J, Mani A, Zhao X, Edwards D, Adachi M, Soboleva T, et al. Properties of Nafion® NR-211 membranes for PEMFCs. Journal of Membrane Science 2010;356:44–51.

[4] DuPont, Nafion®; PFSA Membranes N115, N117, N1110 Datasheet, www2.dupont.com/FuelCells/en_US/assets/downloads/dfc101.pdf.

[5] DuPont, Nafion®; PFSA Membranes NR211 and NR212 Datasheet, www2.dupont.com/FuelCells/en_US/assets/downloads/dfc201.pdf.

[6] Hamrock S. Membranes and MEAs for Dry, Hot Operating Conditions, DOE Hydrogen and Fuel Cells Program. FY2011 Annual Progress Report 2011:662–6.

[7] Zawodzinski Jr TA, Deroiun C, Radzinski S, Sherman RJ, Springer T, Gottesfeld S. Water Uptake By and Transport Through Nafion 117 Membranes. Journal of the Electrochemical Society 1993;vol. 140:1041.

[8] Zawodzinski Jr TA, Lopez C, Jestel R, Valerio J, Gottesfeld S. A Comparative Study of Water Uptake By and Transport Through Ionomeric Fuel Cell Membranes. Journal of the Electrochemical Society 1993;vol. 140:1981.

[9] Cleghorn S, Kolde J, Liu W. Catalyst Coated Composite Membranes. In: Vielstich W, Lamm A, Gastegier HA, editors. Handbook of Fuel Cells: Fundamentals, Technology, and Applications. Fuel Cell Technology and Applications, vol. 3. New York: John Wiley & Sons; 2003. p. 566–75.

[10] Zawodzinski Jr TA, Neeman M, Sillerud L, Gottesfeld S. Determination of Water Diffusion Coefficients in Perfluorosulfonate Ionomeric Membranes. Journal of Physical Chemistry 1991;vol. 95:6040.

[11] Springer TE, Zawodzinski TA, Gottesfeld S. Polymer Electrolyte Fuel Cell Model. Journal of the Electrochemical Society 1991;vol. 138(No. 8):2334–42.

[12] Zawodzinski, TA, Membranes Performance and Evaluation, NSF Workshop on Engineering Fundamentals of Low Temperature PEM Fuel Cells (Arlington, VA, November 2001) (http://electrochem.cwru.edu/NSF/presentations/NSF_Tom-Zawodzinsk_Membranes.pdf).

[13] La Conti AB, Fragala AR, Boyack JR. Solid Polymer Electrolyte Electrochemical Cells: Electrode and Other Materials Considerations. In: McIntyre JDE, Srinivasan S, Will FG, editors. Electrode Materials and Processes for Energy Conversion and Storage, in Proc. the Electrochemical Society. NJ: Pennington; 1977. p. 354. PV 77-6.

[14] Fuller T, Newman J. Experimental Determination of the Transport Number of Water in Nafion 117 Membrane. Journal of the Electrochemical Society 1992;vol. 139:1332.

[15] Zawodzinski Jr TA, Springer T, Uribe F, Gottesfeld S. Characterization of Polymer Electrolytes for Fuel Cell Applications. Solid State Ionics 1993;vol. 60:199.

[16] Yeo SC, Eisenberg A. Physical Properties and Supermolecular Structure of Perfluorinated Ion-containing (Nafion) Polymers. Journal of Applied Polymer Science 1977;vol. 21:875.

[17] Eisman GA. The Physical and Mechanical Properties of a New Perfluorosulfonic Acid Ionomer for Use as a Separator/Membrane in Proton Exchange Processes. In: Van Zee JW, White RE, Kinoshita K, Burney HS, editors. Diaphragms, Separators, and Ion

Exchange Membranes, in Proc. the Electrochemical Society. NJ: Pennington; 1986. p. 156–71. PV 86-13.

[18] Verbruge M. Methanol Diffusion in Perfluorinated Ion-Exchange Membranes. Journal of the Electrochemical Society 1989;vol. 136:417.

[19] Slade RCT, Hardwick A, Dickens PG. Investigation of H^+ Motion in NAFION Film by Pulsed ^1H NMR and A.C. Conductivity Measurements. Solid State Ionics 1983;vols. 9–10:1093. Part 2.

[20] Newman J. Electrochemical Systems. Englewood Cliffs, NJ: Prentice Hall; 1991.

[21] Motupally S, Becker AJ, Weidner JW. Diffusion of Water in Nafion 115 Membranes. Journal of the Electrochemical Society 2000;vol. 147:3171.

[22] Nguyen TV, White RE. A Water and Heat Management Model of Proton-Exchange-Membrane Fuel Cells. J. Electrochem. Soc. 1993;140:2178–86.

[23] Husar A, Higier A, Liu H. In situ Measurements of Water Transfer Due to Different Mechanisms in a Proton Exchange Membrane Fuel Cell. Journal of Power Sources 2008;183:240–6.

[24] Buchi FN, Scherer GG. Investigation of the Transversal Water Profile in Nafion Membranes in Polymer Electrolyte Fuel Cells. Journal of the Electrochemical Society 2000;vol. 148(No. 3):A181–8.

[25] Janssen GJM, Overvelde MLJ. Water Transport in the Proton-Exchange-Membrane Fuel Cell: Measurements of the Effective Drag Coefficient. Journal of Power Sources 2001;vol. 101:117–25.

[26] Jeo RS, McBreen J. Transport Properties of Nafion Membranes in Electrochemically Regenerative Hydrogen/Halogen Cells. Journal of the Electrochemical Society 1979;vol. 126:1682.

[27] Ogumi Z, Takehara Z, Yoshizawa S. Gas Permeation in SPE Method. Journal of the Electrochemical Society 1984;vol. 131:769.

[28] Bernardi DM, Verbrugge MW. A Mathematical Model of the Solid-Polymer-Electrolyte Fuel Cell,. Journal of the Electrochemical Society 1992;vol. 139:2477.

[29] Sakai TH, Takeraka Torikai E. Gas Diffusion in the Dried and Hydrated Nafions. Journal of the Electrochemical Society 1986;vol. 133(No. 1):88–92.

[30] Gubler L, Sherer GG. Trends for Fuel Cell Membrane Development. Desalination 2010;250:1034–7.

[31] Lin J-C, Kunz HR, Fenton JM. Membrane/Electrode Additives for Low-Humidification Operation. In: Vielstich W, Gasteiger HA, Lamm A, editors. Handbook of Fuel Cells: Fundamentals, Technology, and Applications, Volume 3. John Wiley & Sons; 2003. p. 456–63.

[32] Jones DJ, Rozière J. Inorganic/Organic Composite Membranes. In: Vielstich W, Gasteiger HA, Lamm A, editors. Handbook of Fuel Cells: Fundamentals, Technology, and Applications, vol 3. John Wiley & Sons; 2003. p. 447–55.

[33] Chalkova E, Pague MB, Fedkin MV, Wesolowski DJ, Lvov SN. J. Electrochem. Soc. 2005;152:A1035–40.

[34] Schmidt TJ, Baurmeister J. Electrochem. Soc. Trans. 2006;3(1):861–9.

[35] Xiao L, Zhang H, Scanlon E, Ramanathan LS, Choe E-W, Rogers D, et al. Chem. Mater 2005;17:5328–33.

[36] Paganin VA, Ticianelli EA, Gonzales ER. Development and Electrochemical Studies of Gas Diffusion Electrodes for Polymer Electrolyte Fuel Cells. Journal of Applied Electrochemistry 1996;vol. 26:297–304.

[37] Ralph TR, Hogarth MP. Catalysis for Low Temperature Fuel Cells, Part I: The Cathode Challenges,. Platinum Metals Review 2002;vol. 46(No. 1):3–14.

[38] Gasteiger HA, Gu W, Makharia R, Mathias MF. Catalyst Utilization and Mass Transfer Limitations in the Polymer Electrolyte Fuel Cells. Orlando, FL: Electrochemical Society Meeting; September 2003.

[39] Uribe F, Zawodzinski T, Valerio J, Bender G, Garzon F, Saab A, et al. Fuel Cell Electrode Optimization for Operation on Reformate and Air. In: Proc. 2002 Fuel Cells Lab R&D Meeting. Golden, CO: DOE Fuel Cells for Transportation Program; May 9, 2002.

[40] Qi Z, Kaufman A. Low Pt Loading High-Performance Cathodes for PEM Fuel Cells. Journal of Power Sources 2003;vol. 113:37–43.

[41] Sasikumar G, Ihm JW, Ryu H. Dependence of Optimum Nafion Content in Catalyst Layer on Platinum Loading. Journal of Power Sources 2004;vol. 132:11–7.

[42] Debe MK, Steinbach AJ, Hendricks SM, Kurkowski MJ, Vernstrom GD, Hester AE, et al. Advanced Cathode Catalysts and Supports for PEM Fuel Cells, DOE Hydrogen and Fuel Cells Program. FY2011 Annual Progress Report 2011:699–707.

[43] Debe MK, Steinbach AJ, Vernstrom GD, Hendricks SM, Kurkowski MJ, Atanasoski RT, et al. Extraordinary Oxygen Reduction Activity of Pt3Ni7. Journal of Electrochemical Society 2011;158(8):B910–8.

[44] Adzic R, Vukmirovic M, Sasaki K, Wang J, Shao-Horn Y, O'Malley R. Contiguous Platinum Monolayer Oxygen Reduction Electrocatalysts on High-Stability Low-Cost Supports, DOE Hydrogen and Fuel Cells Program. FY2011 Annual Progress Report 2011:729–33.

[45] Wu G, More KL, Johnston CM, Zelenay P. High-Performance Electrocatalysts for Oxygen Reduction Derived from Polyaniline, Iron, and Cobalt. Science 2011;332:443–7.

[46] Burheim OS. Thermal Signature and Thermal Conductivities of PEM Fuel Cells. Ph.D. Thesis. Trondheim: Norwegian University of Science and Technology; 2009.

[47] Mathias MF, Roth J, Fleming J, Lehnert W. Diffusion Media Materials and Characterization. In: Vielstich W, Lamm A, Gastegier HA, editors. Handbook of Fuel Cells, Fundamentals, Technology and Applications. Fuel Cell Technology and Applications, vol. 3. New York: John Wiley & Sons; 2003. p. 517–37.

[48] U.S. Department of Energy, Transportation Fuel Cell Power Systems. 2000 Annual Progress Report. Washington, DC: U.S. Department of Energy; 2000.

[49] Barbir F, Braun J. Development of Low Cost Bi-Polar Plates for PEM Fuel Cells. In: Proc. Fuel Cell 2000 Research & Development. Philadelphia: Strategic Research Institute Conference; September 2000.

[50] Wilkinson DP, Lamont GJ, Voss HH, Schwab C. Method of Fabricating an Embossed Fluid Flow Field Plate. U.S. Patent #5 1996;527:363.

[51] MatWeb Materials Properties Data, www.matweb.com (accessed April 2012).

[52] Smits FM. Measurement of Sheet Resistivities with the Four-Point Probe. Bell System Technical Journal May 1958:711–8.

[53] Barbir F, Braun J, Neutzler J. Properties of Molded Graphite Bi-Polar Plates for PEM Fuel Cells. International Journal on New Materials for Electrochemical Systems 1999;(No. 2):197–200.

[54] Mishra V, Yang F, Pitchumani R. Electrical Contact Resistance Between Gas Diffusion Layers and Bi-Polar Plates in a PEM Fuel Cell. In: Proc. 2nd International Conference on Fuel Cell Science, Engineering, and Technology; 2004. Rochester, NY.

[55] Majumdar A, Tien CL. Fractal Network Model for Contact Conductance. Philadelphia, PA: Joint ASME/AIChE National Heat Transfer Conference; 1989. 1–9.

CHAPTER FIVE

Fuel Cell Operating Conditions

Fuel cell operating conditions include:
- Pressure
- Temperature
- Flow rates of reactant gases
- Humidity of reactant gases

Typical PEM fuel cell operating conditions are listed in Table 5-1. It should be mentioned that the values listed in Table 5-1 do not represent rigid boundaries but rather the typical values.

These parameters are discussed in the following sections.

5.1 OPERATING PRESSURE

A fuel cell may be operated at ambient pressure or it may be pressurized. As we have already learned, a fuel cell gains some potential when the operating pressure is increased (Figure 5-1, for example [1]), thus generating more power. The gain is not linear with pressure, but it is proportional to the logarithm of pressure ratio. However, pressurization of reactant gases requires power, which, as illustrated in Chapter 9, may offset the gain. Whether there will be any net power gain depends on fuel cell polarization curves, efficiency of the compression devices, system configuration, and so on and therefore must be evaluated for each fuel cell system. The issue of pressurization is also related to the issue of water management, which in turn

Table 5-1 Typical PEM Fuel Cell Operating Conditions

	H₂/air: Ambient to 400 kPa
Pressure	H_2/O_2: up to 1,200 kPa
Temperature	50°C to 80°C
Flow rates	H_2: 1 to 1.2
	O_2: 1.2 to 1.5
	Air: 2 to 2.5
Humidity of reactants	H_2: 0 to 125%
	O_2/Air: 0 to 100%

PEM Fuel Cells
ISBN 978-0-12-387710-9

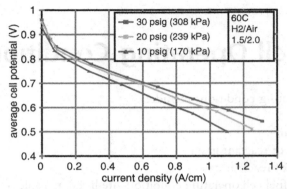

Figure 5-1 Fuel cell performance at various operating pressures [1].

is related to the operating temperature. This issue, therefore, must be addressed from a system perspective.

For a hydrogen/oxygen fuel cell, when both reactants are already stored in pressurized tanks there is no power penalty for compression, so such fuel cells usually operate at elevated pressures, typically from 3 bars up to 10–12 bars. (Note that because of logarithmic dependency the voltage/power gain in going from 10 to 100 bars would be the same as going from 1 to 10 bars.) However, for hydrogen/air fuel cells, pressurization of air requires a mechanical device, a blower or a compressor, which adds to the complexity of the system and which requires power. For such systems the choice is to operate at ambient conditions at the fuel cell exhaust or to operate at elevated pressure, typically up to 3 bars.

When a fuel cell is fed the reactant gases from a pressurized tank, its pressure is controlled by a backpressure regulator placed at the outlet (Figure 5-2a). This pressure regulator keeps the desired preset pressure at the fuel cell outlet. Very often in laboratory settings the inlet pressure is not even recorded. The inlet pressure is always higher because of inevitable pressure drop from inlet to outlet due to reactants' passage through tiny channels inside the fuel cell. However, when the reactant gas (for example, air) is fed to a fuel cell by a mechanical device, a blower or a compressor, which is the case in any practical system, it is the inlet pressure that matters (Figure 5-2b). The compressor or the blower must be capable of delivering the required flow rate at that pressure. The backpressure regulator may still be used to pressurize the cell or, if no backpressure regulator is used, the gas leaves the cell at atmospheric pressure. Note that atmospheric pressure may vary, too, depending on weather conditions or elevation. (Many fuel cell experiments

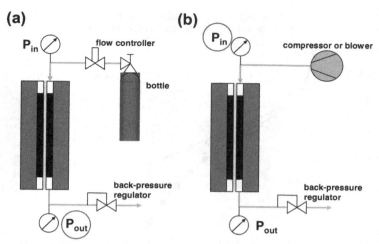

Figure 5-2 Fuel cell operating pressure as a function of reactant gas supply: (a) supply from a high-pressure tank, (b) supply by a mechanical device, a compressor or blower.

reported by the Los Alamos National Laboratory are actually conducted at substandard pressure because of elevation.)

5.2 OPERATING TEMPERATURE

In general, a higher operating temperature results in higher cell potential; however, for each fuel cell design there is an optimal temperature. Figure 5-3 shows an example of how an optimum operating temperature exists [2]. For that particular fuel cell, the optimum operating temperature appears to be between 75°C and 80°C. Operation above 80°C results in diminished performance.

A PEM fuel cell does not have to be heated up to the operating temperature to become operational. As already shown in Figure 3-19, a fuel cell can operate even at freezing conditions, but it cannot reach its full rated power. The car companies have put a lot of effort into investigating fuel cell survival and cold-start capabilities for temperatures as low as −30°C.

The upper limit of operating temperature is determined by the membrane. Because the main function of the proton exchange membrane depends on its state of hydration, the typical PSA membranes cannot operate above 100°C without a danger of being dehydrated and eventually damaged. This temperature is already close to the glass transition temperature of the polymer. For that reason PEM fuel cells with PSA or similar

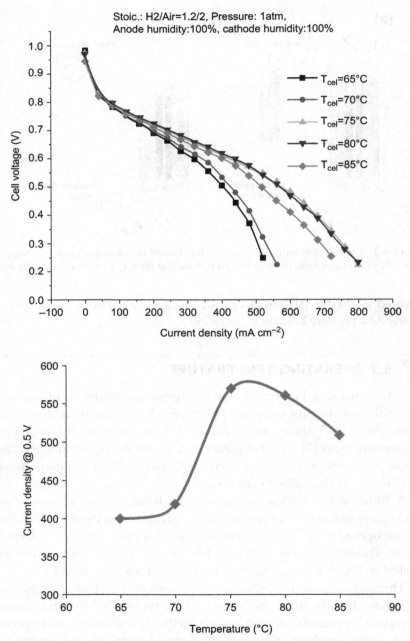

Stoic.: H2/Air=1.2/2, Pressure: 1atm,
Anode humidity:100%, cathode humidity:100%

- ■ T_{cel}=65°C
- ● T_{cel}=70°C
- ▲ T_{cel}=75°C
- ▼ T_{cel}=80°C
- ◆ T_{cel}=85°C

Figure 5-3 Effect of operating temperature on fuel cell performance [2].

membranes are rarely operated at temperatures above 90°C but more typically up to 80°C. Different types of membrane (for example, phosphoric acid, H3PO4, doped polybenzimidazole, or PBI membranes), in which water as the proton solvent is replaced by the acid, allow operation at 140°C or higher.

The operating temperature of practical fuel cells, similarly to operating pressure, must be selected from the system perspective, taking into account not only the cell performance but also the system requirements, particularly the size and parasitic power requirements of the heat management subsystem. A fuel cell generates heat as a byproduct of the electrochemical reaction. To maintain the desired temperature, heat must be taken away from a fuel cell. Some heat dissipates from the outer surface of the fuel cell and some must be taken away with a cooling system. The medium that takes away the heat may be air, water, or a special coolant. The inner design of the fuel cell must allow the heat transfer to occur. Sometimes small fuel cells need a heater to reach the operating temperature. In these fuel cells so much heat is being taken away from the outer surface that an additional heater is required. This, of course, is not very practical, but it is sometimes necessary for testing fuel cells at a desired temperature.

The following is the fuel cell heat balance:

$$Q_{gen} + Q_{react,in} = Q_{reac,out} + Q_{dis} + Q_{cool} \qquad (5\text{-}1)$$

In other words, the heat generated in the fuel cell plus the heat brought into the cell with reactant gases is taken away from the cell by the reactant gases leaving the cell, by heat dissipation from the cell surface to the surrounding, and by the coolant.

The temperature inside a fuel cell may not be uniform; it may vary from inlet to outlet, from inside out, or from cathode to anode. So which temperature is the cell temperature? The cell temperature may be approximated by the following temperatures, which are much easier to measure than the cell temperature:

• Surface temperature
• Temperature of air leaving the cell
• Temperature of coolant leaving the cell

Because of finite temperature differences needed for heat transfer inside a fuel cell, none of these is exactly the cell operating temperature. The surface temperature is clearly lower than the temperature inside a fuel cell in a case in which the fuel cell is heating itself, and it is actually higher than the inside temperature if the fuel cell is heated with the heating pads on its surface. Because most of the losses in the fuel cell may be associated with the cathode reaction, the temperature of air exiting the fuel cell is a good approximation of the cell operating temperature, although again the temperature inside a fuel cell must be at least slightly higher than the air temperature. In a case in which the cell temperature is maintained by the flow of coolant through the cell, the coolant outlet temperature may be used as the operating temperature. The accuracy of these approximations depends on thermal conductivity of the cell materials and air and coolant flow rates.

5.3 REACTANT FLOW RATES

The reactants' flow rate at the inlet of a fuel cell must be equal to or higher than the rate at which those reactants are being consumed in the cell. The rates (in mol s^{-1}) at which hydrogen and oxygen are consumed and water is generated are determined by Faraday's law:

$$\dot{N}_{H2} = \frac{I}{2F} \tag{5-2}$$

$$\dot{N}_{O2} = \frac{I}{4F} \tag{5-3}$$

$$\dot{N}_{H2O} = \frac{I}{2F} \tag{5-4}$$

where:
$\dot{N}$ = consumption rate (mol s^{-1})
I = current (A)
F = Faraday's constant (C mol^{-1})
The mass flow rates of reactants' consumption (in $g\ s^{-1}$) are then:

$$\dot{m}_{H2} = \frac{I}{2F}M_{H2} \tag{5-5}$$

$$\dot{m}_{O2} = \frac{I}{4F}M_{O2} \tag{5-6}$$

The mass flow rate of water generation (in $g\ s^{-1}$) is:

$$\dot{m}_{H2O} = \frac{I}{2F}M_{H2O} \tag{5-7}$$

Most often, the flow rates of gases are expressed in volumetric units, that is, normal liters per minute (Nl min^{-1}) or normal liters per second (Nl s^{-1}), or normal cubic meters per minute ($Nm^3 min^{-1}$) or normal cubic meters per hour ($Nm^3 h^{-1}$). Normal liter or normal cubic meter is a quantity of gas that would occupy 1 liter or 1 cubic meter, respectively, of volume at normal conditions, namely atmospheric pressure, 101.3 kPa, and 0°C. Often in practice and in technical literature, standard conditions are used, such as standard liter per minute (slpm), standard cubic foot per minute (scfm), or standard cubic foot per hour (scfh), but there is a great deal of confusion about standard conditions that vary from source to source: values of 15°C, 15.6°C (60°F), 20°C (68°F), and 21.1°C (70°F) are being used for standard temperature, and values of 101 kPa (1 atm or 14.696 psi), 1 bar (0.987 atm or 14.5 psi), or 30 inHg (1.06 bar or 14.73 psi) have been used for standard pressure. Note that in atmospheric science standard atmosphere is defined at sea level at a temperature of 15°C (59°F) and pressure of 101.3 kPa, whereas most chemical handbooks and textbooks refer to 25°C as standard or reference temperature. Therefore, to avoid confusion about standard temperature and pressure, it is better to use units such as normal liter (Nl) or normal cubic meter (Nm^3), which is an SI unit. Also note that a liter is not an SI unit, but it has been accepted by the International Committee for Weights and Measures (CIPM). The symbol for liter is the letter *1*, but to avoid the risk of confusion between the letter *l* and the number 1, the General Conference on Weights and Measures (CGPM) adopted the alternative symbol for the liter, *L*, which is more widely used in the United States [3]. Thus, both *l* and *L* are internationally accepted symbols for the liter. The script letter ℓ or abbreviation *lit* are not approved symbols for the liter.

For any ideal gas, mols and volumes are directly related by the equation of state:

$$PV = NRT \qquad (5\text{-}8)$$

Molar volume is:

$$v_m = \frac{V}{N} = \frac{RT}{P} \qquad (5\text{-}9)$$

At normal conditions, that is, atmospheric pressure 101.3 kPa and 0°C, molar volume is:

$$v_m = \frac{RT}{P} = \frac{8.314 \times 273.15}{101,300} = 0.022420 \ m^3 \ mol^{-1} = 22.42 \ l \ mol^{-1}$$
$$= 22,420 \ cm^3 \ mol^{-1}$$

The volumetric flow rates of reactants consumption (in normal liters per minute, or Nl min^{-1}) are:

$$\dot{V}_{H2} = 22.42 \times 60 \times \frac{I}{2F} \qquad (5\text{-}10)$$

$$\dot{V}_{O2} = 22.42 \times 60 \times \frac{I}{4F} \qquad (5\text{-}11)$$

Consumption of the reactants, hydrogen and oxygen, and water generation in the fuel cell are summarized in Table 5-2.

Table 5-2 Reactants' Consumption and Water Generation (per Amp and per Cell)

	Hydrogen Consumption	Oxygen Consumption	Water Generation (liq.)
mol/s	5.18×10^{-6}	2.59×10^{-6}	5.18×10^{-6}
g/s	10.4×10^{-6}	82.9×10^{-6}	93.3×10^{-6}
Nl min^{-1}	6.970×10^{-3}	3.485×10^{-3}	N/A
Nm^3 h^{-1}	0.418×10^{-3}	0.209×10^{-3}	N/A

The reactants may, and in some cases must, be supplied in excess of consumption. For example, this is always necessary on the cathode side where water is produced and must be carried out from the cell with excess flow. The ratio between the actual flow rate of a reactant at the cell inlet and the consumption rate of that reactant is called the *stoichiometric ratio*, S:

$$S = \frac{\dot{N}_{act}}{\dot{N}_{cons}} = \frac{\dot{m}_{act}}{\dot{m}_{cons}} = \frac{\dot{V}_{act}}{\dot{V}_{cons}} \qquad (5\text{-}12)$$

Hydrogen may be supplied at the exact rate at which it is being consumed, in so-called *dead-end mode* (Figure 5-4a). If hydrogen is available at elevated pressure, such as in a high-pressure storage tank, the dead-end mode does not require any controls, that is, hydrogen is being supplied as it is being consumed. In a dead-end mode, $S = 1$. If hydrogen loss due to crossover permeation or internal currents is taken into account, then the hydrogen flow rate at the fuel cell inlet is slightly higher than the consumption rate corresponding to the electrical current being generated:

$$S_{H2} = \frac{\dot{N}_{H2,cons} + \dot{N}_{H2,loss}}{\dot{N}_{H2,cons}} > 1 \qquad (5\text{-}13)$$

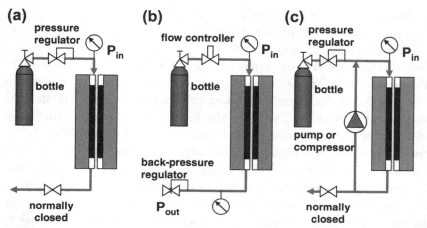

Figure 5-4 Modes of reactant supply: (a) dead-end mode, (b) flow-through mode, and (c) recirculation mode.

Fuel utilization, that is, a ratio between the fuel consumed in the electrochemical reaction and fuel supplied to the fuel cell, is the reverse of the previously defined stoichiometric ratio:

$$\eta_{fu} = \frac{1}{S_{H2}} \qquad (5\text{-}14)$$

Thus for a dead-end operation the fuel utilization is:

$$\eta_{fu} = \frac{\dot{N}_{H2,cons}}{\dot{N}_{H2,cons} + \dot{N}_{H2,loss}} \qquad (5\text{-}15)$$

Even in a dead-end mode, hydrogen has to be periodically purged because of accumulation of inert gases or water that may be present in the feed hydrogen or that can permeate the polymer membrane. The frequency and duration of purges depend on purity of hydrogen, rate of nitrogen permeation of the membrane, and water net transport through the membrane. In calculating the fuel cell efficiency, the loss of hydrogen due to purging must be taken into account through fuel utilization.

$$\eta_{fu} = \frac{\dot{N}_{H2,cons}}{\dot{N}_{H2,cons} + \dot{N}_{H2,loss} + \dot{N}_{H2,prg}\,\tau_{prg}\,f_{prg}} \qquad (5\text{-}16)$$

where:

$\dot{N}_{H2,cons}$ = rate of hydrogen consumption (mol s^{-1})
$\dot{N}_{H2,loss}$ = rate of hydrogen loss (mol s^{-1})

$\dot{N}_{H2,prg}$ = rate of hydrogen purge (mol s^{-1})

τ_{prg} = duration of hydrogen purge (s)

f_{prg} = frequency of purges (s^{-1})

Instead of purging, hydrogen may be supplied in excess ($S > 1$) in so-called *flow-through mode* (Figure 5-4b). In that case fuel utilization is given by Equation (5-14). Air is almost always supplied in a flow-through mode, with stoichiometry about $S = 2$ or higher. In the case of pure reactants (hydrogen and/or oxygen), a recirculation mode may be utilized (Figure 5-4c). In this case the unused gas is returned to the inlet by a pump or compressor, or sometimes a passive device such as an ejector (based on a Venturi tube) may be employed. Note that in case of recirculation, a cell may operate at a stoichiometric ratio much higher than 1, but because unused reactant (hydrogen or oxygen) is not wasted but returned for consumption back to the cell inlet, fuel or oxidant utilization on a system level is high (close to 1). However, periodic purging may still be necessary to get rid of the inert gases that may accumulate in anode and cathode compartments. In that case Equation (5-16) may be used to calculate the fuel or oxygen utilization.

In general, higher flow rates result in better fuel cell performance. This is particularly true when either hydrogen or oxygen is not pure. Although pure hydrogen may be supplied in a dead-end mode ($S = 1$) or with a stoichiometry slightly higher than 1 (1.05 to 1.2), hydrogen in a mixture of gases (such as that coming out of a fuel processor) must be supplied with higher stoichiometries (1.1 to 1.5). The exact flow rate is actually a design variable. If the flow rate is too high, the efficiency will be low (because hydrogen would be wasted), and if the flow rate is too low, the fuel cell performance may suffer.

Similarly, for pure oxygen flow rate, the required stoichiometry is between 1.2 and 1.5, but when air is used, the typical stoichiometry is 2 or higher. Although higher air flow rates result in better fuel cell performance, as shown in Figure 5-5, the air flow rate is also a design variable. Air is supplied to the cell by means of a blower or a compressor (depending on operating pressure) whose power consumption is directly proportional to the flow rate. Therefore, at higher air flow rates the fuel cell may perform better, but power consumption of a blower or particularly of a compressor may significantly affect system efficiency. This idea is discussed in Chapter 9 in greater detail.

There are at least two reasons that fuel cell performance improves with excess air flow rate:

- Higher flow rate helps remove product water from the cell.
- Higher flow rate keeps oxygen concentration high.

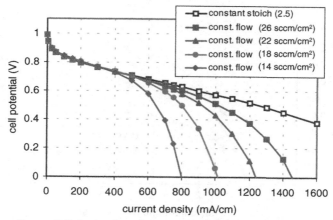

Figure 5-5 Fuel cell performance at different air-flow rates.

Because oxygen is being consumed in the cell, its concentration at the cell outlet depends on the flow rate. If air is supplied at the exact stoichiometric ratio ($S = 1$), all the oxygen in the supplied air will be consumed in the fuel cell, that is, oxygen concentration in air exhaust will be zero. The higher the flow rate, the higher the oxygen concentration at the outlet and throughout the cell, as shown in Figure 5-6.

If oxygen volume or molar fraction at the fuel cell inlet is $r_{O2,in}$, then the oxygen volume or molar fraction at the outlet is:

$$r_{O2,out} = \frac{S - 1}{\dfrac{S}{r_{O2,in}} - 1} \tag{5-17}$$

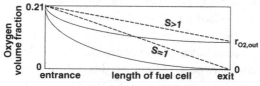

Figure 5-6 Oxygen volume or molar fraction through a fuel cell. Dashed lines represent an ideal case where the rate of oxygen consumption is constant; solid lines are more realistic because the oxygen consumption rate, (i.e., current generation), is a function of oxygen concentration and thus as oxygen content decreases so does the oxygen consumption rate.

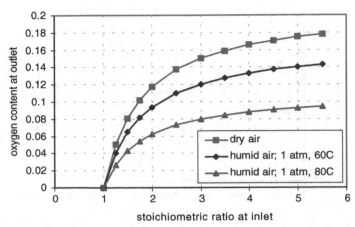

Figure 5-7 Oxygen content (by volume) at the fuel cell outlet as a function of stoichiometric ratio.

From Figure 5-7 it is clear that oxygen content at the outlet rapidly diminishes with stoichiometric ratios below 2. Because air at the outlet is almost always saturated with water vapor, oxygen content is even lower than it would be in dry air. For example, oxygen content at the fuel cell outlet at atmospheric pressure and 80°C operating with a stoichiometric ratio of 2 would be only 6%. An increase of stoichiometric ratio to 3 would result in oxygen content increasing to 8%. Above the stoichiometric ratio of 3 the curve levels off, and little gain in oxygen concentration and fuel cell performance is achieved.

5.4 REACTANT HUMIDITY

Because the membrane requires water to maintain protonic conductivity, as shown in Chapter 4, both reactant gases typically must be humidified before entering the cell. In some cases they have to be saturated, but in some cases excess humidity is needed on the anode side, and less than saturated conditions may be sufficient on the cathode side.

The *humidity ratio* is a ratio between the amount of water vapor present in a gas stream and the amount of dry gas. The *humidity mass ratio* (grams of water vapor/grams of dry gas) is:

$$x = \frac{m_v}{m_g} \qquad (5\text{-}18)$$

The *humidity molar ratio* (mols of water vapor/mols of dry gas) is:

$$\chi = \frac{N_v}{N_g} \tag{5-19}$$

The relationship between mass and molar humidity ratios is:

$$x = \frac{M_w}{M_g} \chi \tag{5-20}$$

Molar ratio of gases is the same as a ratio of partial pressures:

$$\chi = \frac{p_v}{p_g} = \frac{p_v}{P - p_v} \tag{5-21}$$

where P is the total pressure and p_v and p_g are the partial pressures of vapor and gas, respectively.

Relative humidity is a ratio between the water vapor partial pressure, p_v, and saturation pressure, p_{vs}, which is the maximum amount of water vapor that can be present in gas for given conditions:

$$\varphi = \frac{p_v}{p_{vs}} \tag{5-22}$$

Saturation pressure is a function of temperature only. The values of saturation pressure may be found in thermodynamic tables. *ASHRAE Fundamentals* [4] provides an equation that allows us to calculate the saturation pressure (in Pa) for any given temperature between 0°C and 100°C:

$$p_{vs} = e^{aT^{-1}+b+cT+dT^2+eT^3+f\ln(T)} \tag{5-23}$$

where a, b, c, d, e, and f are the coefficients:

$a = -5800.2206$
$b = 1.3914993$
$c = -0.048640239$
$d = 0.41764768 \times 10^{-4}$
$e = -0.14452093 \times 10^{-7}$
$f = 6.5459673$

The humidity ratios may be expressed in terms of relative humidity, saturation pressure, and total pressure by combining Equations (5-20), (5-21), and (5-22):

$$x = \frac{M_w}{M_a} \frac{\varphi p_{vs}}{P - \varphi p_{vs}} \tag{5-24}$$

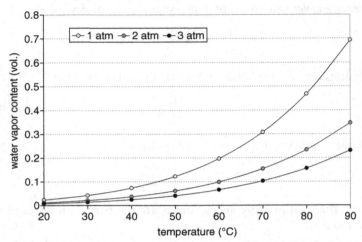

Figure 5-8 Water vapor content in a gas at different pressures and temperatures.

and

$$\chi = \frac{\varphi p_{vs}}{P - \varphi p_{vs}} \tag{5-25}$$

Figure 5-8 shows the water vapor content in gas at different temperatures and pressures. As it follows from Equation (5-25), at lower pressure a gas can contain more water vapor, and as it follows from Equations (5-23) and (5-25), water content in gas increases exponentially with temperature. At 80°C and ambient pressure, water content in air is close to 50%. Water vapor content by volume is:

$$r_{H2O,v} = \frac{\chi}{\chi + 1} = \frac{\varphi p_{vs}}{P} \tag{5-26}$$

Enthalpy of dry gas is:

$$h_g = c_{p,g} t \tag{5-27}$$

where:

h_g = enthalpy of dry gas, J g^{-1}
c_{pg} = specific heat of gas, J g^{-1} K^{-1}
t = temperature in °C

Note that this equation allows the use of degrees Celsius by assuming that a reference zero state is at 0°C (i.e., $h_0 = 0$), so 1 degree of temperature difference on the Celsius scale is equal to 1 Kelvin.

Enthalpy of water vapor is [5]:

$$h_v = c_{p,v}\, t + h_{fg} \tag{5-28}$$

where h_{fg} = heat of evaporation = 2500 J g^{-1} at 0°C.

Enthalpy of humid gas is then [5]:

$$h_{vg} = c_{p,g}t + x(c_{p,v}t + h_{fg}) \tag{5-29}$$

and the unit is Joules per gram of dry gas.

Enthalpy of liquid water is:

$$h_w = c_{p,w}t \tag{5-30}$$

If the gas contains both water vapor and liquid water, such as may be the case at the fuel cell outlet, its enthalpy is [5]:

$$h_{vg} = c_{p,g}t + x_v(c_{p,v}t + h_{fg}) + x_w c_{p,w}t \tag{5-31}$$

where x_v = water vapor content (in grams of vapor per gram of dry gas) and x_w = liquid water content (in grams of liquid water per gram of dry gas). The total water content is:

$$x = x_v + x_w \tag{5-32}$$

Note that when $x_w = 0$, then $x = x_v$; and when $x_w > 0$, then $x_v = x_{vs}$. (When there is liquid water present in gas, gas is already saturated with vapor.)

The processes with humid gases are best seen in an *h-x diagram*, or a so-called *Mollier diagram* (Figure 5-9) [5], where the *x*-axis is tilted by:

$$\left(\frac{dh}{dx}\right)_{t=0°\,C} = h_{fg} \tag{5-33}$$

The saturation line divides the diagram into two distinct regions: the unsaturated region above the saturation line and the fog (mist) region below the saturation line (Figure 5-9). The state of humid gas is determined by its temperature and relative humidity, or temperature and water content, or temperature and dew point. The dew point temperature is the temperature at which all the water vapor present in gas would condense.

The reactant gases in PEM fuel cells are typically humidified. Most commonly, both reactant gases are required to be saturated at the cell operating temperature, although there are cell and MEA designs that require either subsaturated conditions or oversaturation. The process of humidification may be as simple as water or steam injection. In either case, to get

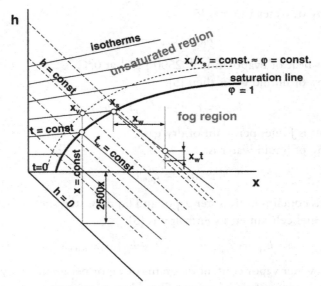

Figure 5-9 Mollier h-x diagram for moist air.

from a dry gas or from ambient–temperature air to fully saturated gas at cell operating temperature, both water and heat are required. Injection of water in relatively dry gas would result in saturation at a temperature lower than that of starting air and water (as shown in Figure 5-10).

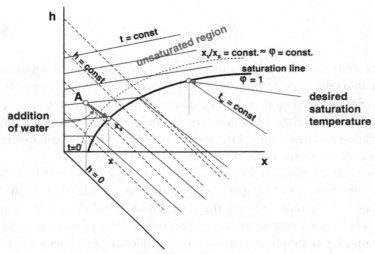

Figure 5-10 Illustration of humidification process in an h-x diagram.

The amount of heat required for humidification may be quite significant, especially if ambient pressure air is to be saturated at relatively high temperatures.

Example

A fuel cell with a 300-cm^2 active area operates at 0.6 A/cm^2 and 0.65 V. Air is supplied at a stoichiometric ratio of 2 and at a pressure of 1.15 bar, and it is humidified by injecting hot water (60°C) just before the stack inlet. Ambient air conditions are 1 bar, 20°C, and 60% RH. The requirement is to saturate the air at cell operating temperature of 60°C. Calculate the air flow rate, the amount of water required for 100% humidification of air at the inlet, and heat required for humidification.

Oxygen consumption is (Equation 5-3):

$$\dot{N}_{O2,cons} = \frac{I}{4F} = \frac{0.6 \, Acm^{-2} \times 300 \, cm^2}{4 \times 96,485} = 0.466 \times 10^{-3} \, \text{mol s}^{-1}$$

Oxygen flow rate at cell inlet is (Equation 5-12):

$$\dot{N}_{O2,act} = S_{O2}\dot{N}_{O2,cons} = 2 \times 0.466 \times 10^{-3} = 0.933 \times 10^{-3} \, \text{mol s}^{-1}$$

Air flow rate at cell inlet is:

$$\dot{N}_{air,in} = \dot{N}_{O2,in}\frac{1}{r_{O2}} = \frac{0.933 \times 10^{-3}}{0.21} = 4.44 \times 10^{-3} \, \text{mol s}^{-1}$$

$$\dot{m}_{air,in} = \dot{N}_{air,in}M_{air} = 4.44 \times 10^{-3} \, \text{mol s}^{-1} \times 28.85 \, \text{g mol}^{-1}$$
$$= 0.128 \, \text{g s}^{-1} \quad \text{Answer}$$

where M_{air} = molecular weight of air = $0.21 \times 32 + 0.79 \times 28$ = 28.85 g mol^{-1}.

The amount of water in the air at the cell inlet (saturated at 1.15 bar and 60°C) is (Equation 5-18):

$$\dot{m}_{H2O,in} = x_s\dot{m}_{air,in}$$

where x_s is water content in air at saturation, that is, $\varphi = 1$ (Equation 5-24):

$$x_s = \frac{M_{H2O}}{M_{air}} \frac{p_{vs}}{P - p_{vs}}$$

where p_{vs} is the saturation pressure at 60°C (Equation 5-23) and P is total pressure, 1.15 bar = 115 kPa.

$$p_{vs} = e^{aT^{-1}+b+cT+dT^2+eT^3+f \, \ln(T)} = 19.944 \, \text{kPa (for T} = 333.15 \, \text{K)}$$

$$x_s = \frac{M_{H2O}}{M_{air}} \frac{p_{vs}}{P - p_{vs}} = \frac{18}{28.85} \frac{19.944}{115 - 19.944} = 0.131 \, g_{H_2O}/g_{air}$$

$$\dot{m}_{H2O,in} = x_s \dot{m}_{air,in} = 0.131 \text{ g}_{H2O}/\text{g}_{air} \times 0.128 \text{ g}_{air} \text{ s}^{-1}$$
$$= 0.0168 \text{ g}_{H2O} \text{ s}^{-1}$$

Ambient air already has some water in it (60% relative humidity at 20°C; $p_{vs} = 2,339$ kPa):

$$x_s = \frac{M_{H2O}}{M_{air}} \frac{\varphi p_{vs}}{P - \varphi p_{vs}} = \frac{18}{28.85} \frac{0.6 \times 2.339}{100 - 0.6 \times 2.339} = 0.00888 \text{ g}_{H2O}/\text{g}_{air}$$

$$\dot{m}_{H2O,amb} = x_s \dot{m}_{air,in} = 0.00888 \text{ g}_{H2O}/\text{g}_{air} \times 0.128 \text{ g}_{air} \text{ s}^{-1}$$
$$= 0.0011 \text{ g}_{H2O} \text{ s}^{-1}$$

Therefore, the amount of water needed for humidification of air at cell inlet is:

$$\dot{m}_{H2O} = 0.0168 - 0.0011 = 0.0157 \text{ g}_{H2O} \text{ s}^{-1} \quad \text{Answer}$$

Heat required for humidification may be calculated from the heat balance:

$$H_{air,amb} + H_{H2O} + Q = H_{air,in} \Rightarrow Q = H_{air,in} - H_{air,amb} - H_{H2O}$$

Enthalpy of humid air is (Equation 5-29):

$$h_{vair} = c_{p,air} \, t + x(c_{p,v} \, t + h_{fg})$$

Humidified air: $h_{vair,in} = 1.01 \times 60 + 0.131 = (1.87 \times 60 + 2500) = 402.8 \text{ J g}^{-1}$
Ambient air: $h_{vair,amb} = 1.01 \times 20 + 0.00888 = (1.87 \times 20 + 2500) = 42.73 \text{ J g}^{-1}$
Water: $h_{H2O} = c_{p,w} \, t = 4.18 \times 60 = 250.8 \text{ J g}^{-1}$

$$Q = 402.8 \text{ J g}^{-1} \times 0.128 \text{ g/s} - 42.73 \text{ J g}^{-1} \times 0.128 \text{ g s}^{-1} - 250.8 \text{ J g}^{-1}$$
$$\times 0.0157 \text{ g s}^{-1}$$

$$= 51.56 \text{ W} - 5.47 \text{ W} - 3.94 \text{ W} = 42.15 \text{ W} \quad \text{Answer}$$

Just for comparison:
Cell electricity generation: $W_{el} = I \times V = 0.6 \text{Acm}^{-2} \times 300 \text{cm}^2 \times 0.65 \text{ V} = 117 \text{ W}$
Cell efficiency: $\eta = V/1.482 = 0.65/1.482 = 0.439$
Heat generation by the cell: $Q = 117/0.439 - 117 = 149.5 \text{ W}$
Water generation rate:

$$\dot{m}_{H2O,gen} = \frac{I}{2F} M_{H2O} = \frac{0.6 \times 300}{2 \times 96,485} 18 = 0.0168 \text{ g s}^{-1}$$

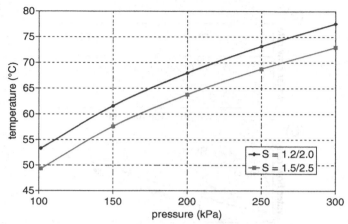

Figure 5-11 Temperatures and pressures at which a fuel cell generates enough water to humidify both hydrogen and air inlet streams.

The fuel cell from this example thus generates more than enough heat and just about enough water needed for humidification of incoming air. With a clever system design it would be possible to capture both heat and water generated by the cell and use it for humidification of incoming air.

Figure 5-11 shows the conditions at which a fuel cell generates enough water for humidification of both air and hydrogen (assuming that both gases are completely dry before humidification). Above the line for given stoichiometry ratios, the need for humidification of reactant gases is greater than the amount of water generated in the stack.

A logical question arises: If a fuel cell generates enough water on the cathode side, why does air have to be humidified before entering the cell? In general, air humidification is needed to prevent drying the portion of the membrane near the air inlet. Figure 5-12a shows that although there is enough water generated in the cell, the air in most of the cell is undersaturated. However, the conditions in Figure 5-12a are not very realistic:

- Air is assumed to enter the cell dry and be heated at the cell operating temperature.
- The cell is isothermal.
- There is no pressure drop.
- Water generation rate (i.e., reaction rate) is constant.

When more realistic conditions are applied, the water profiles in the cell change dramatically, as shown in Figure 5-12b. Conditions are selected so that the product water is sufficient to saturate air at the fuel cell exit. Air

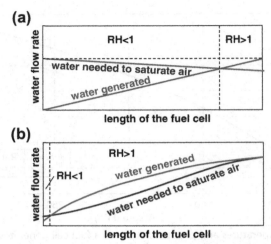

Figure 5-12 Water profiles in a fuel cell: (a) assuming uniform current density distribution and isothermal conditions; (b) assuming realistic current density distribution and air temperature increase from inlet to outlet.

enters the cell at ambient conditions, relatively dry. At low temperatures (20°C to 30°C), small amounts of water are needed to saturate the air, and the product water is more than sufficient. As the air is heated up and its pressure decreases, it needs more and more water. With careful design of air passages and heat transfer inside the fuel cell, it is possible to match the two water profiles even more closely.

Figure 5-13 shows the inlet air water content, shown as the dew point temperature, required to saturate air at the fuel cell outlet for typical air flows. Conditions above the line would result in liquid water at the outlet, whereas conditions below the line would result in undersaturated air at the outlet.

Berning [6] suggested that the dew point temperature of the exhaust gases may be used as a criterion for selecting operating conditions. The dew point temperature at the anode outlet depends on the molar gas flow rates. The molar stream of hydrogen leaving the cell is:

$$\dot{N}_{H2,out} = (S_{H2} - 1)\,\dot{N}_{H2,cons} = (S_{H2} - 1)\,\frac{I}{2F} \qquad (5\text{-}34)$$

The molar flow rate of water leaving the cell is calculated using the definition of the net drag coefficient r_d, defined as [7]:

$$r_d = \frac{\dot{N}_{H2OinH2in} - \dot{N}_{H2OinH2out}}{I/F} \qquad (5\text{-}35)$$

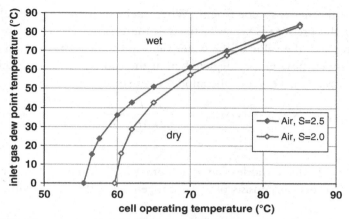

Figure 5-13 Water content in air at inlet (specified as dew point) required to result in saturated conditions at the fuel cell outlet.

Net drag coefficient is a difference between the water flux due to electroosmotic drag and water back diffusion. It is defined the same as the electroosmotic drag, that is, the number of water molecules per proton. It is positive when the net flux is from the anode to the cathode.

For dry hydrogen inlet ($\dot{N}_{H2OinH2in} = 0$) this leads to:

$$\dot{N}_{H2OinH2out} = -r_d\frac{I}{F} \tag{5-36}$$

Molar ratio of water vapor and dry hydrogen gas at the anode outlet is thus:

$$\chi_{H2OinH2out} = \frac{\dot{N}_{H2OinH2out}}{\dot{N}_{H2,out}} = \frac{-r_d\dfrac{I}{F}}{(S_{H2}-1)\dfrac{I}{2F}} = \frac{-2r_d}{(S_{H2}-1)} \tag{5-37}$$

Note that there will be water in the hydrogen gas at the outlet only when the net drag is negative, that is, when the net flux is from the cathode to the anode. The water vapor partial pressure at the anode outlet can then be calculated by combining Equation (5-37) and Equation (5-21):

$$p_{H2OinH2out} = P_{H2out}\frac{2r_d}{2r_d - (S_{H2}-1)} \tag{5-38}$$

From Equation (5-38) it follows that for a given pressure at the anode outlet, the dew point temperature depends only on the stoichiometric flow ratio and the net drag coefficient. When fuel cells operate at an elevated pressure, the pressure is typically fixed at the outlet so that the preceding equation yields the value of the water vapor pressure at the fuel cell outlet.

The molar flow rate of depleted air at the cathode exhaust (neglecting the crossover of nitrogen to the anode) is given by:

$$\dot{N}_{Air,out} = \dot{N}_{Air,in} - \dot{N}_{O2,cons} = \frac{S_{O2}}{0.21}\frac{I}{4F} - \frac{I}{4F} = \frac{I}{4F}\left(\frac{S_{O2}}{0.21} - 1\right)$$

(5-39)

where S_{O2} is the cathode side stoichiometric flow ratio. This equation assumes that dry air consists of 79% nitrogen and 21% oxygen on a molar basis.

The molar water vapor stream leaving the cathode side, assuming the air was dry at the inlet, is:

$$\dot{N}_{H2OinAirout} = \dot{N}_{H2O,gen} - \dot{N}_{H2OinH2out} = \frac{I}{2F} + r_d\frac{I}{F} = \frac{I}{2F}(1 + 2r_d)$$

(5-40)

Note that only if $r_d < 0$ some of the product water will end up on the anode side.

The molar ratio of water vapor and dry gas at the cathode outlet is:

$$\chi_{H2OinAirout} = \frac{\dot{N}_{H2OinAirout}}{\dot{N}_{Air,out}} = \frac{\frac{I}{2F}(1 + 2r_d)}{\frac{I}{4F}\left(\frac{S_{O2}}{0.21} - 1\right)} = \frac{2(1 + 2r_d)}{\left(\frac{S_{O2}}{0.21} - 1\right)}$$

(5-41)

The partial pressure of water vapor in the gas at the cathode outlet finally can be obtained by combining Equations (5-41) and (5-21):

$$P_{H2OinAirout} = P_{Air,out}\frac{2(1 + 2r_d)}{2(1 + 2r_d) + \frac{S_{O2}}{0.21} - 1}$$

(5-42)

Equations (5-38) and (5-42) can be used to calculate the water vapor pressure at the anode and cathode side outlets, respectively, and the dew point temperatures can then be calculated using the following equation, which is a reverse of Equation (5-23):

$$T_{dew} = -2.581 \times 10^{-18}(p_{H2O})^4 + 6.4056 \times 10^{-13}(p_{H2O})^3$$
$$-5.8916 \times 10^{-8}(p_{H2O})^2 + 2.8427 \times 10^{-3}(p_{H2O}) + 22.455$$

(5-43)

where p_{H2O} is the water vapor partial pressure in Pa calculated from Equations (5-38) and (5-42), respectively. Equation (5-43) is an excellent curve-fit between 40°C and 90°C [6].

The dew point temperature depends on the net drag coefficient r_d. However, when the cell is operated on dry reactant gases, the net drag coefficient is typically within a very narrow range of $-0.1 < r_d < 0$ [8]. The dew point temperature depends strongly on the operating pressure, and an increase in cell pressure will lead to an increase in the dew point temperature and consequently increase the danger of condensation and cell flooding.

The preceding equations allow for the construction of charts that show the anode and cathode dew point temperatures T_{dew} as functions of the net drag coefficient r_d and the stoichiometric flow ratio. Due to the dependence of the dew point temperatures on the pressure, different charts have been created for each operating pressure. Figure 5-14 shows exemplarily the calculated dew point temperatures at the anode (left) and cathode (right) for an ambient operating pressure (top), 1.5 bar (middle), and 2.0 bar (bottom).

Combining the preceding analysis of the exhaust gases' dew point with modeling of the processes inside a fuel cell, Berning [6] has suggested that the ideal operating temperature of a fuel cell should be a few degrees above the cathode condensation temperature, which typically ranges between 60°C and 80°C, depending on the pressure and stoichiometric flow ratio. Operating the fuel cell around 10°C above the dew point temperature leads to a dry membrane. When the anode side operates at a stoichiometric flow ratio as low as $\xi = 1.05$, the predicted net drag falls within a very narrow regime and is quite independent of the current density, which greatly helps determine ideal operating conditions. Because the anode-side dew point temperature is very sensitive to rd and increases quickly to values above 80°C, it is advisable to have the anode outlet be the hot end of the fuel cell when operating in counter-flow mode. This can be facilitated by running the coolant counter-flow to the cathode.

Therefore, under certain conditions, operation with dry gases is possible. The criterion for operation with dry gases is that the amount of product water should be sufficient to saturate the outlet gases [9], that is:

$$1 \geq \left[(S_{H2} - 1) + \frac{1}{2}\left(\frac{S_{O2}}{0.21} - 1\right) \right] \frac{p_s(T_{cell})}{P - p_s(T_{cell})} \qquad (5\text{-}44)$$

Figure 5-15 [9] gives the critical gas stoichiometry for typical cell temperatures of 60°C and 80°C and pressure 1.5 and 3 bar. However,

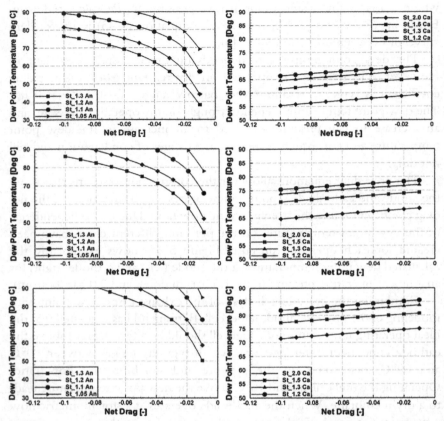

FIGURE 5-14 Dew point temperatures at the anode (left) and the cathode (right) exhaust at ambient operating pressure (top) 1.5 bar (middle) and 2.0 (bottom) for a fuel cell operated with dry gases at the inlet [6].

satisfaction of the preceding equation does not guarantee that the local dehydration will not occur.

Indeed, Tolj et al. [10] have shown what happens inside a fuel cell fed with ambient unhumidified air if the cell hardware is kept at a constant temperature (Figure 5-16). Although the product water is sufficient to humidify the air at the exhaust, the air stream, quickly after entering the cell, heats up, which causes relative humidity to drop below 25%. By the end of the channel the air stream gets fully saturated, but throughout the cathode channel the air is practically dry. This can be avoided if a temperature profile along the cathode channel is established that would prevent air from heating up quickly and that would ensure relative humidity close to 100%

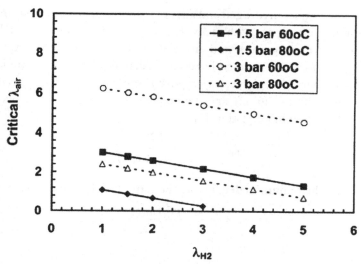

FIGURE 5-15 Maximum air stoichiometry that for a given hydrogen stoichiometry still fulfils Equation (5-44); each point corresponds to a combination of stoichiometries for which the amount of product water is just sufficient to saturate the outlet gas [7].

throughout the channel. This can be accomplished by flowing the coolant counter-flow to the cathode and/or with carefully designed heat removal from the cell. If neither of these measures is sufficient to avoid drying conditions in the fuel cell, humidification of air before it inlets the fuel cell may be necessary.

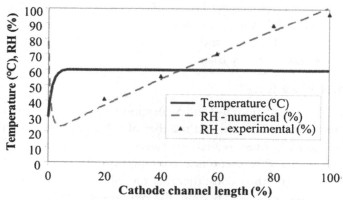

FIGURE 5-16 Temperature and relative humidity along the cathode channel of a fuel cell operated at 60°C with ambient air at the inlet [10].

When both gases (hydrogen and air) are saturated at cell temperature, it is still possible to end up with dehydration on either anode or cathode, depending on the magnitude and direction of the net water drag. A limited positive drag does not necessarily result in dehydration of the anode gas, whereas a substantial negative drag is possible before the cathode gas shows dehydration. The maximum drag before the anode dehydration occurs is given by the following equation [7]:

$$r_{d,\max} = \frac{1}{2} \frac{p_s(T_{cell})}{P - p_s(T_{cell})} \qquad (5\text{-}45)$$

Similarly, the minimum drag before the cathode dehydration occurs is [7]:

$$r_{d,\min} = -\frac{1}{4} \frac{p_s(T_{cell})}{P - p_s(T_{cell})} - \frac{1}{2} \qquad (5\text{-}46)$$

5.5 FUEL CELL MASS BALANCE

Fuel cell mass balance requires that the sum of all mass inputs must be equal to the sum of all mass outputs. The inputs are the flows of fuel and oxidant plus water vapor present in those gases. The outputs are the flows of unused fuel and oxidant plus water vapor present in those gases, plus any liquid water present in either fuel or oxidant exhaust:

$$\sum (\dot{m}_i)_{in} = \sum (\dot{m}_i)_{out} \qquad (5\text{-}47)$$

where i are the species, namely H_2, O_2, N_2, $H_2O(g)$, and $H_2O(l)$. In cases when hydrogen is fed from a fuel processor, other species, such as CO_2 and small amounts of CO, CH_4, and others, may also be present and must be accounted for in the energy balance.

5.5.1 Inlet Flow Rates

As shown earlier (Equations 5-2 and 5-3), consumption of reactants in a fuel cell is proportional to current and number of cells. Furthermore, the stoichiometric ratio is defined as a ratio between the actual flow of reactant at the fuel cell inlet and its theoretical consumption. Therefore, all the flow rates of reactants and their constituents at the inlet are also proportional to current and number of cells. Because the cell power output is:

$$W_{el} = n_{cell} I V_{cell} \qquad (5\text{-}48)$$

all the flows are also proportional to power output and inversely proportional to cell voltage:

$$n_{cell}I = \frac{W_{el}}{V_{cell}} \tag{5-49}$$

Hydrogen mass flow rate $(g\ s^{-1})$ is:

$$\dot{m}_{H2in} = S_{H2}\dot{m}_{H2,cons} = S_{H2}\frac{M_{H2}}{2F}I \cdot n_{cell} \tag{5-50}$$

If hydrogen is in a gas mixture, with r_{H2} volumetric and molar fraction, then the mixture flow rate is:

$$\dot{m}_{fuel} = \frac{S_{H2}}{r_{H2}}\frac{M_{fuel}}{2F}I \cdot n_{cell} \tag{5-51}$$

Oxygen mass flow rate $(g\ s^{-1})$ is:

$$\dot{m}_{O2in} = S_{O2}\dot{m}_{O2,cons} = S_{O2}\frac{M_{O2}}{4F}I \cdot n_{cell} \tag{5-52}$$

Air mass flow rate $(g\ s^{-1})$ is then:

$$\dot{m}_{Airin} = \frac{S_{O2}}{r_{O2}}\frac{M_{Air}}{4F}I \cdot n_{cell} \tag{5-53}$$

Nitrogen (which makes 79% of air by volume at the fuel cell inlet) mass flow rate $(g\ s^{-1})$ is:

$$\dot{m}_{N2in} = S_{O2}\frac{M_{N2}}{4F}\frac{1 - r_{O2in}}{r_{O2in}}I \cdot n_{cell} \tag{5-54}$$

Water vapor $(g\ s^{-1})$ present in hydrogen inlet with relative humidity (φ_{an}) is:

$$\dot{m}_{H2OinH2in} = S_{H2}\frac{M_{H2O}}{2F}\frac{\varphi_{an}P_{vs(T_{an,in})}}{P_{an} - \varphi_{an}P_{vs(T_{an,in})}}I \cdot n_{cell} \tag{5-55}$$

If hydrogen is supplied not as a pure gas but in a gas mixture, with r_{H2} volumetric or molar fraction, water vapor in fuel inlet $(g\ s^{-1})$ is:

$$\dot{m}_{H2O,fuelin} = \frac{S_{H2}}{r_{H2}}\frac{M_{H2O}}{2F}\frac{\varphi_{an}P_{vs(T_{an,in})}}{P_{an} - \varphi_{an}P_{vs(T_{an,in})}}I \cdot n_{cell} \tag{5-56}$$

Water vapor $(g\ s^{-1})$ present in an oxygen inlet with relative humidity (φ_{ca}) is:

$$\dot{m}_{H2OinO2in} = S_{O2}\frac{M_{H2O}}{4F}\frac{\varphi_{ca}P_{vs(T_{ca,in})}}{P_{ca} - \varphi_{ca}P_{vs(T_{ca,in})}}I \cdot n_{cell} \tag{5-57}$$

If air is used as an oxidant, water vapor in the air inlet $(g\ s^{-1})$ is:

$$\dot{m}_{H2OinAirin} = \frac{S_{O2}}{r_{O2}} \frac{M_{H2O}}{4F} \frac{\varphi_{ca}P_{vs(T_{ca,in})}}{P_{ca} - \varphi_{ca}P_{vs(T_{ca,in})}} I \cdot n_{cell} \qquad (5\text{-}58)$$

5.5.2 Outlet Flow Rates

The equations for the outlet mass flow rates must account for reactants' consumption, water generation, and water net transport across the membrane.

The unused hydrogen flow rate is:

$$\dot{m}_{H2out} = (S_{H2} - 1)\ \frac{M_{H2}}{2F} I \cdot n_{cell} \qquad (5\text{-}59)$$

The water content in hydrogen exhaust is equal to water brought into the cell with the hydrogen inlet minus the net water transport across the membrane. As discussed in Chapter 3, water gets "pumped" from anode to cathode because of electroosmotic drag. At the same time, some water diffuses back because of the water concentration gradient and because of the pressure differential. The net water transport is then the difference between these two fluxes. Water balance in the cathode side is then:

$$\dot{m}_{H2OinH2out} = \dot{m}_{H2OinH2in} - \dot{m}_{H2OED} + \dot{m}_{H2OBD} \qquad (5\text{-}60)$$

Electroosmotic drag is proportional to current, just like any other flow in or out of the fuel cell (Equation 4-5). The proportionality constant, ζ_D, the so-called electroosmotic drag coefficient, represents a number of water molecules per proton. When $\zeta_D = 1$, each proton is accompanied by one water molecule to form H_2O^+:

$$\dot{m}_{H2OED} = \zeta_D \frac{M_{H2O}}{F} I \cdot n_{cell} \qquad (5\text{-}61)$$

Back diffusion of water depends on water concentration on both sides of the membrane, water diffusivity through the membrane, and membrane thickness. Because water concentration is not uniform, it is not easy to explicitly calculate back diffusion for the entire cell or a stack of cells. For the sake of mass balance, back diffusion may be expressed as a fraction, β, of electroosmotic drag. When $\beta = 1$, back diffusion is equal to electroosmotic drag, that is, there is no net water transport across the membrane. The coefficient β may be determined experimentally by carefully condensing and measuring water content in both anode and cathode exhaust streams:

$$\dot{m}_{H2OBD} = \beta \dot{m}_{H2OED} = \beta \zeta_D \frac{M_{H2O}}{F} I \cdot n_{cell} \qquad (5\text{-}62)$$

The relationship between the previously defined net water transfer coefficient, r_D (Equation 5-35), the electro-osmotic drag coefficient, ζ_D (defined by Equation 5-61), and coefficient β (defined by Equation 5-62) is:

$$r_D = \zeta_D(1 - \beta) \qquad (5\text{-}63)$$

Depending on the hydrogen flow rate, that is, stoichiometry, and conditions at the outlet (temperature and pressure), water at the hydrogen exhaust may be present as vapor only, or liquid water may be present after the gas is saturated with water vapor. The water vapor content/flux at the anode outlet is the smaller of the total water flux at the anode outlet (calculated by Equation 5-60) and the maximum amount of vapor the exhaust gas can carry (saturation):

$$\dot{m}_{H2OinH2out,V} = \min\left[(S_{H2} - 1)\frac{M_{H2O}}{2F}\frac{P_{vs(T_{out,an})}}{P_{an} - \Delta P_{an} - P_{vs(T_{out,an})}} \right.$$

$$\left. \times I \cdot n_{cell}, \dot{m}_{H2OinH2out} \right] \qquad (5\text{-}64)$$

where ΔP_{an} is the pressure drop on the anode side, that is, the difference in pressure between the inlet and the outlet. In other words, if there is liquid water in the exhaust, the gas is already saturated.

The amount of liquid water, if any, is the difference between the total water present at the exhaust and water vapor:

$$\dot{m}_{H2OinH2out,L} = \dot{m}_{H2OinH2out} - \dot{m}_{H2OinH2out,V} \qquad (5\text{-}65)$$

A similar set of equations may be applied for the cathode exhaust. Oxygen flow rate at the outlet, that is, unused oxygen, is equal to oxygen supplied at the inlet minus oxygen consumed in the fuel electrochemical reaction:

$$\dot{m}_{O2out} = (S_{O2} - 1)\frac{M_{O2}}{4F} I \cdot n_{cell} \qquad (5\text{-}66)$$

Nitrogen flow rate at the exit is the same as the flow rate at the inlet because nitrogen does not participate in the fuel cell reaction:

$$\dot{m}_{N2out} = \dot{m}_{N2in} = S_{O2}\frac{M_{N2}}{4F}\frac{1 - r_{O2in}}{r_{O2in}} I \cdot n_{cell} \qquad (5\text{-}67)$$

The depleted air flow rate is then simply a sum of oxygen and nitrogen flow rates:

$$\dot{m}_{Airout} = \left[(S_{O2} - 1)M_{O2} + S_{O2}\frac{1 - r_{O2in}}{r_{O2in}}M_{N2}\right]\frac{I \cdot n_{cell}}{4F} \qquad (5\text{-}68)$$

Note that the oxygen volume fraction at the outlet is inevitably lower than the inlet volume fraction (Equation 5-16):

$$r_{O2,out} = \frac{S_{O2} - 1}{\dfrac{S_{O2}}{r_{O2,in}} - 1} \qquad (5\text{-}69)$$

Both $r_{O2,in}$ and $r_{O2,out}$ refer to volume or molar fractions of oxygen in dry air. The actual volume fraction in humid air is lower (see Figure 5-7):

$$r^{*}_{O2,out} = \frac{S_{O2} - 1}{\dfrac{S_{O2}}{r_{O2,in}} - 1}\left(1 - \frac{\varphi p_{vs}}{P}\right) \qquad (5\text{-}70)$$

Water content in the cathode exhaust is equal to the amount of water brought in the cell by humid air at the inlet plus water generated in the cell, plus the net water transport across the membrane, that is, the difference between electroosmotic drag and water back diffusion:

$$\dot{m}_{H2OinAirout} = \dot{m}_{H2OinAirin} + \dot{m}_{H2Ogen} + \dot{m}_{H2OED} - \dot{m}_{H2OBD} \qquad (5\text{-}71)$$

Depending on the oxygen/air flow rate, that is, the stoichiometry, and conditions at the outlet (temperature and pressure), water at the cathode exhaust may be present as vapor only, or liquid water may be present after the gas is saturated with water vapor. The water vapor content/flux at the cathode outlet is the smaller of the total water flux at the cathode outlet and the maximum amount the exhaust gas can carry (saturation):

$$\dot{m}_{H2OinAirout,V} = \min\left[\frac{S_{O2} - r_{O2in}}{r_{O2in}}\frac{M_{H2O}}{4F}\frac{P_{vs(T_{out,ca})}}{P_{ca} - \Delta P_{ca} - P_{vs(T_{out,ca})}}\right.$$

$$\left. \times I \cdot n_{cell}, m_{H2OinAirout}\right] \qquad (5\text{-}72)$$

where ΔP_{ca} is the pressure drop on the cathode side, that is, the difference in pressure between the inlet and outlet, so $P_{ca}, - \Delta P_{ca}$ is the pressure at the outlet.

The amount of liquid water, if any, is the difference between the total water present at the exhaust and water vapor:

$$\dot{m}_{H2OinAirout,L} = m_{H2OinAirout} - \dot{m}_{H2OinAirout,V} \qquad (5\text{-}73)$$

The previous set of Equations (5-47 through 5-73) represents the fuel cell mass balance. For a given set of inlet conditions (temperatures, pressures, flow rates, relative humidities) and some known or estimated stack performance characteristics (such as current, pressure drop, temperature difference, electroosmotic drag, and back diffusion), it allows one to calculate the flow rates and particularly water conditions at the outlet, or it may be used to tailor the inlet conditions so that desired conditions at the outlet are obtained, as demonstrated in an example at the end of this chapter.

5.6 FUEL CELL ENERGY BALANCE

Fuel cell energy balance requires that the sum of all energy inputs must be equal to the sum of all energy outputs:

$$\sum (H_i)_{in} = W_{el} + \sum (H_i)_{out} + Q \qquad (5\text{-}74)$$

The inputs are the enthalpies of all the flows into the fuel cell, namely fuel and oxidant, plus enthalpy of water vapor present in those gases. The outputs are:

- Electric power produced
- Enthalpies of all the flows out of the fuel cell, namely unused fuel and oxidant, plus enthalpy of water vapor present in those gases, plus enthalpy of any liquid water present in either fuel or oxidant exhaust
- Heat flux out of the fuel cell, both controlled through a cooling medium and uncontrolled because of heat dissipation (radiation and convection) from the fuel cell surface to the surroundings

For each flow in and out of the fuel cell there is an associated enthalpy, which can be calculated from Equations (5-75) through (5-79).

For each dry gas or a mixture of dry gases, the enthalpy (in J s^{-1}) is:

$$H = \dot{m}c_p t \qquad (5\text{-}75)$$

where:

$\dot{m}$ = mass flow rate of that gas or mixture (g s^{-1})

c_p = specific heat (J g^{-1} K^{-1})

t = temperature in °C

Note that the use of degrees Celsius implies that $0°C$ has been selected as a reference state for all enthalpies.

If a gas is combustible, that is, it has a heating value, its enthalpy is then:

$$H = \dot{m}\left(c_p t + h^0_{HHV}\right) \tag{5-76}$$

where h^0_{HHV} is the higher heating value of that gas ($J g^{-1}$) at $0°C$. Typically, heating values are reported and tabulated at $25°C$. The difference between the heating value at $25°C$ and $0°C$ is the difference between the enthalpies of reactants and products at those two temperatures. For hydrogen it is:

$$h^0_{HHV} = h^{25}_{HHV} - \left(c_{p,H2} + \frac{1}{2}\frac{M_{O2}}{M_{H2}}c_{p,O2} - \frac{M_{H2O}}{M_{H2}}c_{p,H2O(l)}\right) \cdot 25 \tag{5-77}$$

Enthalpy of water vapor is (from Equation 5-28):

$$H = \dot{m}_{H2O(g)}\left(c_{p,H2O(g)}t + h^0_{fg}\right) \tag{5-78}$$

Enthalpy of liquid water is:

$$H = \dot{m}_{H2O(l)}c_{p,H2O(l)}t \tag{5-79}$$

Some properties of the species commonly found in fuel cell inlets and outlets are listed in Table 5-3.

The use of mass and energy balance equations is demonstrated in the following example.

Table 5-3 Properties of Some Gases and Liquids

	Molecular Weight (g mol^{-1})	Specific Heat (J g^{-1} K^{-1})	Higher Heating Value (J g^{-1})
Hydrogen, H_2	2.0158	14.2	141,900
Oxygen, O_2	31.9988	0.913	
Nitrogen, N_2	28.0134	1.04	
Air	28.848	1.01	
Water vapor, H_2O (g)	18.0152	1.87	
Water, H_2O (l)	18.0152	4.18	
Carbon monoxide	28.0105	1.1	10,100
Carbon dioxide	44.0099	0.84	
Methane	16.0427	2.18	55,500
Methanol (l)	32.04	2.5	22,700

Example

A hydrogen/air fuel cell may be cooled through evaporative cooling by injecting liquid water at the air inlet and having the air flow rate sufficiently high. For a fuel cell that generates 1 kW at 0.7 V, find the required oxygen stoichiometric ratio and injected liquid water flow rate (g/s) so that the outlet air is fully saturated at 65°C (and there is no liquid water at the outlet) and that no additional cooling or heating of the fuel cell is required. Air is supplied at room temperature (20°C) with 70% relative humidity. Hydrogen is supplied in a dead-end mode at room temperature (20°C) and dry. Assume that the net water transport through the membrane is sufficient to maintain saturated conditions in the anode chamber at 65°C. Inlet pressure is 120 kPa (for both air and hydrogen); air outlet pressure is atmospheric, that is, 101.3 kPa.

To solve this problem with two unknowns, namely water injection rate, $\dot{m}_{H2OInject}$, and oxygen stoichiometric ratio, S_{O2}, two independent equations are needed. They may be obtained from water and energy balance.

Water balance is:

$$\dot{m}_{H2OinAirIn} + \dot{m}_{H2OInject} + \dot{m}_{H2Ogen} = \dot{m}_{H2OinAirOut,V}$$

Water in air in is (Equation 5-58):

$$\dot{m}_{H2OinAirin} = \frac{S_{O2}}{r_{O2}} \frac{M_{H2O}}{4F} \frac{\varphi_{ca} P_{vs(T_{ca,in})}}{P_{ca} - \varphi_{ca} P_{vs(T_{ca,in})}} I \cdot n_{cell}$$

$r_{O2} = 0.2095$
$M_{H2O} = 18.015$ (from Table 5-3)
$F = 96,485$ A s mol^{-1}
$\varphi_{ca} = 70\%$
$p_{vs} = e^{aT^{-1}+b+cT+dT^2+eT^3+f\ln(T)}$ (Equation 5-23); T = 293.15;
$p_{vs} = 2.339$ kPa
$I \cdot n_{cell} = W/V = 1000$ W/0.7 V $= 1428.6$ Amps
$\dot{m}_{H2OinAirin} = 0.0044 S_{O2}$ g s^{-1}

Water generated is (Equation 5-7):

$$\dot{m}_{H2O,gen} = \frac{I}{2F} M_{H2O} = \frac{1,428.6}{2 \times 96,485} 18.015 = 0.1334 \text{ g s}^{-1}$$

Water vapor in air out is (Equation 5-72):

$$\dot{m}_{H2OinAirout,V} = \left[\frac{S_{O2} - r_{O2in}}{r_{O2in}} \frac{M_{H2O}}{4F} \frac{P_{vs(T_{out,ca})}}{P_{ca} - \Delta P_{ca} - P_{vs(T_{out,ca})}} I \cdot n_{cell} \right]$$

$$p_{vs} = e^{aT^{-1}+b+cT+dT^2+eT^3+f\ln(T)}; T = 65°C = 338.15 \text{ K}; p_{vs}$$
$$= 25.039 \text{ kPa}$$

$$P_{ca} - \Delta P_{ca} = 101.3 \text{ kPa}$$

$$\dot{m}_{H2OinAirout,V} = 0.1045S_{O2} - 0.02189 \text{ g s}^{-1}$$

Water balance is therefore:

$$0.004403S_{O2} + \dot{m}_{H2OInject} + 0.1334 = 0.1045S_{O2} - 0.02189$$

or, after rearranging:

$$\dot{m}_{H2OInject} - 0.1001S_{O2} + 0.1553 = 0 \qquad \text{(Equation E1)}$$

Energy balance is:

$$H_{H2in} + H_{Airin} + H_{H2OinAirin} + H_{H2Oinject}$$
$$= H_{Airout} + H_{H2OinAirOut,V} + W_{el}$$

The flow rates are:

Hydrogen in (Equation 5-50): $\dot{m}_{H2in} = S_{H2}\dfrac{M_{H2}}{2F}I \cdot n_{cell}$

$S_{H2} = 1$ (dead-end mode):

$$\dot{m}_{H2in} = 1\frac{2.0158}{2 \times 96,485}1,428.6 = 0.01492 \text{ g s}^{-1}$$

Air in (Equation 5-53): $\dot{m}_{Airin} = \dfrac{S_{O2}}{r_{O2}}\dfrac{M_{Air}}{4F}I \cdot n_{cell}$

$$\dot{m}_{Airin} = \frac{S_{O2}}{0.2095}\frac{28.848}{4 \times 96,485}1,428.6 = 0.5097S_{O2}$$

Water vapor in air in (from water balance above): $\dot{m}_{H2OinAirin} = 0.0044S_{O2} \text{ g s}^{-1}$

Air out (Equation 5-68):

$$\dot{m}_{Airout} = \dot{m}_{O2Out} + \dot{m}_{N2Out} \quad \dot{m}_{O2out} = [(S_{O2} - 1)M_{O2}]$$

$$\frac{I \cdot n_{cell}}{4F} = 0.1184S_{O2} - 0.1184 \quad \dot{m}_{N2out} = \left[S_{O2}\frac{1 - r_{O2in}}{r_{O2in}}M_{N2}\right]$$

$$\frac{I \cdot n_{cell}}{4F} = 0.3913S_{O2}$$

Water vapor in air out (from water balance above):

$$\dot{m}_{H2OinAirout,V} = 0.1045S_{O2} - 0.02189 \text{ g s}^{-1}$$

Energy flows (enthalpies) are:

Hydrogen in (Equation 5-76): $H_{H2in} = \dot{m}_{H2in}(c_{p,H2} t_{in} + h^0_{HHV})$

Hydrogen higher heating value at $0°C$ is (Equation 5-77):

$$h_{HHV}^0 = h_{HHV}^{25} - \left(c_{p,H2} + \frac{1}{2}\frac{M_{O2}}{M_{H2}}c_{p,O2} - \frac{M_{H2O}}{M_{H2}}c_{p,H2O(l)} \right) \cdot 25$$

$$h_{HHV}^0 = 141,900 - \left(14.2 + \frac{31.9988/2}{2.0158}0.913 - \frac{18.0152}{2.0158}4.18 \right) \cdot 25$$

$$h_{HHV}^0 = 142,298 \text{ J g}^{-1} \text{ K}^{-1}$$

$$H_{H2in} = 0.0149(14.2 \times 20 + 142,298) = 2127.8 \text{ (W)}$$

Air in (Equation 5-75): $H_{Airin} = \dot{m}_{Airin}\, c_{p,Air}\, t_{in}$

$$H_{Airin} = 0.5097 S_{O2} \times 1.01 \times 20 = 10.302 S_{O2}$$

Water vapor in air in (Equation 5-78):

$$H_{H2OinAirin} = \dot{m}_{H2OinAirin}\left(c_{p,H2O(g)}\, t_{in} + h_{fg}^0 \right)$$

$$H_{H2OinAirIn} = 0.0044 S_{O2} (1.87 \times 20 + 2500) = 11.171 S_{O2} \text{ g s}^{-1}$$

Water injected (Equation 5-79):

$$H_{H2OInject} = \dot{m}_{H2OInject} \times 4.18 \times 20 = 83.6 \text{ g s}^{-1}$$

Air out: $H_{Air,out} = \dot{m}_{O2,Out}\, c_{p,O2}\, t_{out} + \dot{m}_{N2,In}\, c_{p,N2}\, t_{out}$

$$= (0.1184 S_{O2} - 0.1184) \times 0.913 \times 65 + 0.3913 S_{O2} \times 1.04$$

$$\times 65 = 33.478 S_{O2} - 7.0291 \text{ g s}^{-1}$$

Water vapor in air out (Equation 5-78):

$$H_{H2OinAirOut} = \dot{m}_{H2OinAirOut}\left(c_{p,H2O(g)}\, t_{out} + h_{fg}^0 \right)$$

$$= (0.1045 S_{O2} - 0.02189) \times (1.87 \times 65 + 2500) = 273.97 S_{O2} - 57.3967$$

Electricity generated: $W_{el} = 1000 \text{ W}$
Energy balance is:

$$2127.8 + 10.302 S_{O2} + 11.171 S_{O2} + 83.6\, \dot{m}_{H2OInject}$$

$$= 33.478 S_{O2} - 7.0291 + 273.97 S_{O2} - 57.3967 + 1000$$

After rearranging the energy balance equation becomes:

$$83.6\, \dot{m}_{H2OInject} - 285.98 S_{O2} + 1192,19 = 0 \qquad \text{(Equation E2)}$$

By combining with the water balance and solving the two equations, E1 and E2, with two unknowns, the result is:

$$SO_2 = 4.25 \text{ and (Answer)}$$

$$\dot{m}_{H2OInject} = 0,27 \text{ g s}^{-1} \text{ (Answer)}$$

Therefore, by injecting $0.27\ g\ s^{-1}$ of liquid water in ambient air at fuel cell inlet with the stoichiometry of 4.25, it would be possible to maintain the desired operating temperature of 65°C and have just enough water at the air outlet to avoid either flooding with liquid water or drying with dry air. Note that such a fuel cell would have to be thermally insulated. Also, operation with dry hydrogen probably would not be possible, that is, additional water may be needed to humidify hydrogen. Nevertheless, such a fuel cell would result in an extremely simple system: no need for air humidification and no need for stack cooling, probably the two bulkiest supporting system components.

PROBLEMS

1. A fuel cell generates 200 amps at 0.6 V. Hydrogen flow rate in the fuel cell is $1.8\ Nl\ m^{-1}$; air flow rate is $8.9\ Nl\ m^{-1}$. Calculate:
   a. Hydrogen stoichiometric ratio
   b. Oxygen stoichiometric ratio
   c. Oxygen concentration at the outlet (neglect water present)

2. If both gases in Problem 1 are 100% saturated at 60°C and 120 kPa, calculate:
   a. The amount of water vapor present in hydrogen (in g/s)
   b. The amount of water vapor present in air (in g/s)
   c. The amount of water generated in the fuel cell reaction (in g/s)

3. In Problem 2, calculate the amount of liquid water at the cell outlet (assuming zero net water transport through the membrane). Both air and hydrogen at the outlet are at ambient pressure and at 60°C.
   a. In hydrogen outlet
   b. In air outlet

4. For Problem 3, calculate:
   a. The fuel cell efficiency
   b. Rate of heat generated (W)

5. An H_2/Air fuel cell operates at 80°C and 170 kPa. Hydrogen is supplied in a dead-end mode. If the net water drag is 0.25 molecules of H_2O per proton, calculate the required relative humidity of hydrogen at the inlet so that there is neither water accumulation nor drying of the membrane on the hydrogen side of the fuel cell. Explain the physical meaning of your result.

6. An H_2/Air fuel cell operates with hydrogen utilization of 84% and oxygen stoichiometric ratio of 2. Both hydrogen and air must be fully saturated at the cell operating pressure and temperature. Which of the following is true?
   a. Air needs approximately 4 times more water
   b. Air needs approximately 2 times more water
   c. Hydrogen needs approximately 2 times more water
   d. Cannot say which reactant needs more water because temperature and pressure are not given

7. Air is fully saturated with water vapor at the fuel cell inlet and at the fuel cell outlet (actually there is also liquid water dripping at the outlet). The air temperature is the same at the inlet and at the outlet. The fuel cell operates with an oxygen stoichiometric ratio of 2. Air passes through a single-channel serpentine flow field. Is the water vapor content (in mols of water vapor per mol of dry gas) at the outlet:
   a. Higher than at the inlet
   b. Lower than at the inlet
   c. The same as at the inlet

Explain!

QUIZ

1. The stoichiometric ratio is:
   a. The ratio between hydrogen and oxygen at the fuel cell inlet
   b. The ratio between hydrogen and oxygen consumed in the fuel cell
   c. The ratio between an actual flow rate of a reactant at the fuel cell inlet and the rate of consumption of that same reactant

2. In the case of an air-breathing fuel cell, a higher oxygen stoichiometric ratio:
   a. Means a higher concentration of O_2 at the outlet
   b. Means a higher concentration of O_2 at the inlet
   c. Has nothing to do with concentration

3. Dead-end operation means:
   a. No hydrogen is leaving the fuel cell
   b. The fuel cell is "dead," that is, produces no current
   c. Consumption is proportional to cell voltage

4. In a dead-end operation, the stoichiometric ratio at the inlet is:
   a. The same as at the outlet
   b. 1
   c. 0

5. Pressure at a fuel cell air inlet:
   a. Is lower than at outlet
   b. Is higher than at the outlet
   c. You cannot say if it is higher or lower because it depends on the current

6. A certain quantity of a gas at higher temperature can contain:
   a. Less water vapor than at a lower temperature
   b. More water vapor than at a lower temperature
   c. The same amount of water vapor as at a lower temperature

7. Relative humidity is:
   a. The percentage of partial pressure of water vapor relative to saturation pressure
   b. The ratio between the amount of water vapor and the amount of dry air
   c. The percentage of water in air

8. If the air flow rate at the inlet is kept proportional to the current (for example, at 2 × stoichiometry), then:
   a. Concentration of O_2 is constant throughout the cell, regardless of current
   b. O_2 concentration decreases along the length of the fuel cell
   c. O_2 concentration reaches zero at certain current

9. If back diffusion of water is equal to the electroosmotic drag and both gases are saturated at the inlet:
   a. There is no net water transport across the membrane
   b. The fuel cell will flood on the air (cathode) side
   c. The fuel cell will dry out on the air (cathode) side

10. The form of water at the cathode (air) exit of the fuel cell depends on:
    a. Operating pressure and temperature only
    b. Stoichiometric ratio and temperature only
    c. Stoichiometric ratio, temperature, pressure, and humidity of air at the inlet

REFERENCES

[1] Barbir F, Fuchs M, Husar A, Neutzler J. Design and Operational Characteristics of Automotive PEM Fuel Cell Stacks. Fuel Cell Power for Transportation 2000;SAE SP-1505:63–9. SAE, Warrendale, PA.

[2] Yan Q, Toghiani H, Causey H. Steady State and Dynamic Performance of Proton Exchange Membrane Fuel Cells (PEMFCs) Under Various Operating Conditions and Load Changes. Journal of Power Sources 2006;Volume 161:492–502. Issue 1.

[3] National Institute of Standards and Technology; http://physics.nist.gov/cuu/Units/outside.html.

[4] ASHRAE Handbook. Fundamentals. Atlanta, GA: ASHRAE; 1981. 1982.

[5] Bosnjakovic F. Technical Thermodynamics. New York: Holt Rinehart and Winston; 1965.

[6] Berning T. The Dew Point Temperature as a Criterion for Optimizing Operating Conditions for Proton Exchange Membrane Fuel Cells, Int. Journal of Hydrogen Energy 2012.

[7] Janssen GJM, Overvelde MLJ. Water Transport in the Proton-Exchange-Membrane Fuel Cell: Measurements of the Effective Drag Coefficient. J. Power Sources 2001;101:117–25.

[8] Berning T, Kær SK. Low Stoichiometry Operation of a Proton-Exchange Membrane Fuel Cell Employing the Interdigitated Fuel Cell: a Modeling Study. J. Power Sources. 2012.

[9] Buchi FN, Srinivasan S. Operating Proton Exchange Membrane Fuel Cells Without External Humidification of the Reactant Gases. J. Electrochemical Society 1997;144:2767–72.

[10] Tolj I, Bezmalinovic D, Barbir F. Maintaining Desired Level of Relative Humidity Throughout a Fuel Cell with Spatially Variable Heat Removal Rates. International Journal of Hydrogen Energy 2011;36:13105–13.

CHAPTER SIX

Stack Design

As discussed in Chapter 3, the fuel cell electrochemical reactions result in theoretical cell potential of 1.23 V, and the actual potential in operation is lower than 1 V. If a single cell were required to generate 1 kW of power, it would need to generate electrical current higher than 1,000 amperes. Such a current could be generated only with a large active area (>1000 cm^2), and it would require very thick cables between the fuel cell and the load to minimize the resistive losses. A more practical solution would be to have multiple cells electrically connected in series. The cell active area and the cross-sectional area of the connecting cables would decrease with a number of cells connected in series. For example, 1 kW power output may be accomplished with as many as 40 cells connected in series, each cell operating at 0.6 V and 1 A/cm^2. Total current would be about 42 amperes at 24 volts.

6.1 SIZING A FUEL CELL STACK

The first step in designing a fuel cell stack is to determine its active area and number of cells in the stack. When a stack is designed for an application, the design inputs come from the application requirements, such as desired power output, desired or preferred stack voltage or voltage range, desired efficiency, and volume and weight limitations. Some of these requirements may conflict with each other, and the stack sizing and design process often results in a compromise solution that meets the key requirements (such as power output) and finds an optimum between the conflicting requirements.

Another design input is the unit performance, best described by the polarization curve. The fuel cell polarization curve is the key for sizing and designing a fuel cell stack. However, as shown in Chapter 5, the fuel cell performance is determined by operational conditions (pressure, temperature, humidity of reactant gases) that must be determined based on the application requirements and constraints.

The stack power output is simply a product of stack voltage and current:

$$W = V_{st} \cdot I \qquad (6-1)$$

PEM Fuel Cells
ISBN 978-0-12-387710-9

The stack potential is simply a sum of individual cell voltages or a product of average individual cell potential and number of cells in the stack:

$$V_{st} = \sum_{i=1}^{N_{cell}} V_i = \overline{V}_{cell} \cdot N_{cell} \qquad (6\text{-}2)$$

The current is a product of current density and cell active area:

$$I = i \cdot A_{cell} \qquad (6\text{-}3)$$

Cell potential and current density are related by the polarization curve:

$$V_{cell} = f(i) \qquad (6\text{-}4)$$

The polarization curve may be defined by a set of data (V_{cell}-i), by any of the Equations (3-43), (3-46), or (3-52), or by linear approximation (Equation 3-68).

The fuel cell stack efficiency may be approximated with a simple Equation (3-61):

$$\eta = V_{cell}/1.482 \qquad (6\text{-}5)$$

which is not valid only for the potentials close to the open circuit potentials.

The previous section represents a set of five equations with eight variables. Additional variables or constraints may be stack volume and weight, both being functions of number of cells in the stack and cell active area, in addition to stack construction and choice of materials. It is therefore a matter of simple arithmetic to calculate the remaining variables if three of them are given—for example, power output, stack voltage, and stack efficiency. (A little complication is the form of the equation $V_{cell} = f(i)$, which may require some iterative process to calculate current density, i, when cell potential, V_{cell}, is given.) If less than three inputs are given, there may be an infinite number of cell active areas and number of cell combinations that satisfy the previous equations, with some limitations (Figure 6-1). Both active area and number of cells in the stack, as well as their combination, have physical and/or technological limits. For example, a large number of cells with very small active area would be difficult to align and assemble. On the other hand, a small number of cells with a large active area would result in a high current/low voltage combination and would result in significant resistive losses in connecting cables.

Very often the number of cells in a stack is limited or determined by required stack voltage. Stacks with active area up to 1000 cm^2 have been demonstrated; more typically, active area is between 50 and 300 cm^2

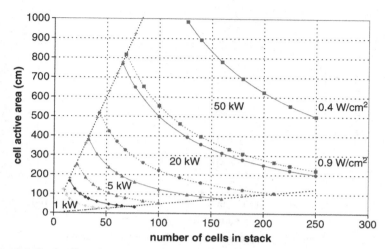

Figure 6-1 Fuel cell stack sizing: number of cells and cell active area for different power outputs (solid lines are for 0.4 W/cm² and dashed lines are for 0.9 W/cm²).

depending on application and desired power output. In larger active areas it is more difficult to achieve uniform conditions, which is one of the key aspects of a successful stack design as it is discussed here. The maximum number of cells in a stack is limited by compression forces, structural rigidity, and pressure drop through long manifolds. Stacks with up to 200–250 cells have been demonstrated.

These limits also apply to the case of the determined system, that is, with three variables given. If the solution for number of cells and cell active area falls outside the limitations, it would be necessary to reconsider and modify the requirements.

Very often, there is a need to optimize the stack efficiency and size. These are two conflicting requirements for a given power output and a given polarization curve. Typical fuel cell polarization curves are shown in Figure 6-2. Most of the published fuel cell stack polarization curves fall within these two lines.

The nominal operating point (cell voltage and corresponding current density) is the operating point at nominal power output, and it may be selected anywhere on the polarization curve. Selection of this point has a profound effect on the stack size and efficiency. Lower selected cell voltage at nominal power results in higher power density and consequently in smaller stack size for any given power output. Figure 6-3 shows the relationship between stack size (total active area per unit power output) and nominal cell potential for the two polarization curves from Figure 6-2. From

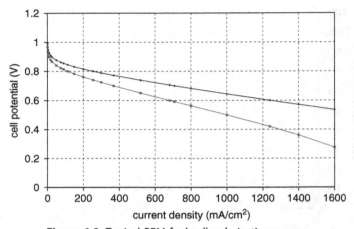

Figure 6-2 Typical PEM fuel cell polarization curves.

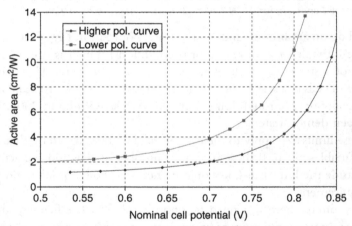

Figure 6-3 Stack size vs. selected nominal cell potential (corresponding to polarization curves in Figure 6-2).

Figure 6-3, it follows that a stack with nominal cell voltage of 0.7 V would require about 40% larger active area than a stack sized at 0.6 V/cell. Selection of 0.8 V/cell at nominal power would result in more than twice the size of a stack sized at 0.7 V/cell. However, higher cell voltage means better efficiency and consequently lower fuel consumption. Currently, most fuel cell manufacturers and developers use between 0.6 V and 0.7 V as voltage at nominal power. However, to reach some system efficiency goals, the fuel cells would have to be rated at 0.8 V/cell or even higher. Indeed, in applications where the efficiency (which translates in reactant consumption)

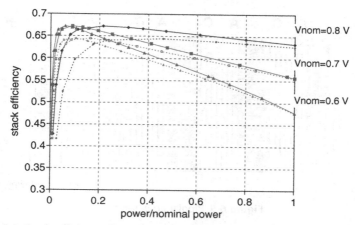

Figure 6-4 Stack efficiency (based on LHV) vs. power output for different selected nominal cell potentials (solid lines are for the higher and dashed lines are for the lower polarization curves in Figure 6-2).

is critical, such as in space applications in which both reactants must be carried on board, the fuel cells are rated and operated above 0.8 V. The extra fuel cell size is negligible compared with the size (and weight) of hydrogen and oxygen saved.

Optimum cell voltage at nominal power should be determined for each application based on the optimization criteria (such as the lowest cost of generated electricity, the least expensive system, or the lowest size and weight), and it should therefore take into consideration many parameters, such as the cost of fuel cell, cost of fuel, lifetime, capacity factor, load profile, system efficiency, and so on [1]. If the stack is operated at partial load (below 20% of nominal power) most of the time, then higher selected nominal cell voltage does not necessarily mean higher operating efficiency. As shown in Figure 6-4, at 20% of nominal power there is no efficiency advantage of higher selected nominal cell voltage. This is due to operation at very low current densities where parasitic losses (including gas permeation through the polymer membrane) may not be negligible.

6.2 STACK CONFIGURATION

A fuel cell stack consists of a multitude of single cells stacked up so that the cathode of one cell is electrically connected to the anode of the adjacent cell. In that way, exactly the same current passes through each of the cells. Note that the electrical circuit is closed, with both electron current passing

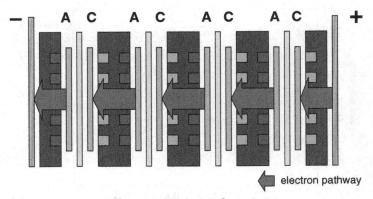

Figure 6-5 Bipolar configuration.

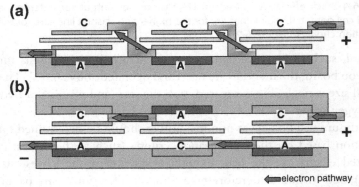

Figure 6-6 Examples of side-by-side stock configurations for smaller fuel cells: (a) zig-zag connections with open air cathode; (b) flip-flop configuration.

through solid parts of the stack (including the external circuit) and ionic current passing through the electrolyte (ionomer), with the electrochemical reactions at their interfaces (catalyst layers).

The bipolar configuration is best for larger fuel cells because the current is conducted through relatively thin conductive plates and thus travels a very short distance through a large area (Figure 6-5). This causes minimum electroresistive losses, even with a relatively bad electrical conductor such as graphite (or graphite polymer mixtures). For small cells it is possible to connect the edge of one electrode to the opposing electrode of the adjacent cell by some kind of a connector [2–4]. This is applicable only to very small active area cells because current is conducted in a plane of very thin electrodes, thus traveling a relatively long distance through a very small cross-sectional area (Figure 6-6).

The main components of a fuel cell stack are the membrane electrode assemblies, or MEAs (membranes with electrodes on each side with a catalyst layer between them), gaskets at the perimeter of the MEAs, bipolar plates, bus plates (one at each end of the active part of the stack) with electrical connections, and the end plates (one at each end of the stack) with fluid connections [5]. Cooling of the stack, that is, its active cells, must be arranged in some fashion as discussed next. The whole stack must be kept together by tie rods, bolts, shroud, or some other arrangement (Figure 6-7).

In some stack configurations, humidification of one or both reactant gases is included in the stack, either in a separate stack section or between the cells. In both cases water is used for both cooling and humidification, and the heat generated in each of the active cells is used for humidification. Figure 6-8 shows a stack configuration in which the reactant gases first pass through a humidification section and then through the active cells, whereas water first passes through the active portion of the stack, gets heated up, and then passes through the humidification section, where heat and water transfer between water and the reactant gases occurs through polymer membranes [6]. The advantage of this stack configuration is in its compactness of the gas humidification system and reduced heat dissipation, but this very advantage is also its main drawback: lack of versatility in controlling water and heat management and inability to separate heat from water management.

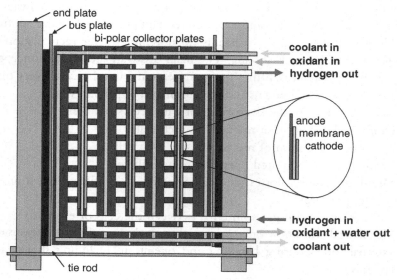

Figure 6-7 Stack schematic [5].

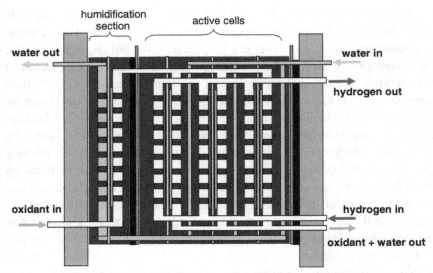

Figure 6-8 Stack configuration with internal humidification.

In another configuration (Figure 6-9), water passes between the cells, which are separated by a porous graphite plate from the reactant gases [7]. The porous plate allows water transport, effectively facilitating water and heat management in each cell. Hydrogen, air, and water pressures must be carefully regulated so that water in the porous plate channels is always at slightly negative pressure compared with reactant pressure. Pore size controls bubble point such that the reactants do not mix. Water recirculation to a radiator provides stack cooling. The main disadvantage of this concept is inability to separate heat from water management, but others are a very tight pressure control requirement and operation at low pressures only.

The following are the key aspects of a fuel cell stack design:

- Uniform distribution of reactants to each cell
- Uniform distribution of reactants inside each cell
- Maintenance of required temperature in each cell
- Minimum resistive losses (choice of materials, configuration, uniform contact pressure)
- No leak of reactant gases (internal between the cells or external)
- Mechanical sturdiness (internal pressure, including thermal expansion; external forces during handling and operation, including shocks and vibrations)

These aspects are discussed in the following sections.

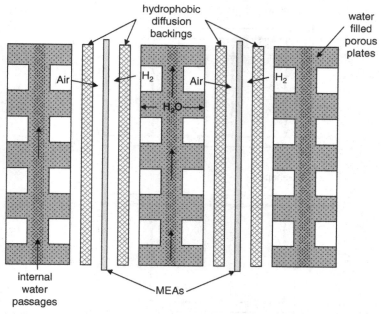

Figure 6-9 Stack configuration with water management through porous plate. *(Adapted from [7].)*

6.3 UNIFORM DISTRIBUTION OF REACTANTS TO EACH CELL

Because fuel cell performance is sensitive to the flow rate of the reactants, it is absolutely necessary that each cell in a stack receive approximately the same amount of reactant gases. Uneven flow distribution would result in uneven performance between the cells. Uniformity is accomplished by feeding each cell in the stack in parallel through a manifold that can be either external or internal. External manifolds can be made much bigger to ensure uniformity. They result in a simpler stack design, but they can only be used in a cross-flow configuration and are, in general, difficult to seal. Internal manifolds are more often used in PEM fuel cell design, not only because of better sealing but also because they offer more versatility in gas flow configuration.

It is important that the manifolds that feed the gases to the cells and the manifolds that collect the unused gases are properly sized. The cross-sectional area of the manifolds determines the velocity of gas flow and the pressure drop. As a rule of thumb, the pressure drop through the manifolds

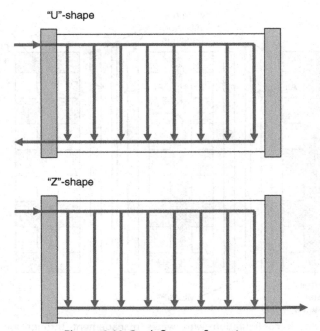

Figure 6-10 Stack flow configurations.

should be an order of magnitude lower than the pressure drop through each cell in order to ensure uniform flow distribution.

The flow pattern through the stack can be either a *U* shape, where the inlet and outlet are at the same side of the stack and the flows in inlet and outlet manifolds are in opposite directions from each other, or a *Z* shape, where the inlets and outlets are on opposite sides of the stack and the flows in inlet and outlet manifolds are parallel to each other (Figure 6-10). If properly sized, both should result in uniform flow distribution to individual cells. Stacks with more than 100 cells have been successfully built.

The procedure to calculate the pressure drop through the manifolds and the entire stack, and the resulting flow distribution, involves a flow network problem consisting of *N*-1 loops [8], where *N* is the number of cells in the stack. The flow in any network must satisfy the basic relations of continuity and energy conservation as follows:

1. The flow into any junction must equal the flow out of it.
2. The flow in each segment has a pressure drop that is a function of the flow rate through that segment.
3. The algebraic sum of the pressure drops around any closed loop must be zero.

The first requirement is satisfied with the following relationships:

$$Q_{in}(i) = Q_{cell}(i) + Q_{in}(i+1) \text{ in inlet manifold} \qquad (6\text{-}6)$$

$$Q_{out}(i) = Q_{cell}(i)$$
$$+ Q_{out}(i+1) \text{ in outlet manifold, "U" configuration}$$
$$(6\text{-}7)$$

$$Q_{out}(i) = Q_{cell}(i)$$
$$+ Q_{out}(i-1) \text{ in outlet manifold, "Z" configuration}$$
$$(6\text{-}8)$$

Note that by definition of the cells and flows numbering shown in Figure 6-11:

$$Q_{out}(1) = Q_{stack} \qquad (6\text{-}9)$$

For a nonoperating stack, that is, one with no species consumption or generation, the flow at the stack outlet is equal to the flow at the inlet:

$$Q_{out}(1) = Q_{in}(1) \text{ for "U" configuration} \qquad (6\text{-}10)$$

$$Q_{out}(N) = Q_{in}(1) \text{ for "Z" configuration} \qquad (6\text{-}11)$$

The pressure drop, derived directly from the Bernoulli equation, in each manifold segment is [8]:

$$\Delta P(i) = -\rho \frac{[u(i)]^2 - [u(i-1)]^2}{2} + f\rho \frac{L}{D_H} \frac{[u(i)]^2}{2} + K_f \rho \frac{[u(i-1)]^2}{2}$$
$$(6\text{-}12)$$

where:

ρ = density of the gas (kg m^{-3})
u = velocity (m s^{-1})
f = friction coefficient
L = length of the segment (m)
D_H = hydraulic diameter of the manifold segment (m)
K_f = local pressure loss coefficient

The first term on the right side of Equation (6-12) represents the energy decrease along the streamline in the manifold due to velocity loss; the second term is the pressure drop due to friction with the walls; and the last term is the pressure loss due to local disturbances described by the geometric coefficient.

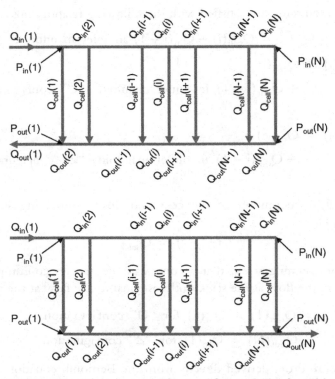

Figure 6-11 Designation of pressure and flow and flow variables for U (above) and Z (below) stack flow configurations.

The loss of pressure due to local disturbances, such as flow branching out or in (tees) or sudden changes in direction (elbows), should be taken into account in calculating the pressure drop in each segment. Typically in fluid-mechanics textbooks, these losses are called *minor losses*; however, in fuel cell manifolds they may not be negligible. Koh et al. [8] showed that the flow distribution largely depends on the local pressure loss coefficients. Though some geometrical pressure loss coefficients are available for various pipe fittings, none fit a specific shape of the fuel cell gas flow manifolds. Koh et al. [8] concluded that the geometrical pressure loss coefficients should be determined experimentally for each stack design.

For laminar flow (Re < 2000), the friction coefficient f for a circular conduit is:

$$f = \frac{64}{Re} \tag{6-13}$$

For turbulent flow, the friction coefficient is primarily a function of the wall roughness. The walls of fuel cell manifolds may be considered "rough" because they are not a smooth pipe but instead consist of the bipolar plates clamped together, often with the gasket material protruding into the manifold. The relative roughness, ε/D, can be as high as 0.1. In that case the friction coefficient (according to Karman [9]) is:

$$f = \frac{1}{\left(1.14 - 2 \, log\dfrac{\varepsilon}{D}\right)^2} \tag{6-14}$$

The pressure drop through the cell, ΔP_{cell}, is discussed in Section 6.4.5, "Pressure Drop Through the Flow Field." In most cases it is a linear function of the flow rate through each cell.

The third requirement is that the algebraic sum of the pressure drops around any closed loop must be zero:

$$\text{For } i = 1 \text{ to } (N-1) \tag{6-15}$$

$$\Delta P_{in}(i+1) + \Delta P_{cell}(i+1) + \Delta P_{out}(i+1) - \Delta P_{cell}(i)$$

$$= 0 \text{ in "U" configuration} \tag{6-16}$$

$$\Delta P_{in}(i+1) + \Delta P_{cell}(i+1) - \Delta P_{out}(i) - \Delta P_{cell}(i)$$

$$= 0 \text{ in "Z" configuration} \tag{6-17}$$

For a given flow rate at the stack entrance and a known pressure either at the stack inlet or at the stack outlet, it is possible to calculate the flow rate through each of the cells using a method of successive approximations, such as the Hardy-Cross method commonly used for pipe networks [9] or the method suggested by Koh et al. [8]. The procedure in either method is similar:

1. It starts with an approximation of the flows in each of the cells $Q_{cell} = Q_{stack}/N_{cell}$.
2. It calculates the pressure drops in individual manifold segments and cells.
3. It checks the sum of pressure drops in each of the loops (it should be zero).
4. It adjusts the flows in each loop by a correction, ΔQ, proportional to the error in sum of the pressure drops, and the process is repeated one loop at a time until the error is negligible.

For the same size conduits, Z configuration usually results in more uniform flow distribution than U configuration. It may be shown that for U

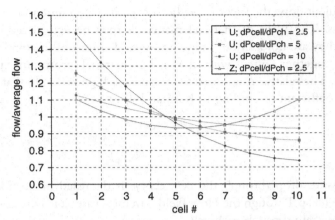

Figure 6-12 Flow distribution in individual cells of a 10-cell stack for U and Z configurations and for various ratios of pressure drop through individual cells and through the inlet manifold (dP_{cell}/dP_{ch}).

configuration, uniform distribution of flows through the individual cells results when the pressure drop through the cell is at least an order of magnitude higher than the pressure drop through the stack inlet manifold (Figure 6-12). The inlet and outlet manifolds must be sized accordingly. Sometimes it is beneficial to have more than one manifold for one reactant, especially for the stacks with larger active area. In that case the pressure drop through each manifold must also be balanced to ensure uniform flow distribution to each manifold. Similar methods as depicted previously may be used for that calculation.

When the stack is in operation, the change in flow rate, gas density, and viscosity due to species consumption and generation must be taken into account, although it should not significantly affect the pressure drop results (<5%).

6.4 UNIFORM DISTRIBUTION OF REACTANTS INSIDE EACH CELL

Once the reactant gases enter the individual cell, they must be distributed over the entire active area. This is typically accomplished through a flow field, which may be in a form of channels covering the entire area in some pattern or porous structures. The following sections describe the key flow field design variables.

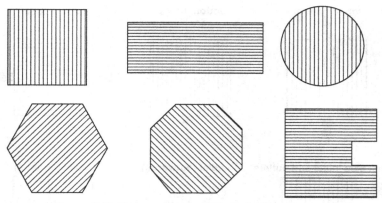

Figure 6-13 Various shapes of the fuel cell active area.

6.4.1 Shape of the Flow Field

The flow fields come in different shapes and sizes. The size comes from the power/voltage requirements as shown in Section 6.1, "Sizing a Fuel Cell Stack." The shape is the result of positioning the inlet and outlet manifolds, flow field design, heat management, and manufacturing constraints. The most common shapes of the flow field are square and rectangular, but circular, hexagonal, octagonal, or irregular shapes have been used or at least tried (Figure 6-13).

6.4.2 Flow Field Orientation

The orientation of the flow field and positions of inlet and outlet manifolds are important only because of gravity's effect on water that may condense inside the flow field (the effect of gravity on the reactant gases is negligible). Condensation may take place either during operation, depending on the choice of operational conditions, or after shutdown. Numerous combinations are possible, some of which are shown in Figure 6-14.

Anode and cathode may be oriented in the same direction, in opposite directions, or in cross-configuration. The position of the anode vs. the cathode may have some effect on fuel cell performance because of varied concentration of reactant gases and water. In some cases the flow fields are oriented so that the cathode outlet is next to the anode inlet, and vice versa, allowing water exchange through the membrane due to the water concentration gradient (i.e., the exiting gas has much higher temperature and water content).

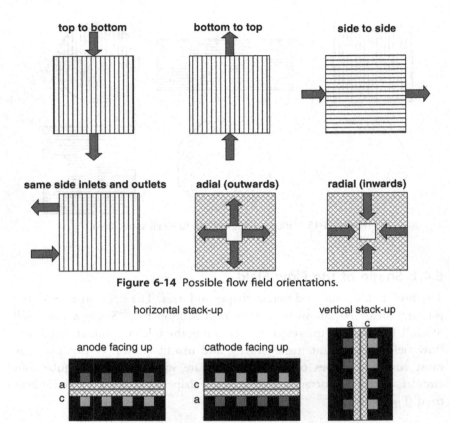

Figure 6-14 Possible flow field orientations.

Figure 6-15 Stack and cell orientation options.

The stack orientation, and so the flow field orientation, may be either vertical or horizontal. In the latter case, either anode or cathode may be facing up (Figure 6-15). Again, this may have some effect on liquid water removal, particularly after shutdown and cooling.

6.4.3 Configuration of Channels

There are many configurations of channels that have been tried in PEM fuel cells, all with the same goal: to ensure uniform reactant gases distribution and product water removal (see Figure 6-16). Some most common designs are as follows:

- *Straight channels with large manifolds.* Although this appears to ensure uniform distribution, it actually does not work in PEM fuel cells. Distribution is indeed uniform but only under ideal conditions. Any

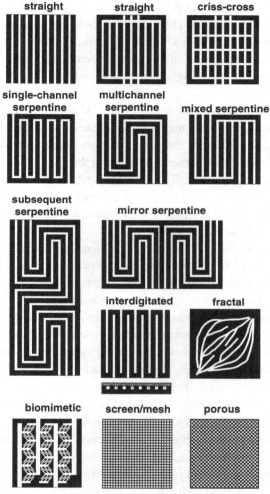

Figure 6-16 Various flow field configurations.

water droplet that develops in a channel would effectively block the entire channel, and the velocity would not be sufficient to push the water out.

- *Straight channels with small manifolds.* This design has the same short-comings and, in addition, has inherent maldistribution of reactant gases because the channels immediately below or above the manifold receive most of the flow. The early fuel cells built with such a flow field exhibited low and unstable cell voltages.

- *Criss-cross configuration.* This flow field attempts to eliminate the short-comings of the straight channel flow field by introducing traversal channels allowing the gas to bypass any "trouble" spot, that is, coalescing water droplets. The problems of low velocities and uneven flow distribution due to positioning of the inlet and outlet manifolds are not reduced with this design.
- *Single-channel serpentine.* As described by Watkins et al. [10], this is the most common flow field for small active areas. It ensures that the entire area is covered, although the concentration of reactants decreases along the channel. There is a pressure drop along the channel due to friction on the walls and due to turns. The velocity is typically high enough to push any water condensing in the channel. Attention must be paid to pressure differentials between the adjacent channels, which may cause significant bypassing of channel portions.
- *Multichannel serpentine.* A single-channel serpentine configuration would not work for the large flow field areas because of a large pressure drop. Although a pressure drop is useful in removing the water, excessive pressure drop may generate larger parasitic energy losses. Watkins et al. [11] proposed a flow field that has a multitude of parallel channels meandering through the entire area in a serpentine fashion. Except for the lower pressure drop, this flow field has the same features, advantages, and shortcomings of the single-channel serpentine. The fact that there are parallel channels means that there is always a possibility that one of the channels may get blocked, as discussed previously with straight channels.
- *Multichannel serpentine with mixing.* As suggested by Cavalca et al. [12], this flow field design allows gases to mix at every turn in order to minimize the effect of channel blocking. This does not reduce the chance of channel blockage, but it limits its effect to only a portion of the channel because the flow field is divided in smaller segments, each with its own connecting channel to both the inlet and the outlet.
- *Subsequent serially linked serpentine.* This flow field also divides the flow field into segments in an attempt to avoid the long, straight channels and relatively large pressure differentials between the adjacent sections, thus minimizing the bypassing effect [13].
- *Mirror serpentine.* This is another design to avoid large pressure differentials in adjacent channels, particularly suited for a larger flow fields with multiple inlets and outlets. These are arranged so that the resulting serpentine patterns in adjacent segments are mirror images of each other,

which results in balanced pressures in adjacent channels, again minimizing the bypassing effect [14].

- *Interdigitated.* First described by Ledjeff [15], advocated by Nguyen [16],
  and successfully employed by Energy Partners in its NG-2000 stack series
  [17,18], this flow field differs from all of the previously described fields
  because the channels are discontinued, that is, they do not connect the
  inlet to the outlet manifolds. This way the gas is forced from the inlet
  channels to the outlet channels through the porous back-diffusion layer.
  Wilson et al. [19] suggested a variation of interdigitated flow field in
  which the channels are made by cutting out the strips of the gas diffusion
  layer. Convection through the porous layer shortens the diffusion path
  and helps remove any liquid water that otherwise may accumulate in the
  gas diffusion layer, resulting in better performance, particularly at higher
  current densities. However, depending on the properties of the gas
  diffusion layer, this flow field may result in higher pressure drops. Due to
  the fact that the most of the pressure drop occurs in the porous media,
  the uniformity of flow distribution between individual channels and
  between individual cells strongly depends on uniformity of gas diffusion
  layer thickness and effective porosity (after being squeezed). One of the
  problems with this flow field is inability to remove liquid water from
  the inlet channels. Issacci and Rehg [20] suggested the porous blocks at
  the end of the inlet channels, allowing water to be removed.
- *Biomimetics.* Suggested by Morgan Carbon [21], this is a further refinement of the interdigitated concept. Larger channels branch to smaller
  side channels, further branching to really tiny channels interweaving
  with outlet channels that are arranged in the same fashion—tiny channels
  leading to larger side channels leading to the large channels. This type of
  branching occurs in nature (leaves or lungs), hence the name *biomimetic.*
- *Fractal.* This flow field suggested by the Fraunhoffer Institute [22] is
  essentially the interdigitated flow field concept, but the channels are not
  straight and they have branches.
- *Mesh.* Metallic meshes and screens of various sizes are successfully being
  used in electrolyzers. The uniformity may greatly be affected by positioning of the inlet manifolds. The researchers at Los Alamos National
  Laboratory successfully incorporated metal meshes in fuel cell design
  [23]. The problems with this design are introduction of another
  component with tight tolerances, corrosion, and interfacial contact
  resistance.

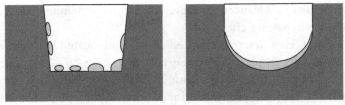

Figure 6-17 The shape of the channel cross-section affects the form of liquid water formation.

- *Porous media flow field* [24]. This is similar to the mesh flow field; the difference is in pore sizes and material. The gas distribution layer must be sufficiently thick and have enough pores sufficiently large to permit a substantially free flow of reactant gas both perpendicular to and parallel to the catalyst layer. Although metallic porous materials (foams) are brittle, carbon-based ones may be quite flexible. This type of flow field may only be applicable for smaller fuel cells because of the high pressure drop.

6.4.4 Channel Shape, Dimensions, and Spacing

The flow field channels may have different shapes, often resulting from the manufacturing process rather than functionality. For example, slightly tapered channels would be very difficult to obtain by machining, but they are essential if the bipolar plate is manufactured by molding. Channel geometry may have an effect on water accumulation. In the round-bottomed channel, condensed water forms a film of water at the bottom, whereas in the channel with tapered walls, condensed water forms small droplets (Figure 6-17). The sharp corners at the bottom of the channel help break the surface tension of the water film, resisting film formation [25].

The shape and size of the water droplets in the channels also depend on hydrophobicity of the porous media and the channel walls. Figure 6-18 shows the possible combinations of hydrophobicity and hydrophilicity of the porous gas diffusion layer and the channel walls and their effect on water droplet shape and size [26].

Typical channel dimensions are around 1 mm but may vary from 0.4 mm to 4 mm. The spacing between the channels is similar. With today's advances in micromanufacturing techniques (MEMS, photolithography) it is possible to produce channels of 0.1 mm and even smaller. The dimensions of the channels and their spacing affect the reactant gas access to the gas diffusion layer and pressure drop as well as electrical current and heat

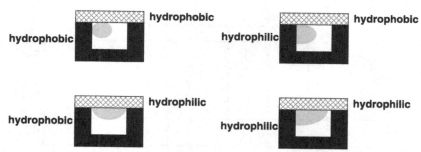

Figure 6-18 Possible combinations of hydrophobicity and hydrophilicity of the porous gas diffusion layer and the channel walls and their effect on droplet size and shape.

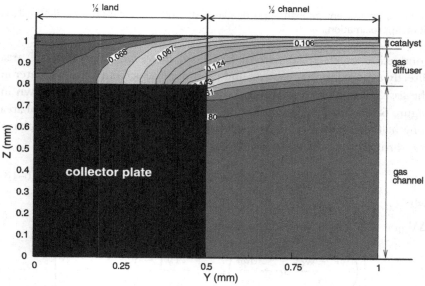

Figure 6-19 Oxygen concentration distribution in and above the channel. *(Adapted from [27].)*

conduction. Wider channels allow more direct contact of the reactants' gas to the gas diffusion layer and also provide wider area for water removal from the gas diffusion layer. Figure 6-19 shows O_2 concentration in a cross-section of an H_2/air fuel cell with serpentine or straight channels [27]. Oxygen concentration, and therefore current density, is higher in the area directly above the channel, and it is significantly lower in the area above the land between the channels.

However, if the channels are too wide there will be no support for the MEA, which will deflect into the channel. Wider spacing enhances

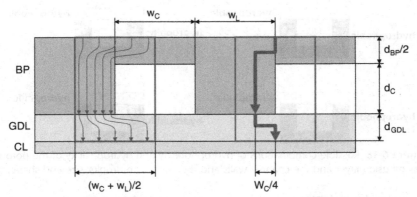

Figure 6-20 Current path through bipolar plate and gas diffusion layer (left, actual; right, approximation).

conduction of electrical current and heat; however, it reduces the area directly exposed to the reactants and promotes the accumulation of water in the gas diffusion layer adjacent to these regions. For a geometry shown in Figure 6-20, and with simplification of the current path in the control area (one half channel and one half spacing between the channels), the voltage loss through the control area is [28]:

$$\Delta V = \Delta V_{BP} + \Delta V_{GDL} + \Delta V_{CR} \qquad (6\text{-}18)$$

where:

ΔV_{BP} is the voltage drop through the bipolar plate:

$$\Delta V_{BP} = \left[\left(\frac{w_L + w_C}{w_L} d_C + \frac{d_{BP}}{2} \right) \rho_{BP,z} + \frac{(w_L + w_C)w_C}{4d_{BP}} \rho_{BP,xy} \right] i \quad (6\text{-}19)$$

ΔV_{GDL} is the voltage drop through the gas diffusion layer:

$$\Delta V_{GDL} = \left[d_{GDL}\rho_{GDL,z} + \frac{(w_L + w_C)w_C}{8d_{GDL}} \rho_{GDL,xy} \right] i \qquad (6\text{-}20)$$

and ΔV_{CR} is the voltage drop due to interfacial contacts:

$$\Delta V_{CR} = R_{CR}\frac{w_L + w_C}{w_C} i \qquad (6\text{-}21)$$

where dimensions w_L, w_C, d_{GDL}, d_C, and d_{BP} are defined in Figure 6-20, and

ρ = resistivity of either bipolar plate (BP) or gas diffusion layer (GDL) in either z-direction (through-plane) or xy-direction (in-plane), Ωcm

R_{CR} = contact resistance between the gas diffusion layer and bipolar plate, Ωcm^2

i = current density at the GDL–catalyst layer interface, A cm^{-2}

Note that the voltage loss through the entire bipolar plate, two gas diffusion layers, and two interfaces is twice as much as what is calculated by Equation (6-18).

In general, as landing width w_L narrows, the fuel cell performance improves until there is either MEA deflection into the channel or the gas diffusion layer crashes because of excessive force applied. The optimum channel size and spacing are therefore a balance between maximizing the open area for the reactant gas access to the gas diffusion layer and providing sufficient mechanical support to the MEA and sufficient conduction paths for electrical current and heat.

Wilkinson and Vanderleeden [25] suggested the use of the following equation to calculate the maximum deflection of the MEA in the channel, d_{max} (mm):

$$d_{max} = \frac{0.032\left(1 - v^2\right)}{t^3 \left(\dfrac{1}{b^4} + \dfrac{1}{L^4}\right)} \frac{p}{E} \qquad (6-22)$$

where:

v = Poisson's ratio

t = MEA thickness, (mm)

b = unsupported channel width, (mm)

L = channel length, (mm)

p = pressure, (kPa)

E = Young's modulus, (kPa)

6.4.5 Pressure Drop Through the Flow Field

Most of the flow fields are arranged as a number of parallel channels (Figure 6-17). In that case the pressure drop along a channel is also the pressure drop in the entire flow field. The pressure drop along a flow field channel may be approximated by the equation for incompressible flow in

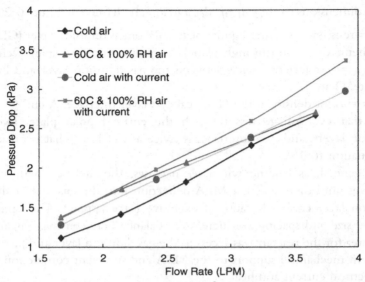

Figure 6-21 Pressure drop of a three-cell, 65-cm² stack as a function of flow rate [33].

pipes and conduits with sufficient accuracy as long as the pressure drop is less than 30% of the inlet pressure:

$$\Delta P = f \frac{L}{D_H} \rho \frac{\bar{v}^2}{2} + \sum K_L \rho \frac{\bar{v}^2}{2} \tag{6-23}$$

where:

f = friction factor
L = channel length, m
D_H = hydraulic diameter, m
ρ = fluid density, $kg\,m^{-3}$
v– = average velocity, $m\,s^{-1}$
K_L = local resistance (for example, in sharp turns)

Hydraulic diameter is defined as four times the channel cross-sectional area divided by its perimeter. For a typical rectangular channel with w_c as width and d_c as depth (Figure 6-21):

$$D_H = \frac{2w_C d_C}{w_C + d_C} \tag{6-24}$$

Channel length is:

$$L = \frac{A_{cell}}{N_{ch}(w_C + w_L)} \tag{6-25}$$

where:

A_{cell} = cell active area, m^2

N_{ch} = number of parallel channels

w_C = channel width, m

w_L = space between the channels, m

The velocity in a fuel cell channel at the entrance of the cell is:

$$\bar{v} = \frac{Q_{stack}}{N_{cell}N_{ch}A_{ch}} \qquad (6\text{-}26)$$

where:

v– = velocity in the channel, $m\,s^{-1}$

Q_{stack} = air flow rate at the stack entrance, $m^3\,s^{-1}$

N_{cell} = number of cells in the stack

N_{ch} = number of parallel channels in each cell

A_{ch} = cross-sectional area of the channel, for a rectangular channel $A_{ch} = bd$, as defined previously

The total flow rate at the stack entrance is (combining Equations 5-42 and 5-47, and dividing by density of humid air):

$$Q_{stack} = \frac{I}{4F}\,\frac{S}{r_{O2}}\,\frac{RT_{in}}{P_{in} - \varphi P_{sat(Tin)}}\,N_{cell} \qquad (6\text{-}27)$$

where:

Q = volumetric flow rate, $m^3\,s^{-1}$

I = stack current, A, $I = iA_{cell}$

F = Faraday constant 96, 485, $As\,mol^{-1}$

S = oxygen stoichiometric ratio

r_{O2} = oxygen content in air, by volume = 0.2095

R = universal gas constant = $8.314\,J\,mol^{-1}\,K^{-1}$

T_{in} = temperature at stack inlet, K

P_{in} = pressure at stack inlet, Pa

φ = relative humidity

P_{sat} = saturation pressure at given inlet temperature (Equation 5-25)

N_{cell} = number of cells in stack

By combining the previous Equations (6-24 through 6-27), the velocity at the stack inlet is:

$$\bar{v} = \frac{i}{4F}\,\frac{S}{r_{O2}}\,\frac{(w_C + w_L)L}{w_C d_C}\,\frac{RT}{P - \varphi P_{sat}} \qquad (6\text{-}28)$$

The *Reynolds number* is important to determine whether the flow in the channel is laminar or turbulent. The Reynolds number at the channel entrance is:

$$Re = \frac{\rho \bar{v} D_H}{\mu}$$

$$= \frac{1}{\mu} \frac{i}{2F} \frac{S}{r_{O2}} \frac{(w_C + w_L)L}{w_C + d_C} \left(M_{Air} + M_{H2O} \frac{\varphi P_{sat(Tin)}}{P_{in} - \varphi P_{sat(Tin)}} \right)$$

$$(6\text{-}29)$$

The flow rate at the stack outlet is somewhat different than the flow at the inlet. It can be lower, equal, or higher, depending on the conditions at the inlet and outlet (flow rate, temperature, pressure, and humidity). Assuming that the outlet flow is saturated with water vapor, the flow rate at the stack outlet is:

$$Q_{stack}^{out} = \frac{I}{4F} \left(\frac{S}{r_{O2}} - 1 \right) \frac{R T_{out}}{P_{in} - \Delta P - P_{sat(Tout)}} N_{cell} \qquad (6\text{-}30)$$

where:

ΔP = pressure drop through the stack

The ratio between the outlet and inlet flow rate, thus velocity, is:

$$\frac{Q_{stack}^{out}}{Q_{stack}^{in}} = \frac{S - r_{O2in}}{S} \frac{T_{out}}{T_{in}} \frac{P_{in} - \varphi P_{vs(Tin)}}{P_{in} - \Delta P - P_{sat(Tout)}} \qquad (6\text{-}31)$$

The first factor is always lower than 1, the second factor is either higher than or equal to 1 (depending on the inlet temperature being lower than or equal to outlet temperature), and the third factor also depends on the inlet temperature and humidity (if the gas at the inlet is saturated at the stack temperature, then this factor is higher than 1). For all practical purposes, the difference between the inlet and outlet flow rates varies within ±5%.

The values for viscosity of common fuel cell gases are shown in Table 6-1. The variation of viscosity with pressure is small for most gases, but viscosity varies with temperature [29]:

$$\mu = \mu_0 \left(\frac{T_0 + C}{T + C} \right) \left(\frac{T}{T_0} \right)^{\frac{3}{2}} \qquad (6\text{-}32)$$

where μ_0 is known viscosity at temperature T_0, from Table 6-1; the coefficient, C, is also listed in Table 6-1.

TABLE 6-1 Viscosity of Fuel Cell Gases (at 25°C)

	Viscosity kg m^{-1}s^{-1}	Coefficient C
Hydrogen	0.92×10^{-5}	72
Air	1.81×10^{-5}	120
Water vapor	1.02×10^{-5}	660

Viscosity of gas mixtures, such as humidified air or humidified hydrogen, can be calculated from the following equation [30]:

$$\mu_{mix} = \frac{\mu_1}{1 + \Psi_1 \dfrac{M_2}{M_1}} + \frac{\mu_2}{1 + \Psi_2 \dfrac{M_1}{M_2}} \tag{6-33}$$

where:

$$\Psi_1 = \frac{\sqrt{2}}{4}\left(1 + \left(\frac{\mu_1}{\mu_2}\right)^{0.5}\left(\frac{r_2}{r_1}\right)^{0.25}\right)^2\left(1 + \frac{r_1}{r_2}\right)^{-0.5} \tag{6-34}$$

$$\Psi_2 = \frac{\sqrt{2}}{4}\left(1 + \left(\frac{\mu_2}{\mu_1}\right)^{0.5}\left(\frac{r_1}{r_2}\right)^{0.25}\right)^2\left(1 + \frac{r_2}{r_1}\right)^{-0.5} \tag{6-35}$$

and

μ_1, μ_2 are the viscosities of components 1 and 2

r_1, r_2 are the volume fractions of components 1 and 2 in the mixture

M_1, M_2 are the molecular weights of components 1 and 2

For steady laminar flow in a channel, the product of the friction factor and the Reynolds number is a constant [31]:

$$Re\, f = \text{constant} \tag{6-36}$$

For a circular channel, $Re\,f = 64$. For rectangular channels, the value of $Re\,f$ depends on the channel aspect ratio, w_C/d_C:

$$Re\, f \approx 55 + 41.5\ \exp\left(\frac{-3.4}{w_C/d_C}\right) \tag{6-37}$$

For a square channel, $Re\, f \approx 56$.

Though some geometrical pressure-loss coefficients, K_L, are available for various bends or elbows, none fits a specific shape of gas flow channels in fuel cells. Values as high as $30\,f$ have been suggested for 90–degree bends and as high as $50\,f$ for close pattern return bends [29].

For fully turbulent flow, the friction coefficient f is independent of the Reynolds number and may be approximated by the Karman's equation

(Equation 6-14). Note that the three walls of the channel are smooth, but the fourth one is the porous gas diffusion layer.

For the porous flow fields, the pressure drop may be determined by Darcy's law [32]:

$$\Delta P = \mu \frac{Q_{cell}}{kA} L \tag{6-38}$$

where:

μ = viscosity of the fluid, $\mathrm{kg\,m^{-1}\,s^{-1}}$

Q_{cell} = volumetric flow rate through a cell, $\mathrm{m^3\,s^{-1}}$

k = permeability, $\mathrm{m^2}$

A = cross-sectional area of the flow field, $\mathrm{m^2}$

L = length of the flow field, m

The same equation may be used to approximate the pressure drop through any flow field as long as the flow is laminar. The permeability factor, k, then refers to the entire flow field and must be determined experimentally.

In most cases the flow through the fuel cell flow field is laminar, which means that the pressure drop is linearly proportional to velocity, that is, to flow rate. However, in a fuel cell channel there are some deviations from the uniform pipe flow:

- Roughness of the GDL is different than that of the channel walls.
- The reactant gas participates in the chemical reaction and the flow rate varies along the channel, although not significantly.
- Temperature may not be uniform along the channel.
- Typically the channel is not straight, but there are numerous sharp turns (90 or 180 degrees).
- Liquid water may be present inside the channel, either in the form of little droplets or as a film, in both cases effectively reducing the channel cross-sectional area.

Figure 6-21 shows a linear relationship between the flow rate and the pressure drop through the cathode side of a three-cell, 65-cm^2 stack when dry air at room temperature is run through the stack and no current, and thus no water, is being generated [33]. The Reynolds number at the entrance of the cathode channels was <250 at the highest flow rate. When humidified air (100% RH at 60°C) is run through the stack, the pressure drop is higher because of condensation in the cold, nonoperating stack; however, as the flow rate is increased, the pressure drop approaches that of the dry air. This may be explained by improved water removal from the stack at higher channel velocities.

When the stack is operational and generates water, the pressure drop is linearly proportional to the flow rate if the incoming air is dry because all the product water gets evaporated in the flow of air. Note that the molar flow rate at the exit is higher than the flow rate at the inlet because each oxygen molecule consumed is replaced by two water vapor molecules. When the incoming air is fully humidified, evaporation of the product water is no longer possible; as a result the pressure drop starts to increase exponentially with the air flow rate (and with current, that is, water generation rate), as shown in Figure 6-21.

Example

Calculate the pressure drop through a cathode flow field of a 60-cell stack with 65-cm^2 cell area. The stack operates at 125 kPa (inlet), 60°C, with saturated air. The flow rate is kept proportional to current at three times stoichiometry. Nominal operating point is 0.4 A cm^{-2} at 0.7 V. The cathode flow field consists of six parallel serpentine channels 0.8 mm wide, 0.8 mm deep, and 0.8 mm apart, with four 90-degree bends.

Solution

The pressure drop is (Equation 6-23):

$$\Delta P = f \frac{L}{D_H} \rho \frac{\bar{v}^2}{2} + \sum K_L \rho \frac{\bar{v}^2}{2}$$

Hydraulic diameter is (Equation 6-24):

$$D_H = \frac{2w_c d_c}{w_c + d_c} = 2 \times 0.08 \times 0.08/(0.08 + 0.08) = 0.08 \text{ cm}$$

Channel length is (Equation 6-25):

$$L = \frac{A_{cell}}{N_{ch}(w_C + w_L)} = \frac{65}{6 \times (0.08 + 0.08)} = 67.7 \text{ cm}$$

The flow rate at the stack entrance is (Equation 6-27):

$$Q_{stack} = \frac{I}{4F} \frac{S}{r_{O2}} \frac{RT_{in}}{P_{in} - \varphi P_{sat(Tin)}} N_{cell}$$

$$= \frac{0.4 \times 65}{4 \times 96,485} \frac{3}{0.21} \frac{8.314 \times (273.15 + 60)}{125,000 - 19,944} 60$$

$$= 0.00152 \text{ m}^3 \text{ s}^{-1} = 1520 \text{ cm}^3 \text{ s}^{-1}$$

where 19,944 Pa is the saturation pressure at 60°C.

The velocity in a fuel cell channel at the entrance of the cell is (Equation 6-26):

$$v = \frac{Q_{stack}}{N_{cell} \, N_{ch} \, A_{ch}} = \frac{1520}{60 \times 6 \times 0.08 \times 0.08} = 660 \text{ cm s}^{-1}$$

The Reynolds number at the channel entrance is (Equation 6-29):

$$Re = \frac{\rho \bar{v} D_H}{\mu}$$

$$\rho = \text{density of humidified air} = \frac{(P - P_{sat})M_{air} + P_{sat}M_{H2O}}{RT}$$

$$= \frac{(125,000 - 19,944) \times 29 + 19,944 \times 18}{8314 \times (273.15 + 60)} = 1.23 \text{ kg m}^3 = 0.00123 \text{ g cm}^3$$

Viscosity of humidified air is (from Table 6-1 and Equations 6-32 through 6-35):

$$\mu = 2 \times 10^{-5} \text{ kg m}^{-1} \text{ s}^{-1} = 0.0002 \text{ g cm}^{-1} \text{ s}^{-1}$$

$$Re = \frac{\rho v D_H}{\mu} = 0.00123 \times 660 \times 0.08/0.0002 = 324.7$$

$$Re f \approx 55 + 41.5 \exp\left(\frac{-3.4}{b/d}\right) = 56$$

Friction factor:

$$f = 56/Re = 56/324.7 = 0.172$$

And finally, the pressure drop is:

$$\Delta P = f \frac{L}{D_H} \rho \frac{\bar{v}^2}{2} + \sum K_L \rho \frac{\bar{v}^2}{2} = 0.172 \frac{0.677}{0.0008} 1.23 \frac{6.6^2}{2}$$

$$+(4 \times 30 \times 0.172) \, 1.23 \frac{6.6^2}{2} = \Delta P = 4452 \text{ Pa} \quad \text{Answer}$$

This pressure drop has been calculated based on inlet conditions. The velocity at the outlet is somewhat lower, so the average velocity throughout the channel would be lower and, consequently, the pressure drop would be somewhat lower. The exact solution may be obtained through an iterative process.

The inlet and outlet manifolds of this 60-cell stack should be sized so that the pressure drop through a manifold is at least an order of magnitude lower than the pressure drop through individual cells, that is, less than 445 Pa (the total stack pressure drop would then be about 5300 Pa).

6.5 HEAT REMOVAL FROM A FUEL CELL STACK

To maintain the desired temperature inside the cells, the heat generated as a byproduct of the electrochemical reactions must be taken away from the cells and from the stack. Different heat management schemes (Figure 6-22) may be applied, such as:

1. *Cooling with coolant flowing between the cells.* Coolant may be deionized water, antifreeze coolant, or air. Cooling may be arranged between each cell, between each pair of cells (in such a configuration, one cell has the cathode and the other cell has the anode next to the cooling arrangement), or between a group of cells (this is feasible only for low–power densities because it results in higher temperatures in the center cells). Equal distribution of coolant may be accomplished by the manifolding arrangement similar to that of reactant gases. If air is used as a coolant, equal distribution may be accomplished by a plenum.

2. *Cooling with coolant at the edge of the active area (with or without fins).* The heat is conducted through the bipolar plate and then transferred to the cooling fluid, typically air. To achieve relatively uniform temperature distribution within the active area, the bipolar plate must be a very good thermal conductor. In addition, the edge surface may not be sufficient for heat transfer and fins may need to be employed. This method results in a much simpler fuel cell stack and fewer parts, but it has heat transfer limitations and is typically used for low-power outputs.

3. *Cooling with phase change.* Coolant may be water or another phase-change medium. Use of water simplifies the stack design because water is already used in both anode and cathode compartments.

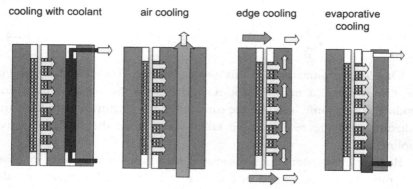

Figure 6-22 Different cell/stack cooling options.

6.5.1 Stack Heat Balance

There are several ways to set the fuel cell stack energy balance. In general, energy of fuel (higher heating value) is converted into either electricity or heat, or:

$$\frac{I}{2F} H_{HHV} n_{cell} = Q_{gen} + IV_{cell} n_{cell} \qquad (6\text{-}39)$$

Heat generated in a fuel cell stack is then:

$$Q_{gen} = (1.482 - V_{cell})In_{cell} \qquad (6\text{-}40)$$

The previous equation assumes that all the product water leaves the stack as liquid at 25°C, which may be the case if the inlet is fully saturated at the stack operating temperature. If all of the product water leaves the stack as vapor, then the following equation is more appropriate:

$$Q_{gen} = (1.254 - V_{cell})In_{cell} \qquad (6\text{-}41)$$

The previous Equations (6-39 through 6-41) are just approximations. A complete stack energy balance (such as shown in Equations 5-62 through 5-67) should take into account the heat (enthalpy) brought into the stack with reactant gases as well as the heat of the unused reactant gases leaving the stack, including both latent and sensible heat of water at the stack inlet and stack outlet:

Enthalpy of reactant gases in = Electricity generated + Enthalpy of unused reactant gases including heat of product water + Heat dissipated to the surrounding + Heat taken away from the stack by active cooling
or

$$\sum Q_{in} = W_{el} + \sum Q_{out} + Q_{dis} + Q_c \qquad (6\text{-}42)$$

On closer examination of this energy balance, it is clear that some of the heat generated in the stack is carried away by reactant gases and product water, some is lost to the surroundings by natural convection and radiation, and the rest must be taken away from the stack by active cooling.

Because heat generation is associated with the (voltage) losses in a fuel cell, most heat is generated in the catalyst layers, predominantly on the

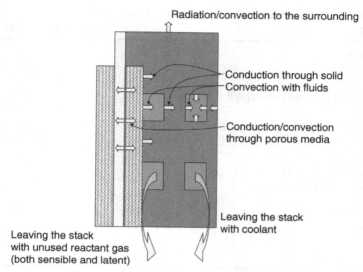

Figure 6-23 Heat paths in a fuel cell segment.

cathode side, then in the membrane due to ohmic losses and in the electrically conductive solid parts of the fuel cell (also due to ohmic losses). This heat is first carried by heat conduction through solid parts of the fuel cell, namely porous electrode structures, including the gas diffusion layer and bipolar plates (Figure 6-23). Some heat is transferred to the reactant gases (depending on their temperature), some is transferred to the cooling medium through convection, and some is conducted to the edge of the stack, where it is transferred to the surrounding air through radiation and natural convection (or, in some cases, forced convection when this is the primary way of stack temperature control).

6.5.2 Heat Conduction

The rate of heat transferred by conduction in the x-direction through a finite cross-sectional area A is, according to the Fourier law of conduction, proportional to the temperature difference:

$$Q_x = kA \frac{dT}{dx} \tag{6-43}$$

where k is the thermal conductivity, $W\ m^{-1}\ k^{-1}$.

TABLE 6-2 Thermal Conductivity of Some Fuel Cell Materials

Material	Thermal Conductivity[a] $W\,m^{-1}\,K^{-1}$
Aluminum	237
Copper	401
Nickel	91
Nickel alloys (Inconel, Hasteloy)	12
Titanium	22
Stainless steel 316	13
Platinum	71
Graphite	98
Graphite/polymer mix	~20[b]
Carbon fiber paper	1.7[b]
Teflon	0.35
Liquid water	0.611
Water vapor	0.0198
Air	0.0267
Hydrogen	0.198

[a]At 300 K; [b]through-plane.

The values of thermal conductivity, k, for some typical fuel cell materials are given in Table 6-2.

More generally, the steady-state heat conduction is governed by the equation (easily obtained by differentiating Equation 6-43):

$$\frac{d^2T}{dx^2} = 0 \qquad (6\text{-}44)$$

which can also easily be extended to three-dimensional steady-state heat conduction:

$$\nabla \cdot (\nabla T) = 0 \qquad (6\text{-}45)$$

To solve this equation, two boundary conditions (for each direction) must be given that describe the behavior of T at the system boundaries (constant T or prescribed flux).

When the heat is conducted through two adjacent materials with different thermal conductivities, the third boundary condition comes from a requirement that the temperature at the interface is the same for both materials (in cases when the contact resistance may be neglected) or there is a discontinuity in temperature distribution at the interface described by either contact resistance R_{tc} or thermal contact coefficient h_{tc}, defined by:

$$Q = h_{tc}A\Delta T \tag{6-46}$$

$$R_{tc} = \frac{1}{h_{tc}A} \tag{6-47}$$

For the case involving internal heat generation due to electrical resistance, Equation (6-44) becomes:

$$\frac{d^2T}{dx^2} + \frac{q_{int}}{k} = 0 \tag{6-48}$$

where:

$$q_{int} = \text{rate of heat generation per unit volume}$$

In the fuel cell, internal heat generation results from electrical and ionic resistance:

$$q_{int} = i^2 \rho_s \tag{6-49}$$

In porous media, an effective thermal conductivity is used that takes into account porosity of the media, ε:

$$k_{eff} = -2k_s + \left[\frac{\varepsilon}{2k_s + k} + \frac{1 - \varepsilon}{3k_s} \right]^{-1} \tag{6-50}$$

Example

A fuel cell operates at 0.6 V and 1 A cm^{-2}. Calculate the temperature distribution through a gas diffusion layer/bipolar plate sandwich on the cathode side. Assume a constant heat flux at the gas diffusion/catalyst layer interface equal to all the fuel cell losses except those due to ionic and electrical resistance. Assume that half of the resistive losses apply to the anode side and half to the cathode. Ionic resistance through the membrane is 0.1 Ohm-cm^2. At the outer edge of the bipolar plate, assume that heat is removed by a cooling fluid at 60°C, with heat transfer coefficient $h = 1600$ W m^{-2} K^{-1}. Electrical resistivity of the gas diffusion layer and bipolar plate is 0.08 Ohm-cm and 0.06 Ohm-cm, respectively. There is a contact resistance of 0.005 Ohm-cm^2 between the gas diffusion layer and bipolar plate. Effective thermal conductivity of GDL and bipolar plate is $1.7\ \mathrm{W\,m^{-1}\,K^{-1}}$ and $20\ \mathrm{Wm^{-1}\,K^{-1}}$, respectively. There is a thermal contact resistance between these two layers equal to $1°\mathrm{C/W}$ thickness of GDL and the bipolar plate is 0.38 mm and 3.3 mm, respectively.

Solution

Assume one-dimensional steady-state conduction.

Let T_1 be the temperature at the bipolar plate–cooling fluid interface, T_2 be the temperature of the bipolar plate facing the gas diffusion layer, T_3 be the temperature of the gas diffusion layer facing the bipolar plate, and T_4 be the temperature of the gas diffusion layer facing the catalyst layer.

Voltage loss through GDL is: $V = IR = i\ \rho_{GDL}d_{GDL} = 1 \times 0.08 \times 0.038 = 0.00304\ \mathrm{V}$

Voltage loss through bipolar plate $= i\ \rho_{BP}d_{BP} = 1 \times 0.06 \times 0.33 = 0.00198\ \mathrm{V}$

Voltage loss due to contact resistance $= 1 \times 0.005 = 0.005\ \mathrm{V}$

Total electrical resistance (one side) $= 0.00304 + 0.00198 + 0.005 = 0.01\ \mathrm{V}$

Total electrical resistance (both sides) $= 0.01 \times 2 = 0.02\ \mathrm{V}$

Voltage loss due to ionic resistance $= 1 \times 0.1 = 0.1\ \mathrm{V}$

Total resistive losses $= 0.12\ \mathrm{V}$

Fuel cell voltage corrected for resistance $= 0.6 + 0.12 = 0.72\ \mathrm{V}$

Heat generation in fuel cell (except resistance) $= (1.254 - 0.72) \times 1^2 = 0.534\ \mathrm{W\,cm^{-2}}$

Heat flux at the catalyst–gas diffusion layer interface (including $1/2$ of the ionic resistance):

$$Q_4 = 0.534 + 1^2 \times 0.1/2 = 0.584\ \mathrm{W\,cm^{-2}}$$

Heat generation due to resistance in GDL $= i \times \Delta V = 1 \times 0.00304 = 0.00304\ \mathrm{W\,cm^{-2}}$

Heat generation due to resistance in bipolar plate $= 1 \times 0.00198 = 0.00198\ \mathrm{W\,cm^{-2}}$

Heat generation due to contact resistance $= 1 \times 0.005 = 0.005\ \mathrm{W\,cm^{-2}}$

Heat flux at the bipolar plate–cooling fluid interface:

$$Q_1 = 0.584 + 0.00304 + 0.00198 + 0.005 = 0.594\ \mathrm{W\,cm^{-2}}$$

Heat flux at the bipolar plate–gas diffusion layer interface (on the bipolar plate side):

$$Q_2 = 0.594 - 0.00198 = 0.592\ \mathrm{W\,cm^{-2}}$$

Heat flux at the bipolar plate–gas diffusion layer interface (on the GDL side):

$$Q_3 = 0.592 - 0.005 = 0.587\ \mathrm{W\,cm^{-2}}$$

Convection heat flux from the bipolar plate surface to the cooling fluid (Equation 6-52):

$$Q_1 = h(T_F - T_1)$$

The temperature T_1 is then:

$$T_1 = T_F + Q_1/h = 60 + .594/0.16 = 63.71$$

The governing equation for heat flux inside the bipolar plate is (Equation 6-48):

$$\frac{d^2T}{dx^2} + \frac{q_{int}}{k} = 0$$

After two integrations, it becomes:

$$T = \frac{q_{int}}{k}\frac{x^2}{2} + C_1 x + C_2$$

The integration constants, C_1 and C_2, can be determined from the boundary conditions:

$$At\ x = 0 \quad T = T_1$$

$$At\ x = d_{BP} \quad k\frac{dT}{dx} = Q_2$$

therefore:

$$C_2 = T_1$$

and

$$C_1 = \frac{Q_2}{k} + \frac{q_{int}}{k} d_{BP}$$

Temperature distribution in the bipolar plate is:

$$T = T_1 + \frac{Q_2}{k_{BP}}x + \frac{q_{int,BP}}{k_{BP}}\left(d_{BP}x - \frac{x^2}{2}\right)$$

where:

$$q_{int,BP} = 0.00198/0.33 = 0.006\ W\ cm^{-3}$$

The temperature T_2 (at $x = d_{BP}$) is:

$$T_2 = T_1 + \frac{Q_2}{k_{BP}} d_{BP} + \frac{q_{int,BP}}{k_{BP}}\frac{d_{BP}^2}{2} = 63.71 + \frac{0.592}{0.19}0.33 + \frac{0.006}{0.19}\frac{0.33^2}{2}$$

$$= 64.74°C$$

There is a discontinuity in temperature at the gas diffusion layer/bipolar plate interface due to contact resistance. Temperature T_3 is then:

$$T_3 = T_2 + R_{tc}\ Q_3 = 64.74 + 1 \times 0.587 = 65.33°C$$

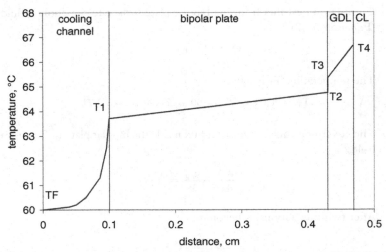

Figure 6-24 Temperature distribution across the bipolar plate/gas diffusion layer "sandwich."

The temperature distribution inside the GDL is (applying the same equation as for the bipolar plate):

$$T = T_3 + \frac{Q_4}{k_{GDL}}x + \frac{q_{int,GDL}}{k_{GDL}}\left(d_{GDL}x - \frac{x^2}{2}\right)$$

where:

$$q_{int,GDL} = 0.00304/0.038 = 0.080 \text{ W cm}^{-3}$$

The temperature T_4 (at $x = d_{GDL}$) is:

$$T_4 = T_3 + \frac{Q_4}{k_{GDL}}\,d_{GDL} + \frac{q_{int,GDL}}{k_{GDL}}\frac{d_{GDL}^2}{2}\; 65.33 + \frac{0.584}{0.017}\,0.038 + \frac{0.08}{0.017}\frac{0.038^2}{2}$$

$$= T_4 = 66.6 - 64°C$$

The resulting temperature distribution across the bipolar plate/gas diffusion layer sandwich is shown in Figure 6-24.

6.5.3 Active Heat Removal

From the heat removal point of view, a fuel cell stack may be considered a heat exchanger with internal heat generation. The walls of the fuel cell cooling channels may be neither at constant temperature nor have the constant heat flux, but these two cases are often used as boundary cases for heat transfer analyses.

The heat to be removed by active cooling (Equation 6-42) is:

$$Q_c = \sum Q_{in} - W_{el} - \sum Q_{out} - Q_{dis} \qquad (6\text{-}51)$$

The same heat, Q_c, has to be transferred to the cooling fluid:

$$\frac{dQ_c}{dA_c} = h(T_S - T_C) \qquad (6\text{-}52)$$

or integrated over the entire heat exchange surface, A_c, just as in a heat exchanger:

$$Q_c = UA_c\text{LMTD} \qquad (6\text{-}53)$$

where:

h = local heat transfer coefficient, $Wm^{2\circ}C$

U = overall heat transfer coefficient, $Wm^{2\circ}C$

A_c = heat exchange area = surface area of the cooling channels, m^2

LMTD = logarithmic mean temperature difference, °C, defined as

$$\text{LMTD} = \frac{(T_S - T_C)_{in} - (T_S - T_C)_{out}}{\ln\dfrac{(T_S - T_C)_{in}}{(T_S - T_C)_{out}}} \qquad (6\text{-}54)$$

The temperature difference between the stack body, T_S, and the cooling fluid, T_C, may be constant (constant thermal flux case), or it may vary from one side of the stack to the other, depending on the position of coolant inlets and outlets vs. reactant inlets and outlets, internal coolant and reactant passages configuration, and current density configuration.

The same heat, Q_c, has to be "absorbed" by the cooling fluid and carried out of the fuel cell stack:

$$Q_c = \dot{m}c_p(T_{c,out} - T_{c,in}) \qquad (6\text{-}55)$$

The temperature difference $\Delta T_c = (T_{c,out} - T_{c,in})$ is a design variable that has to be selected in conjunction with the coolant flow rate. As usual in fuel cell design, in selecting the temperature difference between coolant outlet and inlet there are conflicting requirements. To achieve uniform temperature distribution through the stack, ΔT_c should be selected as small as practically possible, unless larger temperature gradients are required by stack design (for example, to facilitate water management). However, small ΔT_c would result in large coolant flow rate, which would increase parasitic power and reduce system efficiency. On the other side, larger ΔT_c would result in lower temperature to which the coolant must be cooled, and there may be practical limits imposed by the ambient temperature and by the characteristics and performance of the heat rejection device.

The coefficient of convection heat transfer, h, depends on the *Nusselt number*, that is, properties of the coolant, geometry of the passages, and flow characteristics:

$$h = Nu \frac{k}{D_H} \quad \text{or} \quad h = Nu_L \frac{k}{L} \quad \text{or} \quad \overline{h} = \overline{Nu}_L \frac{k}{L} \qquad (6\text{-}56)$$

The Nusselt number represents the ratio of convection heat transfer for fluid in motion to conduction heat transfer for a motionless layer of fluid [34]. Common expressions for the Nusselt number and for various flow characteristics (i.e., developed/undeveloped, laminar/turbulent) are summarized in Table 6-3 [34], and the thermal properties of some fluids commonly found in fuel cells are given in Table 6-4.

TABLE 6-3 Nusselt Number for Various Internal Flow Conditions (Adapted from [34])

Condition	Equation
Laminar flow	
Hydraulically fully developed,	
$x/D > 0.05\, Re\, Pr$	
Thermally fully developed	
$x/D > 0.05\, Re\, Pr$	
Uniform wall temperature	
Square tube	$Nu = 2.98$
Circular tube	$Nu = 3.66$
Thermal entry	
Uniform wall temperature	
Uniform wall temperature	
$x/D < 0.01\, Re\, Pr$	
Uniform wall heat flux $x/D < 0.01\, Re\, Pr$	
Hydraulic and thermal entry x/D	
$< 0.01\, Re\, Pr$	
Turbulent flow	
Hydraulically fully developed	
Thermally fully developed	
	where $f = 4/(1.58 \ln Re - 3.28)^2$
Thermal entry $x/D < 60$	
Hydraulic and thermal entry $x/D < 60$	
Transitional turbulent flow	$Nu = C_{tr}Nu_{L2} + (1 - C_{tr})Nu_{T8}$
	$Nu_{L2} = Nu$ for $Re - 2000$
	$Nu_{T8} - Nu$ for $Re - 8000$
	$C_{tr} = 1.33 - Re/6000$

TABLE 6-4 Properties of Some Fuel Cell Gases and Coolant Mediums (Upper Number Is at 20°C and Lower Is at 60°C)

Gas	Density kg m^{-3}	Viscosity kg m^{-1} s^{-1}	Thermal Conductivity W m^{-1} K^{-1}	Specific Heat kJ kg^{-1} K^{-1}
Water	998	0.001	0.602	4.18
	984	0.000466	0.654	4.18
Ethylene glycol	1116	0.0214	0.249	2.38
	1088	0.0052	0.260	2.56
Propylene glycol	1036	0.054	0.200	2.47
	1010	0.0075		2.72
Air	1.21	18×10^{-6}	0.0257	1.005
	1.06	20×10^{-6}	0.0287	1.008
Hydrogen	0.0841	8.8×10^{-6}	0.178	14.2
	0.0741	9.6×10^{-6}	0.198	14.4

Example

Coolant (50% water and 50% ethylene glycol) flows with the velocity of 1 m/s through a 1-mm-square channel at 80°C. Calculate the heat transfer coefficient for hydraulically and thermally fully developed flow under uniform wall heat flux conditions. If the heat flux is 1 W cm^{-2}, calculate the required temperature difference between the wall and the fluid.

Solution

The heat transfer coefficient is:

$$h = Nu \frac{k}{D_H}$$

$Nu = 3.61$
$D_H = 4 \, A/Pm = 4 \times 0.001^2/4 \times 0.001 = 0.001$ m
k for water at 80°C = 0.671 W m^{-1} K^{-1}
k for ethylene glycol at 80°C = 0.261
k for 50/50 mixture of water and ethylene glycol = 0.466 W m^{-1} K^{-1}

$$h = 3.61 \times 0.466/0.001 = 1682 \text{ W m}^{-2} \text{ K}^{-1}$$

The average temperature difference between the wall of the cooling channel and coolant required for transfer of 1 W cm^{-2} (or 10,000 W m^{-2}) is:

$$T_s - T_C = \frac{Q}{AK} = \frac{10000 \text{ W m}^{-2}}{1682 \text{ W m}^{-2} \text{ K}} = 6°C$$

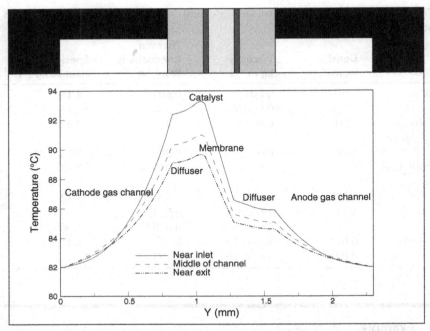

Figure 6-25 Temperature distribution through a fuel cell characteristic cross-section. *(Adapted from [28].)*

Equation (6-45) may be solved numerically for a variety of boundary conditions, such as constant T_S at the walls or constant or prescribed heat flux. Because of complicated three-dimensional heat transfer pathways (shown in Figure 6-23), calculation of heat fluxes and temperature profiles in a fuel cell stack requires 3-D numerical simulation. Figure 6-25 shows temperature distribution in a representative cross-section of a fuel cell obtained by 3-D numerical simulation [28]. From Figure 6-25, it is obvious that there are significant temperature variations inside a fuel cell stack. Because most heat in a fuel cell stack is produced in the cathode catalyst layer, that layer expectedly has the highest temperature.

6.5.4 Heat Dissipation from the Stack by Natural Convection and Radiation

Maximum heat that the stack may lose through natural convection and radiation to the surroundings is:

$$Q_{dis} = \frac{T_S - T_0}{R_{th}} \qquad (6\text{-}57)$$

where:

T_S = stack surface temperature
T_0 = temperature of the surrounding walls
R_{th} = thermal resistance, defined as

$$R_{th} = \cfrac{1}{\cfrac{1}{R_C} + \cfrac{1}{R_R}} \tag{6-58}$$

where:

R_C = convective thermal resistance, defined as

$$R_C = \frac{1}{hA_S} \tag{6-59}$$

and

R_R = radiative thermal resistance defined as

$$R_R = \frac{1}{\sigma F A_S (T_S + T_0)(T_S^2 + T_0^2)} \tag{6-60}$$

where:

σ = Stefan–Boltzman constant = $5.67 \times 10^{-8}\,\mathrm{W\,m^{-2}\,K^{-4}}$
F = shape factor; for the first approximation it may be assumed as $F = 1$
A_S = stack exposed surface area, $\mathrm{m^2}$

For short stacks, the ratio of exposed (external) surface area and the (internal) active area, A_S/A_{act}, may be too large and the stack cannot reach the operating temperature at all because the rate of heat dissipation is higher than the rate of heat generation. This is often the case with single cells used in laboratories that must use heat pads (or other means of temperature control) to obtain the desired operating temperature.

The heat transfer coefficient, h, in Equation (6-59) is a function of the Nusselt number, Nu:

$$h = \frac{k}{L} Nu_L \tag{6-61}$$

For vertical plates and natural convection, the Nusselt number is some empirical function of Prandtl and Rayleigh numbers, $Nu_L = f(Pr, Ra_L)$, such as, for example:

$$Nu_L = \left\{ 0.825 + \frac{0.387 Ra_L^{1/6}}{\left[1 + \left(\dfrac{0.5}{Pr}\right)^{9/16}\right]^{8/27}} \right\}^2 \tag{6-62}$$

TABLE 6-5 Air Thermal Properties

Property	@300K	@350K
Density, ρ, $kg\,m^{-3}$	1.1774	0.998
Specific heat, c_p, $kJ\,kg^{-1\circ}C^{-1}$	1.0057	1.0090
Thermal conductivity, k, $W\,m^{-1\circ}C^{-1}$	0.02624	0.03003
Thermal diffusivity, α, $m^2\,s^{-1}$	0.2216×10^{-4}	0.2983×10^{-4}
Thermal expansion, β, C^{-1}	0.00333	0.00286
Viscosity, μ, $kg\,m^{-1}\,s^{-1}$	1.846×10^{-5}	2.075×10^{-5}
Kinematic viscosity, ν, $m^2\,s^{-1}$	15.68×10^{-6}	20.76×10^{-6}
Prandtl number, Pr	0.708	0.697

where:

$$Ra_L = \frac{g\beta(T_S - T_0)L^3}{\nu\alpha} \qquad (6\text{-}63)$$

and

$$Pr = \frac{\nu}{\alpha} \qquad (6\text{-}64)$$

where:

g = gravity acceleration ($9.81\,m\,s^{-2}$)

β = thermal expansion coefficient; for gases $\beta = 1/T$

L = characteristic length or length of travel of the fluid in the boundary layer, that is, the height of the stack, m

ν = kinematic viscosity, $m^2\,s^{-1}$

α = thermal diffusivity, $m^2\,s^{-1}$

Fluid properties may be evaluated at the free stream or film (wall) temperature.

For horizontal plate:

$Nu_L = 0.54\,Ra_L^{1/4}$

L = A/Pm (area/perimeter)

Some air properties at 300 and 350 K are listed in Table 6-5.

6.5.5 Alternative Stack Cooling Options

Air Cooling

Air is already passing through the cathode compartment in excess of oxygen exact stoichiometry. Can the same air be used as a coolant? Theoretically, yes, although the flow rate would have to be much higher. How high? This can be found from a simple heat balance: Heat generated by a fuel cell must be equal to heat taken away by the flow of air.

The heat generated, assuming that the product water evaporates and leaves the stack as vapor, which actually should be the case for this cooling scheme, is (Equation 6-41):

$$Q = (1.254 - V_{cell})In_{cell} \qquad (6\text{-}65)$$

The heat transferred to air is:

$$Q = \dot{m}c_p(T_{Air,out} - T_{Air,in}) \qquad (6\text{-}66)$$

The mass flow rate at the stack exit is given by Equation (5-56):

$$\dot{m}_{Airout} = \left[(S_{O2} - 1)M_{O2} + S_{O2} \frac{1 - r_{O2in}}{r_{O2in}} M_{N2} \right] \frac{i \cdot n_{cell}}{4F} \qquad (6\text{-}67)$$

By combining the previous equations, an expression for required stoichiometric ratio is obtained:

$$S_{O2} = \frac{M_{O2} + \dfrac{4F(1.254 - V_{cell})}{c_p \Delta T}}{M_{O2} + \dfrac{1 - r_{O2in}}{r_{O2in}} M_{N2}} \qquad (6\text{-}68)$$

The only variables in the previous equation are the cell potential, V_{cell}, and temperature difference between air at the stack inlet and outlet, ΔT. The cell potential determines the cell efficiency, and at a lower efficiency more heat is generated (which follows directly from Equation 6-65). The air temperature difference is determined by the ambient temperature and the stack operating temperature. Figure 6-26 shows the resulting cathode

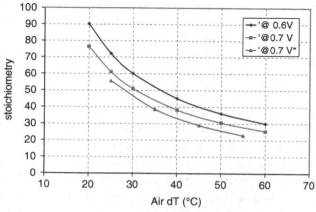

Figure 6-26 Required air stoichiometry for stack cooling. (*This curve takes into account heat dissipation from the stack surface.)

stoichiometric ratio as a function of two typical operating cell potentials (0.6 V and 0.7 V) and as a function of air temperature difference. The two upper curves were calculated by Equation (6-68), and the lower curve took into account the heat dissipation from the stack surface (for a 1 kW stack with the ratio of surface area to active area $A_s/A_{act} = 0.27$). Very large air flow rates are required, with a stoichiometric ratio higher than 20. Product water is not sufficient to saturate these large amounts of air, and therefore this cooling scheme would cause severe drying of the anode and would not be practical. Relative humidity at the outlet may be calculated from the following equation:

$$\varphi = \frac{2r_{O2,in}}{S_{O2} + r_{O2,in}} \frac{P}{P_{vs}} \qquad (6\text{-}69)$$

For a normal operating range of temperatures (from $40°C$ to $80°C$) and for stoichiometric ratios above 20, relative humidity at the outlet would be below 10%, and the cell would experience severe drying.

Evaporative Cooling

If additional liquid water is injected at the stack cathode inlet, it would be possible to prevent drying and at the same time dramatically reduce the air flow rate requirement, because cooling would be achieved by evaporation of injected water. Using the mass and energy balance equation (Chapter 5), it is possible to calculate the exact amount of water and air flow rate needed to achieve saturation conditions at the outlet and provide sufficient cooling of the stack. The example in Chapter 5 shows the calculation for one set of operational conditions. Figure 6-27 shows the resulting water and air flow rate requirements for evaporative cooling over a range of operating conditions (cell potential and temperature) for an atmospheric pressure fuel cell (outlet pressure is atmospheric, and air at the inlet is at $20°C$ and 70% relative humidity). In addition, the conditions in Figure 6-27 assume no net water flux across the membrane, that is, electroosmotic drag is equal to back diffusion. In case that electroosmotic drag is higher than back diffusion, the water injection requirement would be reduced. Indeed, Wilson et al. [35] suggested adiabatic cooling of an atmospheric stack with water introduced on the anode side and then relying on wicking through the gas diffusion layer (using hydrophilic thread) and electroosmotic drag to saturate the ambient air on the cathode side.

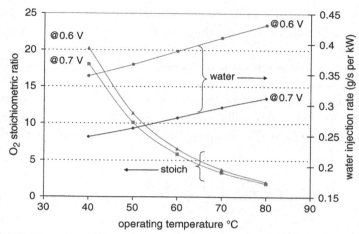

Figure 6-27 Air and water injection rates required for stack cooling and air saturation at the outlet.

Edge Cooling

If a fuel cell flow field is made narrow enough, the heat generated may be removed on the sides of the cells instead of the more conventional way between the cells. In this case, heat is conducted in the plane of the bipolar plate rather than through it. Active cooling may still be needed at the edge of the bipolar plates. To enhance heat transfer, fins may be added and/or high thermal conductivity material may be used.

Obviously, in this case the maximum temperature will be achieved in the center of the flow field. In a narrow and long flow field, the heat transfer may be approximated as one-dimensional (i.e., heat removed through the long sides is significantly larger than the heat removed at the narrow sides).

The equation that describes one-dimensional heat transfer (conduction) in a flat plane with internal heat generation is:

$$\frac{d^2T}{dx^2} + \frac{Q}{kAd_{BP}^{eff}} = 0 \qquad (6\text{-}70)$$

where:

Q = heat generated in the cell (either given by Equation 6-65 or by a detailed energy balance analysis), W

k = bipolar plate in-plane thermal conductivity (in some cases, in-plane conductivity may be significantly different from through-plane thermal conductivity), $Wm^{-1}K^{-1}$

A = cell active area, m^2

d_{BP}^{eff} = effective (or average) thickness of the bipolar plate in the active area, m

Although the heat is not actually generated inside the plate but rather in a thin catalyst layer above the plate, Equation (6-70) may be used with sufficient accuracy because the thickness of the plates is very small compared with the width (i.e., a few millimeters vs. a few centimeters). In that case, the solution of Equation (6-70) for symmetrical cooling on both sides with $T(0)$ = $T(L)$ = T_0 is:

$$T - T_0 = \frac{Q}{kAd_{BP}^{eff}} \frac{L^2}{2} \left[\frac{x}{L} - \left(\frac{x}{L} \right)^2 \right] \tag{6-71}$$

where:

T_0 = the temperature at the edge of the active area

L = width of the active area

The maximum temperature difference is between the edge ($x = 0$ or $x = L$) and the center ($x = L/2$):

$$\Delta T_{max} = \frac{Q}{kAd_{BP}^{eff}} \frac{L^2}{8} \tag{6-72}$$

In addition, the heat must be conducted from the edge of the flow field to the edge of the plate or to the fin over the flow field border with width b. The thickness of the plate at the border is d_{BP}. For this reason, the temperature will further decrease, according to Fourier's law:

$$T_0 - T_b = \frac{Q}{2kA} \frac{L}{d_b} b \tag{6-73}$$

where T_b is the temperature at the edge of the bipolar plate.

Therefore, the total temperature difference between the center of the plate and the edge of the plate, or the base of a fin, is:

$$\Delta T_{max} = \frac{Q}{kA} L \left(\frac{L}{8d_{BP}^{eff}} + \frac{b}{2d_{BP}} \right) \tag{6-74}$$

Figure 6-28 shows the geometry of the narrow flat plate, and Figure 6-29 shows the maximum temperature difference in the active area (temperature in the center – temperature at the edge of the active area) as a function of flow field width and plate thermal conductivity ($Q = 0.279 W/cm^2$,

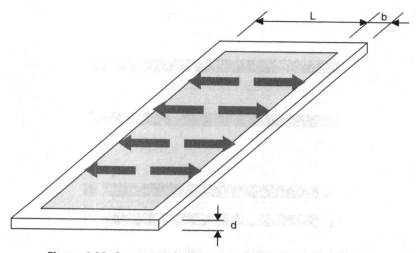

Figure 6-28 Geometry of narrow active area with edge cooling.

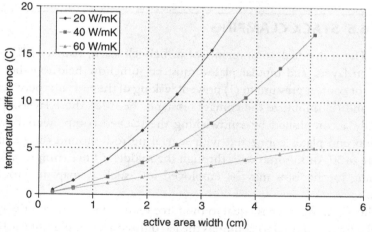

Figure 6-29 Maximum temperature difference in the active area (temperature in the center – temperature at the edge of the active area) as a function of flow field width and plate thermal conductivity ($Q = 0.279$ W/cm², $d_{BP}^{eff} = 1.8$ mm).

$d_{BP}^{eff} = 1.8$ mm). The maximum temperature difference in the active area should be limited to <5–10°C. In that case, edge cooling is limited by the active area width and bipolar plate conductance. Unless a high-conductivity material is used, this kind of cooling may be employed only for low-power densities (below 0.3/cm²) and narrow flow fields (2–3 cm).

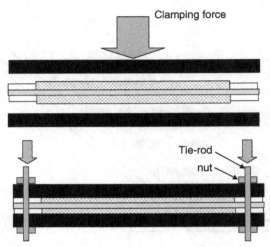

Figure 6-30 Compression of fuel cell components with tie rods.

6.6 STACK CLAMPING

The individual components of a fuel cell stack, namely MEAs, gas diffusion layers, and bipolar plates, must be somehow held together with sufficient contact pressure to (1) prevent leaking of the reactants between the layers and (2) minimize the contact resistance between those layers. This is typically accomplished by sandwiching the stacked components between the two end plates connected with several tie rods around the perimeter (Figure 6-30) or in some cases through the middle. Other compression and fastening mechanisms may be employed too, such as snap-in shrouds or straps.

The clamping force is equal to the force required to compress the gasket plus the force required to compress the gas diffusion layer, plus internal force (for example, the internal operating pressure).

The pressure required to prevent the leak between the layers depends on the gasket material and design. Various materials, ranging from rubber to proprietary polymer configurations, are used for fuel cell gaskets. The designs also vary from manufacturer to manufacturer, including flat or profile gaskets or gaskets as individual components or molded on bipolar plates or on or around the gas diffusion layer. A so-called seven-layer MEA includes the catalyzed membrane, two gas diffusion layers—one on each side—and the gasket that keeps the entire MEA together.

The torque on the bolts necessary to achieve the required force may be calculated from the following equation:

$$T = \frac{FK_b D_b}{N_b} \tag{6-75}$$

where:

  T = tightening torque, Nm
  F = clamping force, N
  K_b = friction coefficient (0.20 for dry and 0.17 for lubricated bolts)
  D_b = bolt nominal diameter, m
  N_b = number of bolts

Too much force on the perimeters may cause bending of the end plates (as shown in Figure 6-31), which has an adverse effect on the compression over the active area. Compression distribution inside the cell may be monitored by pressure-sensitive films (which register only the highest force applied) or by pressure-sensitive electronic pads, which, connected to a monitor, allow inspection of compression force distribution in real time throughout the assembly process. Because of the possibility of bending, the end plates must be designed with sufficient stiffness. Alternatively, end plates with a hydraulic or pneumatic piston that applies a uniform force throughout the active area may be used. Another alternative is to put the tie rods through the center of the plate and then design the flow field around it.

Figure 6-32 shows the compressive force distribution in a cell using a pressure-sensitive film with two different end plate designs: one with inadequate stiffness that resulted in insufficient force (i.e., very bad contacts)

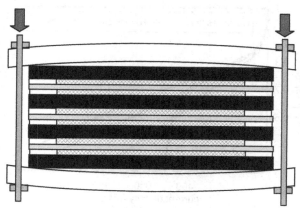

Figure 6-31 Bending of the end plates if too much force is applied on tie rods around the perimeter.

Figure 6-32 Compressive force distribution in a cell with inadequate stiffness of the end plates (left) and in a cell with a hydraulic piston in one of the end plates (right).[36]

inside the active area and one with a hydraulic piston that resulted in very uniform compressive force distribution inside the active area [36].

As discussed in Chapter 4 (Figure 4-22), a pressure of 1.5 to 2.0 MPa is required to minimize the contact resistance between a gas diffusion layer and a bipolar plate. The gas diffusion layer is compressible (Figure 4-17), and the required "squeeze" may be determined by the cell design, that is, by carefully matching the thicknesses of the gas diffusion layer, gaskets, and hardstops or recesses on the bipolar plate. If the gas diffusion layer is compressed too much, it will collapse and will lose its main function—gas and water permeability, as is very graphically illustrated in Figure 6-33. As the

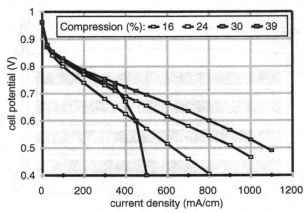

Figure 6-33 Effect of too much compression of the gas diffusion media on cell performance.

compression (expressed as a percentage of gas diffusion media squeeze) is increased, the cell performance improves because of the reduced interfacial resistance. If the compression is too much, the polarization curve exhibits severe mass transport problems. Optimum compression must be experimentally determined for any gas diffusion media. The test cells may be specially designed to allow change of compression, even during cell operation [37,38].

The stack design must also ensure that adequate clamping force is maintained during cell operation. As a result of different coefficients of thermal expansion for different materials used in a fuel cell stack, the compression force may increase or decrease when the stack reaches its operating temperature. This must be compensated for by the use of coil, disc, or polyurethane springs.

PROBLEMS

1. Determine the required number of cells and cell active area for a fuel cell stack that has to generate 50 kW at 120 volts. The polarization curve may be approximated by:

$$V_{cell} = 0.85 - 0.2i \text{ (where i is in A cm}^{-2})$$

The stack should have efficiency of 0.5.

2. For an H_2/O_2 fuel cell polarization curve determined by the following parameters: $T = 60°C$, $P = 101.3 \text{ kPa}$, $i_0 = 0.002 \text{ A cm}^{-2}$, $R = 0.21 \text{ Ohm-cm}^2$, $i_L = 2 \text{ A cm}^{-2}$, $i_{loss} = 1.2 \text{ m A cm}^{-2}$, determine the efficiency at 0.6 V and 0.7 V. Also determine the efficiency at 20% of nominal power for both $V_{nom} = 0.6 \text{ V}$ and $V_{nom} = 0.7 \text{ V}$.

3. A fuel cell has 100-cm² active area covered by six parallel channels on the cathode. Each channel is 0.8 mm wide and 0.8 mm deep, with equal spacing between the channels of 0.8 mm. The fuel cell generates 0.8 A/cm². Air at the inlet is fully saturated at 60°C. Pressure at the inlet is 200 kPa, and there is a 15–kPa pressure drop through the flow field. Oxygen stoichiometric ratio is 2.5. Calculate the velocity and Reynolds number at the air inlet and outlet (neglect liquid water at the outlet).

4. For a fuel cell from Problem 3, calculate the heat generated at 0.65 V and 1 A cm^{-2} using Equation (6–41) and by doing a detailed mass and heat balance analysis (assume that hydrogen is supplied in dead–end mode

saturated at $60°C$ and that net water transport through the membrane is such that there is no water accumulation on the anode side). Explain the difference in results.

5. Calculate the amount of liquid water produced in the cell from Problem 3. This liquid is probably dispersed in numerous droplets. However, assuming that a liquid water film is formed at the bottom of the channel and that water in the film is moving with a velocity that is $^1/_3$ of the air velocity, calculate the depth of the film at the fuel cell exit, or the percentage of the channel's cross-sectional area filled with liquid water.

6. Calculate the temperature at a center of the long, narrow flow field (2.2 cm) of a fuel cell operating at 0.75 V and 0.33 A cm^{-2}. Cooling is obtained at the edge of the cell by flowing air at $25°C$ (h $= 50$ W m^{-2} K^{-1}). The bipolar plate is made out of graphite/polymer mixture with k $= 19$ W m^{-1} K^{-1}, and it is 2 mm thick in the active area and 3 mm thick at the border. The border around the active area is 8 mm wide.

QUIZ

1. A flow field is:
   a. Flow channels that feed each cell in the fuel cell stack
   b. A "maze" of channels on the surface of the bipolar plate
   c. A plane in which the fuel cell electrochemical reaction takes place
2. The velocity in the flow field channel does not depend on:
   a. Channel length
   b. Channel depth
   c. Channel width
3. One of the most important features of the flow field is:
   a. It allows the flow of the reactant gases through each cell with minimum pressure drop
   b. It has a uniform supply of reactants over the entire active area
   c. It allows significant heat removal by circulating excess air
4. Nonuniform distribution of gases over the active area results in:
   a. "Starving" regions
   b. High pressure drop
   c. "Hiccups"
5. The flow in the fuel cell channels is typically:
   a. Laminar

    b. Turbulent

    c. Mixed/transient

6. A characteristic of laminar flow is:

    a. Pressure drop is minimal

    b. Pressure drop is higher than in turbulent flow

    c. Pressure drop is directly proportional to the flow rate

7. If all the product water in a fuel cell is in liquid form, its volume compared with the volume of air exiting the fuel cell would be:

    a. An order of magnitude less

    b. About the same

    c. Several orders of magnitude less

8. As oxygen is consumed along the channel, O_2 content in air:

    a. Linearly decreases

    b. Decreases faster near the entrance

    c. Decreases faster near the exit

9. Removal of heat from the fuel cell stack is intended to:

    a. Keep the stack at a desired temperature

    b. Maintain a desired efficiency

    c. Maintain some liquid water in the membrane

10. "Squeezing" the gas diffusion layer:

    a. Always improves the fuel cell performance because it reduces the interfacial resistance

    b. Improves the performance only up to a certain "squeeze"; if the squeeze is excessive, it has an adverse effect

    c. It has nothing to do with performance—it only prevents overboard leaks

REFERENCES

[1] Barbir F. PEM Fuel Cell Stack Design Considerations. In: Proc. Fuel Cell Technology: Opportunities and Challenges. New Orleans, LA: AIChE Spring National Meeting; March; 2002. p. 520–30.

[2] Barbir F. Development of an Air-Open PEM Fuel Cell, SBIR Phase I Final Technical Report. A report by Energy Partners, Inc. to U.S. Army Research Laboratory; 1995. contract DAAL01-95-C-3511.

[3] Ledjeff K, Nolte R. New SPFC-Technology with Plastics. In: Savadogo O, Roberge PR, Veziroglu TN, editors. New Materials for Fuel Cell Systems. Montreal: Editions de l'Ecole Politechnique de Montreal; 1995. p. 128–34.

[4] Cisar A, Weng D, Murphy OJ. Monopolar Fuel Cells for Nearly Passive Operation. In: Proc. 1998 Fuel Cell Seminar. CA: Palm Springs; November 1998. p. 376–8.

[5] Barbir F. Progress in PEM Fuel Cell Systems Development, in Hydrogen Energy System. In: Yurum Y, editor. Utilization of Hydrogen and Future Aspects, NATO ASI Series E-295. Dordrecht, The Netherlands: Kluwer Academic Publishers; 1995. p. 203–14.

[6] Chow CY, Wozniczka BM. Electrochemical Fuel Cell Stack with Humidification Section Located Upstream from the Electrochemically Active Section; 1995. U.S. Patent 5,382,478.

[7] Reiser CA, Sawyer RD. Solid Polymer Electrolyte Fuel Cell Stack Water Management System; 1988. U.S. Patent 4,769,297.

[8] Koh J-H, Seo HK, Lee CG, Yoo Y-S, Lim HC. Pressure and Flow Distribution in Internal Gas Manifolds of a Fuel Cell Stack. Journal of Power Sources 2003;115:54–65.

[9] Daugherty RL, Franzini JB. Fluid Mechanics with Engineering Applications. 7th ed. New York: McGraw-Hill; 1977.

[10] Watkins DS, Dircks KW, Epp DG. Novel Fuel Cell Fluid Flow Field Plate; 1991. U.S. Patent 4,988,583.

[11] Watkins DS, Dircks KW, Epp DG. Fuel Cell Fluid Flow Field Plate; 1992. U.S. Patent 5,108,849.

[12] Cavalca C, Homeyer ST, Walsworth E. Flow Field Plate for Use in a Proton Exchange Membrane Fuel Cell; 1997. U.S. Patent 5,686,199.

[13] Rock JA. Serially Linked Serpentine Flow Channels for PEM Fuel Cell; 2001. U.S. Patent 6,309,773.

[14] Rock JA. Mirrored Serpentine Flow Channels for Fuel Cell; 2000. U.S. Patent 6,099,984.

[15] Ledjeff K, Heinzel A, Mahlendorf F, Peinecke V. Die Reversible Membran-Brennstoffzelle. In: Elektrochemische Energiegewinnung, Dechema Monographien, 128; 1993. 103.

[16] Nguyen TV. Gas Distributor Design for Proton-Exchange-Membrane Fuel Cells. Journal of the Electrochemical Society 1996;143(5):L103–5.

[17] Barbir F, Neutzler J, Pierce W, Wynne B. Development and Testing of High-Performing PEM Fuel Cell Stack. In: Proc. 1998 Fuel Cell Seminar. CA: Palm Springs; 1998. p. 718–21.

[18] Barbir F, Fuchs M, Husar A, Neutzler J. Design and Operational Characteristics of Automotive PEM Fuel Cell Stacks, Fuel Cell Power for Transportation. Warrendale, PA: SAE; 2000. SAE SP-150563–69.

[19] Wilson MS, Springer TE, Davey JR, Gottesfeld S. Alternative Flow Field and Backing Concepts for Polymer Electrolyte Fuel Cells. In: S. Gottesfeld, G. Halpert, A. Langrebe, editors. Proton Conducting Membrane Fuel Cells I. The Electrochemical Society Proceedings Series, PV 95-23; 1995. p. 115. Pennington, NJ.

[20] Issacci F, Rehg TJ. Gas Block Mechanism for Water Removal in Fuel Cells; 2004. U.S. Patent 6,686,084.

[21] Chapman AR, Mellor IM. Development of BioMimetic™ Flow-Field Plates for PEM Fuel Cells. In: Proc. 8th Grove Fuel Cell Symposium; September 2003. London.

[22] Tüber K, Oedegaard A, Hermann M, Hebling C. Investigation of Fractal Flow Fields in Portable PEMFC and DMFC. In: Proc. 8th Grove Fuel Cell Symposium; September 2003. London.

[23] Zawodzinski C, Wilson MS, Gottesfeld S. Metal Screen and Foil Hardware for Polymer Electrolyte Fuel Cells. In: Gottesfeld S, Fuller TF, editors. Proton Conducting Membrane Fuel Cells II, Electrochemical Society Proceedings, 98–27; 1999. p. 446–56.

[24] Damiano PJ. Fuel Cell Structure; 1978. U.S. Patent 4,129,685.

[25] Wilkinson DP, Vanderleeden O. Serpentine Flow Field Design. In: Vielstich W, Lamm A, Gasteiger H, editors. Handbook of Fuel Cell Technology: Fundamentals, Technology and Applications, 3. New York: John Wiley & Sons; 2003. p. 315–24. Part 1.

[26] Mathias, M., Design, Engineering, Modeling, and Diagnostics, NSF Workshop on Engineering Fundamentals of Low Temperature PEM Fuel Cells (Arlington, VA,

November 2001; http://electrochem.cwru.edu/NSF/presentations/NSF_Mathias_ Diagnostics.pdf).

[27] Mathias MF, Roth J, Fleming J, Lehnert W. Diffusion Media Materials and Characterization. In: Vielstich W, Lamm A, Gasteiger H, editors. Handbook of Fuel Cell Technology: Fundamentals, Technology and Applications, 3. New York: John Wiley & Sons; 2003. p. 517–37. Part 1.

[28] Liu H, Zhou T, You L. Development of Unified PEM Fuel Cell Models. In: DOE Fuel Cells for Transportation Program. Annual National Laboratory R&D Meeting, Oak Ridge National Laboratory; June 6–8, 2001.

[29] Crane Co. Flow of Fluids Through Valves, Fittings and Pipe. New York: Author; 1982. Technical Paper 410.

[30] Bird RB, Stewart WE, Lightfoot EN. Transport Phenomena. New York: John Wiley & Sons; 1960.

[31] Kee RJ, Korada P, Walters K, Pavol M. A Generalized Model of the Flow Distribution in Channel Networks of Planar Fuel Cells. Journal of Power Sources 2002;109(1):148–59.

[32] Dullien FAL. Porous Media: Fluid Transport and Pore Structure. 2nd ed. San Diego, CA: Academic Press; 1992.

[33] Barbir F, Gorgun H, Wang X. Relationship Between Pressure Drop and Cell Resistance as a Diagnostic Tool for PEM Fuel Cells. Journal of Power Sources 2005 (in press).

[34] Lindon TC. Heat Transfer: Professional Version. Englewood Cliffs, NJ: Prentice Hall; 1993.

[35] Wilson MS, Moeller-Holst S, Webb DM, Zawodzinski C. Efficient Fuel Cell Systems, 1999 Annual Progress Report: Fuel Cells for Transportation. Washington, DC: U.S. Department of Energy; 1999. 80–83.

[36] Wang X, Zhang B. Pressurized Endplates for Uniform Pressure Distributions in PEM Fuel Cells. Storrs, CT: First International Conference on Fuel Cell Development and Deployment; March 2004.

[37] Ihonen J, Jaouen F, Lindbergh G, Sundholm G. A Novel Polymer Electrolyte Fuel Cell for Laboratory Investigations and In-Situ Contact Resistance Measurements. Electrochimica Acta 2001;46(19):2899–911.

[38] Higier A, Husar A, Haberer G, Liu H. Design of a Single Cell, Variable Compression PEM Fuel Cell Test Fixture. In: Proc. 2002 Fuel Cell Seminar. CA: Palm Springs; 2002. p. 45.

CHAPTER SEVEN

Fuel Cell Modeling

Modeling plays a significant and important role in the fuel cell design and development process. Because of its importance, modeling is initiated early in a fuel cell development cycle, as shown in Figure 7-1 [1]. The flowchart illustrates a typical procedure in fuel cell development, where the process begins with a set of requirements. Requirements include power and energy requirements, environmental operating conditions, size and volume limitations, safety specifications, and others. Along with the requirements, knowledge of materials, processes, and material interactions is necessary to properly construct a fuel cell stack. Once several designs are down-selected, modeling of the designs is performed to determine how well the candidate systems satisfy requirements. The modeling helps the designer further down-select designs to fabricate and test. The tests performed on the final designs can result in either a final prototype or an iteration of existing designs for improvement. Reliable diagnostics are needed not only to find out what is wrong with the existing design so that it can be improved but also to calibrate and verify the models and assumptions used in developing them.

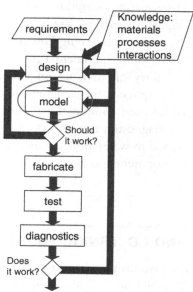

Figure 7-1 Role of modeling and diagnostics in the fuel cell development process.

PEM Fuel Cells
ISBN 978-0-12-387710-9

This design iteration loop can be quite lengthy and frequent, especially when stringent requirements or poor modeling capabilities exist. On the contrary, improvements in modeling capability can help the designer find and improve on existing designs that satisfy stringent requirements. Hence, modeling has a critical role in the fuel cell design and development process. A designer can use an accurate and robust model to design and develop fuel cell stacks more efficiently and often with better performance and lower manufacturing cost.

Using fuel cell modeling as a successful design tool requires the model to be robust, accurate, and able to provide usable answers quickly. In terms of robustness, the model should be able to predict fuel cell performance under a large range of operating conditions. For example, a PEM fuel cell can be operating at different temperatures, humidity levels, and fuel mixtures. A robust model should be able to predict fuel cell performance under varying conditions. The model must also predict fuel cell performance accurately. It should be mentioned here that accuracy does not necessarily mean that a model should accurately predict the absolute value of all the physical phenomena being modeled at any point in space and time. Instead, a model should accurately predict the relative values or the trends (e.g., an increase in value of one parameter should result in either no, low, moderate, or significant increase or decrease of another parameter). Accuracy can be contributed to by using the correct assumptions, using the correct properties and other numerical input parameters, predicting the correct physical phenomena by using the correct governing equations, and being able to match the modeling results with experimental data. However, enhancing model robustness and accuracy often trades off with computational efficiency. To provide answers quickly, the designer must select a model that balances robustness, accuracy, and computational effort.

Models coupled with diagnostics can provide insight in fuel cell inner workings, particularly related to water and heat management. As such, they may be employed in development and application of model-based predictive control strategies.

7.1 THEORY AND GOVERNING EQUATIONS

Physical phenomena occurring within a PEM fuel cell can in general be represented by the solution of conservation equations for mass, momentum, energy, species, and current transport. In addition, equations that deal specifically with phenomena in a fuel cell may be used where applicable, such as:

- Darcy's equation for fluid flow in conduits and porous media
- Fick's law for diffusion
- Stefan–Maxwell equation for multispecies diffusion
- Fourier's law for heat conduction
- Faraday's law for the relationship between electrical current and consumption of reactants in an electrochemical reaction
- Butler–Volmer equation for relationship between electrical current and potential
- Ohm's law of electrical current conduction

In addition, empirical equations, particularly those describing water behavior in the polymer membrane and related phenomena, are often used in modeling in the absence of equations that describe the actual physical phenomena.

Any model is only as good as the assumptions on which it is built. Assumptions are needed to simplify the model. It is important to understand the assumptions in order to understand the model's limitations and to accurately interpret its results. Common assumptions used in fuel cell modeling are:

- Ideal gas properties
- Ideal gas mixtures
- Incompressible flow
- Laminar flow
- Isotropic and homogeneous membrane and electrode structures
- Negligible ohmic potential drop in solid components
- The mass and energy transport through porous structures of porosity ε is modeled from a macro perspective using the volume-averaged conservation equations

7.1.1 Conservation of Mass

The general equation for mass conservation, which is valid for all the processes inside a fuel cell such as fluid flow, diffusion, phase change, and electrochemical reactions, is:

$$\frac{\partial \rho}{\partial t} + \nabla \cdot (\rho v) = 0 \qquad (7\text{-}1)$$

where:

$\rho =$ density, kg m^{-3}

$v =$ velocity vector, ms^{-1}

$\nabla =$ operator, $\dfrac{d}{dx} + \dfrac{d}{dy} + \dfrac{d}{dz}$

The transient term represents accumulation of mass with time, and the second term represents the change in mass flux.

7.1.2 Conservation of Momentum

Momentum conservation is described by the following equation:

$$\frac{\partial(\rho v)}{\partial t} + \nabla \cdot (\rho v v) = -\nabla p + \nabla \cdot (\mu^{eff} \nabla v) + S_m \tag{7-2}$$

where:

p = fluid pressure, Pa

μ^{eff} = mixture average viscosity, $kgm^{-1} s^{-1}$

S_m = momentum source term

The transient term in the momentum conservation equation represents the accumulation of momentum with time, and the second term describes advection momentum flux. The first two terms on the right side of the momentum conservation equation represent momentum imparted due to pressure and viscosity, respectively. The source term is different for different regions of the fuel cell, as follows:

For gas channels:

$$S_m = 0 \tag{7-3}$$

For backing layers and voids of the catalyst layers:

$$S_m = -\frac{\mu}{K} \varepsilon v \tag{7-4}$$

where:

K = permeability of the gas diffusion layer (or the catalyst layer), m^2

ε = porosity of the gas diffusion layer

This source term for the momentum conservation equation represents a pressure drop arising from Darcy's drag force imposed by the pore walls on the fluid.

For water transport in the polymer phase, an additional source term is *electrokinetic permeability*:

$$S_m = -\frac{\mu}{K_p} \varepsilon_m x_m v + \frac{K_\phi}{K_p} c_f n_f F \nabla \phi_m \tag{7-5}$$

where:

ε_m = membrane water porosity

x_m = volume fraction of ionomer in the catalyst layer

K_f = electrokinetic permeability, m^2
K_p = hydraulic permeability of the membrane, m^2
c_f = concentration of fixed charge, mol m^{-3}
n_f = charge number of the sulfonic acid ions
F = Faraday's constant
ϕ_m = ionomer phase potential

7.1.3 Conservation of Energy

Conservation of energy for any domain in a PEM fuel cell is described by:

$$(\rho c_p)_{eff} \frac{\partial T}{\partial t} + (\rho c_p)_{eff}(\mathbf{v} \cdot \nabla T) = \nabla \cdot (k_{eff} \nabla T) + S_e \qquad (7\text{-}6)$$

where:

c_p = mixture-averaged specific heat capacity, J kg^{-1} K^{-1}
T = temperature, K
k = thermal conductivity, W m^{-1} K^{-1}
S_e = energy source term

Subscript *eff* represents effective properties for the porous media [2]:

$$(\rho c_p)_{eff} = (1 - \varepsilon)\rho_s c_{p,s} + \varepsilon \rho c_p \qquad (7\text{-}7)$$

$$k_{eff} = -2k_s + \left[\frac{\varepsilon}{2k_s + k} + \frac{1 - \varepsilon}{3k_s} \right]^{-1} \qquad (7\text{-}8)$$

where ρ_s, $c_{p,s}$, and k_s represent the density, specific heat capacity, and thermal conductivity of the solid matrix, respectively.

The source term in the energy conservation equation may include heat from reactions, ohmic heating, and/or heat of evaporation or condensation in case there is a phase change.

In gas channels, the only possible heat source or sink is phase change, that is, condensation of water vapor present in the gas streams as a heat source and evaporation of liquid water present in the channels (in the form of a film or droplets) as a heat sink.

Evaporation will happen only if both of the following conditions are satisfied:

- There is liquid water present in the gas stream, $x_{H2O(l)} > 0$.
- The gas is not saturated, $x_{H2O(g)} < x_{sat}$ (saturated gas may become unsaturated if either the pressure decreases or the temperature increases).

Condensation will happen only if the gas is already fully saturated, $x_{H2O(g)} > x_{sat}$, and the temperature of the gas drops (i.e., in contact with cooler parts of the fuel cell). Condensation may also happen when the reactant gas or its constituents (hydrogen or oxygen) "disappear" in the electrochemical reaction, although this does not happen in the channels but rather in the catalyst layer.

The maximum rate of condensation (in mols s^{-1}) is:

$$N_{H_2O,cond} = \frac{P - \dfrac{P_{sat}}{y_w}}{P - P_{sat}} N_{H_2O(v)in} \qquad (7\text{-}9)$$

where water vapor molar fraction, y_w, is:

$$y_w = \frac{N_{H_2O(v)in}}{N_{H_2O(v)in} + N_{gas}} \qquad (7\text{-}10)$$

The same equation may be used for evaporation (when $P_{sat}/y_w > P$), assuming there is liquid water present in the gas stream.

$$N_{H_2O,evap} = \frac{\dfrac{P_{sat}}{y_w} - P}{P - P_{sat}} N_{H_2O(v)in} \qquad (7\text{-}11)$$

However, in case of evaporation it may be easier to deal with water content (x or χ), defined as mass or molar ratio of water vapor and dry gas, respectively:

$$N_{H_2O,evap} = \min\left[N_{H_2O(1)}, N_{gas}(\chi_{sat} - \chi)\right] \qquad (7\text{-}12)$$

The actual, local rate of evaporation would depend on local conditions, particularly on the water/gas interface. Bosnjakovic [3] defines the rate of heat exchange during the process of evaporation using the evaporation coefficient, σ, defined as the mass flow rate of dry air that must be saturated in order to evaporate ($x_{sat} - x_{H2O(g)}$) amount of water over 1 m^2 of contact surface. In that case:

$$S_e = -\sigma A_{fg}(x_{sat} - x_{H_2O(g)})\Delta h_{fg} \qquad (7\text{-}13)$$

where:

σ = evaporation coefficient, kgm^{-2} s^{-1}

A_{fg} = phase-change surface area per unit volume, m^{-1}

x_{sat} = maximum mass fraction of water vapor in dry gas (at saturation), kg$_w$ kg$_g^{-1}$

$x_{H2O(g)}$ = fraction of water vapor in dry gas, kg$_w$ kg$_g^{-1}$

Δh_{fg} = heat of evaporation, J kg^{-1}

The evaporation coefficient, σ, may be determined from the dimensionless Lewis factor, $\sigma c_p/\alpha$, which is given by the following equation derived from the analogy between heat and mass transport [3]:

$$\frac{\sigma c_p}{\alpha} = \left[\frac{D\rho c_p}{k(1+x)}\right]^{1-n}\left(\frac{v_h}{v_d}\right)^{m-n}\frac{\dfrac{M_w}{M_g}+x}{x_{sat}-x}\ln\frac{\dfrac{M_w}{M_g}+x_{sat}}{\dfrac{M_w}{M_g}+x} \qquad (7\text{-}14)$$

where:

α = coefficient of heat transfer between water surface and gas, W m^2 K^{-1}
D = diffusion of water vapor through gas, m^2 s^{-1}
ρ = density of humid air, kgm^{-3}
c_p = heat capacity, J kg^{-1} K^{-1}
k = thermal conductivity of humid gas, W m^{-1} K^{-1}
x = fraction of water vapor in dry gas, kg$_w$ kg$_g^{-1}$
v_h = kinematic viscosity of humid gas, m^2 s^{-1}
v_d = kinematic viscosity of dry gas, m^2 s^{-1}
M_w = molecular mass of water, kgk mol^{-1}
M_g = molecular mass of gas, kgk mol^{-1}
m = coefficient, ~0.75 [3]
n = coefficient, ~0.33 [3]

In gas diffusion layers, the possible heat sources are due to ohmic resistance through solid and phase change in pores (again, evaporation is possible only if there is liquid water present, i.e., $x_{H2O(l)} > 0$, and the gas is not saturated, i.e., $x_{H2O(g)} < x_{sat}$):

$$S_e = \frac{i_e^2}{\kappa_s^{eff}} - \sigma A_{fg}(x_{sat} - x_{H2O(g)})(\Delta h_{fg}) \qquad (7\text{-}15)$$

where:

i_e = current density, A m^{-2}
κ_s^{eff} = effective electric conductivity of the gas diffusion layer, S cm^{-1}

In catalyst layers, the source term includes the heat released by the electrochemical reaction, heat generated due to ionic and electronic resistance, and heat of water evaporation (again, only if there is liquid water present and the gas is not saturated):

$$S_e = |j|\left[|\Delta V_{act}| - \frac{T\Delta S}{nF}\right] + \left(\frac{i_m^2}{\kappa_m^{eff}} + \frac{i_e^2}{\kappa_s^{eff}}\right)$$
$$- \sigma A_{fg}(x_{sat} - x_{H2O(g)})(\Delta h_{fg}) \qquad (7\text{-}16)$$

where:

j = transfer current density, A cm^{-3}, defined later by Equations (7-29) and (7-30)

ΔV_{act} = activation overpotential, V

i_m = ionic current density, A cm^{-2}

κ_m^{eff} = effective ionic conductivity of ionomer phase in the catalyst layer, S cm^{-1}

In the membrane, the only heat source is due to ohmic resistance:

$$S_e = \frac{i_m^2}{\kappa_m} \tag{7-17}$$

7.1.4 Conservation of Species

Species conservation equations representing mass conservation for the individual gas phase species are:

$$\frac{\partial(\varepsilon \rho x_i)}{\partial t} + \nabla \cdot (\mathbf{v} \varepsilon \rho x_i) = \nabla \cdot (\rho D_i^{eff} \nabla x_i) + S_{s,i} \tag{7-18}$$

where:

x_i = mass fraction of gas species, $i = 1, 2, \ldots, N$ (for example, $i = 1$ for hydrogen, $i = 2$ for oxygen, $i = 3$ for water vapor, etc.; liquid water may be treated as a separate species)

$S_{s,i}$ = source or sink terms for the species

The first two terms in the species conservation equation represent species accumulation and advection terms, and the first term on the right side represents Fickian diffusion of species in a porous medium. In the porous medium, $D_{i,eff}$ is a function of the porosity, ε, and tortuosity, τ. Because there is considerable lack of information for gas transport in PEM porous media, one common relationship used in literature is given by the Bruggman model [4], where $\tau = 1.5$:

$$D_{i,eff} = D_i \varepsilon^\tau \tag{7-19}$$

where:

D_i = the free-stream mass diffusion coefficient

The Stefan–Maxwell equations for a multicomponent species system may be added [5] to define the gradient in mole fraction of the components:

$$\nabla y_i = RT \sum_j \frac{y_i N_j - y_j N_i}{p D_{ij}^{eff}} \tag{7-20}$$

where:

y_i = gas phase mole fraction of species i

N_i = superficial gas-phase flux of species i averaged over a differential volume element, which is small with respect to the overall dimensions of the system but large with respect to the pore size [5]

$D_{i,j}^{eff}$ = effective binary diffusivity of the pair i,j in the porous medium, which may be calculated for any temperature and pressure from the following equation [6]:

$$D_{ij}^{eff} = \frac{a}{p}\left(\frac{T}{\sqrt{T_{c,i}\,T_{c,j}}}\right)^b \left(p_{c,i}p_{c,j}\right)^{1/3}\left(T_{c,i}\,T_{c,j}\right)^{5/12}\left(\frac{1}{M_i}+\frac{1}{M_j}\right)^{1/2}\varepsilon^{1.5}$$

(7-21)

where:

T_c and p_c = the critical temperature and pressure of the species i and j

M = molecular weight of species i and j

a = 0.0002745 for diatomic gases, H_2, O_2, and N_2, and a = 0.000364 for water vapor

b = 1.832 for diatomic gases, H_2, O_2, and N_2, and b = 2.334 for water vapor

The source term in the species conservation equation, $S_{s,i}$, is equal to zero everywhere except in the catalyst layers where the species are consumed or generated in the electrochemical reactions. In that case the source terms, $S_{s,i}$, for hydrogen, oxygen, water vapor, and liquid water, respectively, are:

$$S_{s,H_2} = -j_a \frac{M_{H_2}}{2F}$$

(7-22)

$$S_{s,O_2} = -j_c \frac{M_{O_2}}{4F}$$

(7-23)

$$S_{s,H_2O(g)} = \sigma A_{fg}\left(x_{sat} - x_{H_2O(g)}\right)$$

(7-24)

$$S_{s,H_2O(l)} = +j_c \frac{M_{H_2O}}{2F} - \sigma A_{fg}\left(x_{sat} - x_{H_2O(g)}\right)$$

(7-25)

In the water source term, it is assumed that water is generated as liquid and then it evaporates if the neighboring air or oxygen is not saturated.

7.1.5 Conservation of Charge

Current transport is described by a governing equation for conservation of charge:

$$\nabla \cdot (\kappa_s^{\text{eff}} \nabla \phi_s) = S_{\phi s} \qquad (7\text{-}26)$$

for electrical current, and

$$\nabla \cdot (\kappa_m^{\text{eff}} \nabla \phi_m) = S_{\phi m} \qquad (7\text{-}27)$$

for ionic current, where:

κ_s^{eff} = electrical conductivity in the solid phase, S cm^{-1}

κ_m^{eff} = ionic conductivity in the ionomer phase (including the membrane), S cm^{-1}

ϕ_s = solid phase potential, V

ϕ_m = electrolyte phase potential, V

S_ϕ = source term representing volumetric transfer current

at the anode catalyst layer $S_{\phi s} = -j_a$ and $S_{\phi m} = j_a$;

at the cathode catalyst layers $S_{\phi s} = j_c$, $S_{\phi m} = -j_c$; and

$S_\phi = 0$ elsewhere.

An additional complication is that the ionomer phase conductivity, κ_m, strongly depends on temperature and water content inside the ionomer, λ, which in turn is a function of conditions (relative humidity) outside the membrane, as discussed in Chapter 4 (Equations 4-3 and 4-2, respectively).

In any volume on either anode or cathode, the electronic and ionic currents generated are equal, that is, $S_{\phi s} = S_{\phi m}$. Also, the total current (of either electrons or ions) generated in the anode catalyst layer must be equal to the total current "consumed" in the cathode catalyst layer (and also must be equal to the total current going through the membrane):

$$\int_{V_a} j_a dV = \int_{V_c} j_c dV \qquad (7\text{-}28)$$

The transfer currents, j, are the result of the electrochemical reactions that take place on the catalyst surface and are driven by the surface overpotential ΔV_{act}, which is the potential difference between the solid phase and the electrolyte membrane:

$$\Delta V_{\text{act}} = \phi_s - \phi_m - V_{\text{ref}} \qquad (7\text{-}29)$$

where the subscripts s and m indicate the solid phase and electrolyte membrane–ionomer phase, respectively. The reference potential of the electrode is zero on the anode side, and it equals the theoretical cell potential side at given temperature and pressure on the cathode (combination of Equations 2-18 and 2-37):

$$V_{ref} = E_{T,P} = -\left(\frac{\Delta H}{nF} - \frac{T\Delta S}{nF}\right) + \frac{RT}{nF}\ln\left[\left(\frac{P_{H_2}}{P_0}\right)\left(\frac{P_{O_2}}{P_0}\right)^{1/2}\right]$$

(7-30)

If the actual reactants' concentrations, that is, partial pressures, at the catalyst surface are used in the last term of Equation (7-30), V_{ref} then also includes the concentration polarization losses:

$$V_{ref} = E_{T,P} = -\left(\frac{\Delta H}{nF} - \frac{T\Delta S}{nF}\right) + \frac{RT}{nF}\ln\left[\left(\frac{P_{H_2}^{eff}}{P_0}\right)\left(\frac{P_{O_2}^{eff}}{P_0}\right)^{1/2}\right]$$

(7-31)

Referring to Figure 3.9, the overpotential defined by Equation 7-29 is positive on the anode and negative on the cathode. The cell potential is then a difference between the cathode and anode solids at the two ends of the cell:

$$V_{cell} = \phi_{s,c} - \phi_{s,a}$$

(7-32)

The relationship between the surface overpotential and transfer current density is given by the Butler–Volmer equation. For the anode (from Equation 3-16):

$$j_a = a i^+_{0,a}\left\{\exp\left[\frac{-\alpha_a F \Delta V_{act,a}}{Rt}\right] - \exp\left[\frac{(1-\alpha_a)F\Delta V_{act,a}}{RT}\right]\right\}$$

(7-33)

where:

j_a = transfer current density, A m^{-3}

a = electrocatalytic surface area per unit volume, m^{-1}

$i^+_{0,a}$ = anode exchange current density per unit of electrocatalytic (Pt) surface area, A m^{-2}

and for the cathode (from Equation 3-17):

$$j_c = a i^+_{0,c}\left\{\exp\left[\frac{-\alpha_c F \Delta V_{act,c}}{Rt}\right] - \exp\left[\frac{(1-\alpha_c)F\Delta V_{act,c}}{RT}\right]\right\}$$

(7-34)

The exchange current density at different pressures and temperatures (as per Equation 3-15) is defined as [7]:

$$i^+{}_0 = i_0{}^{ref} \left(\frac{P_r}{P_r^{ref}} \right)^\gamma \exp\left[\frac{E_C}{RT} \left(1 - \frac{T}{T_{ref}} \right) \right] \qquad (7\text{-}35)$$

The average current density is the total current generated in a fuel cell divided by the geometric area:

$$i_{avg} = \frac{1}{A} \int_{V_a} j_a dV = \frac{1}{A} \int_{V_c} j_c dV \qquad (7\text{-}36)$$

7.2 MODELING DOMAINS

The previous set of equations is applied to a computational domain using finite difference, finite volume, or finite element methods. The input is either the cell voltage or the average current density. In addition, the flow rates and conditions at the inlet must also be prescribed, as must the boundary conditions at the outside walls. The boundary conditions depend on the model domain. Several typical computational domains may be of interest, namely:

- 1-D model through the membrane, z-direction as defined in Figure 7-2, for analysis of fluxes, concentrations, temperatures, and potentials in the catalyst layer and the membrane for any given conditions in the channel(s).
- 2-D model of a partial cross-section, yz-direction as defined in Figure 7-2, for analysis of fluxes and concentrations in the gas diffusion and catalyst layers. This is similar to the 1-D model, but it is extended in two dimensions to include the effect of the "ribs" or "lands" between the channels. This domain may include only one side (either cathode or anode) or both sides.
- 2-D model of along-the-channel cross-section, xz-direction (Figure 7-2) for analysis of changes along the channel due to reduced concentrations of reactant gases, pressure drop, and increased concentrations of water. This domain may include only one side (typically cathode) or both sides if the effects of co- or counterflow of hydrogen on the other side are of interest.
- 3-D model of a partial cross-section, essentially a combination of the two 2-D models (Figure 7-3).

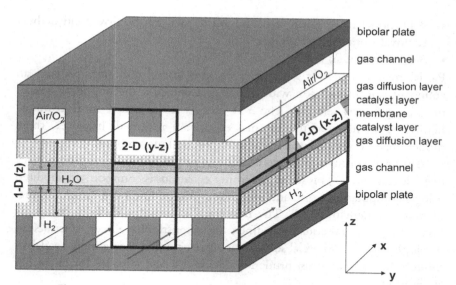

Figure 7-2 The 1-D and 2-D modeling domains for a PEM fuel cell.

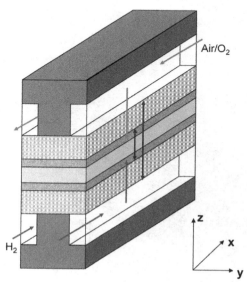

Figure 7-3 The 3-D modeling domain for a PEM fuel cell.

- 3-D model of an entire cell, needed to analyze the flow field or heat removal from the cell (Figure 7-4).
- 3-D model of an entire stack.

Partial models (1-D or 2-D) may be very useful and may provide an abundance of information with sufficient accuracy if the boundary conditions are properly selected. The 3-D models are of course more comprehensive, particularly if the thermal effects are included; however, these models may be very complex and may require long computing times.

A number of numerical methods to solve this system of equations have been documented. This set of governing equations is commonly discredited using a computational grid and solved using finite difference, finite volume, of finite element methods. The numerical solution procedure, computational methods, and equation stiffness characteristics have been described in detail [9−11]. Commercial software is also becoming increasingly available, though the designer must prudently decide the best software for his or her application.

When a fuel designer decides on a solution procedure or software, care must be taken in using the model properly. Strong emphasis needs to be placed on properly validating the model before use. Even though comparison with numerical results obtained elsewhere is acceptable, it is highly recommended to validate model predictions to experimental data as close as possible to the designer's system of interest.

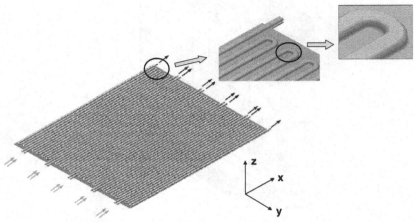

Figure 7-4 Entire flow field as a modeling domain [8].

7.3 MODELING EXAMPLES

Here we use several examples to illustrate the procedure and capability of PEM fuel cell models.

7.3.1 One-Dimensional Through-the-Membrane Model (Bernardi–Verbrugge [5])

One of the first and most often cited PEM fuel cell models is the Bernardi–Verbrugge model [5]. This is a one-dimensional model that treats the cathode gas diffusion electrode bonded to a polymer electrolyte and transport of neutral and charged species within. The model results in a set of differential equations, which once solved allow determination of species concentration profiles, spatial dependence of the pressure and potential drop, and identification of various contributions to the total potential drop. The model is based on simplifying assumptions such as:

- Isothermal conditions throughout the domain
- Gases behave like ideal gases
- Total pressure within the gas diffuser is constant
- The gases are well mixed and of uniform composition
- Water is available on both boundaries of the domain
- The membrane is fully hydrated, as are the pores in the gas diffuser
- The gas phase is in equilibrium with the liquid water phase

The modeling domain includes the membrane, the catalyst layer, and the gas diffusion layer, as shown in Figure 7-5. The conditions in the gas chamber are considered to be given and do not change in z-direction (or in any direction).

The model sets the governing equations and boundary conditions for each of the three regions.

In the membrane, the liquid water transport inside the membrane and catalyst layers is described by a form of Schlögl's equation of motion, which in essence is a special case of Equation (7-2) with source terms as specified in Equation (7-5). According to the equation, the velocity of water inside the MEA is due to potential and pressure gradients:

$$u = \frac{K_\phi}{\mu} z_f c_f F \frac{d\phi_m}{dz} - \frac{K_p}{\mu} \frac{dP}{dz} \tag{7-37}$$

where:

u = velocity in z-direction, m s^{-1}
K_ϕ = electrokinetic permeability, m^2

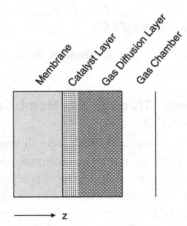

Figure 7-5 Modeling domain of 1-D through-the-membrane model (Bernardi–Verbrugge).

μ = viscosity, $kg\,m^{-1}\,s^{-1}$
z_f = charge number of the sulfonic acid ions inside the membrane, -1
c_f = concentration of sulfonic acid ions inside the membrane, $mol\,m^{-3}$
F = the Faraday constant
ϕ_m = potential in the electrolyte phase, V
K_p = hydraulic permeability, m^2
P = pressure, Pa

The following electroneutrality in the membrane applies:

$$-z_f c_f = \sum_i z_i c_i \qquad (7\text{-}38)$$

Because the only mobile ions in the membrane fluid pore are hydrogen ions:

$$-z_f c_f = c_H^+ \qquad (7\text{-}39)$$

where c_H^+ is proton concentration inside the membrane.

This means that the proton concentration in the membrane can be considered constant, and diffusion is not a mode of proton transport.

The potential in the membrane changes because of conduction and migration:

$$\kappa_m \frac{d\phi_m}{dz} = -i + F c_f u \qquad (7\text{-}40)$$

where:

κ_m = ionic conductivity of the membrane, S cm^{-1}

i = current density in the electrolyte phase, A cm^{-2}

In addition, current conservation in the membrane is expressed as (from Equation 7-27, with $S_{\phi m} = 0$ and κ_m = const.):

$$\frac{di}{dz} = 0 \tag{7-41}$$

The continuity equation for dissolved oxygen species (from Equation 7-18) yields:

$$u \frac{dc_{O_2}}{dz} = D_{O_2}^{eff} \frac{d^2 c_{O_2}}{dz^2} \tag{7-42}$$

The general equation of continuity for incompressible fluid flow (Equation 7-1) eliminates the velocity variable from the set of equations:

$$\frac{du}{dz} = 0 \tag{7-43}$$

In the catalyst layer, the same equations of liquid water (Equation 7-37) and charge transport (Equation 7-40) in the electrolyte phase apply— however, with effective properties K_ϕ^{eff}, K_p^{eff}, and κ^{eff} to account for the porous nature of the catalyst layer where the electrolyte phase occupies only a portion, ε_m, of the total volume.

The continuity equation for water transport in the catalyst layer accounts for water generation:

$$\rho \frac{du}{dz} = -\frac{1}{2F} \frac{di}{dz} \tag{7-44}$$

Similarly, the continuity equation for dissolved oxygen species also accounts for water production and oxygen depletion:

$$u \frac{dc_{O_2}}{dz} = D_{O_2}^{eff} \frac{d^2 c_{O_2}}{dz^2} - \left(\frac{1}{4F} - \frac{1}{2F} \frac{c_{O_2} M_{H_2O}}{\rho_{H_2O(l)}} \right) \frac{di}{dz} \tag{7-45}$$

The standard Butler–Volmer equation describes the relationship between the current generation rate and surface overpotential (Equation 7-34).

Because this model only deals with one side of the fuel cell, it is not necessary to account for the potential increase between the anode and the cathode, and in that case the overpotential is simply a difference between potentials in solid and electrolyte phases:

$$\Delta V_{act,c} = \phi_s - \phi_m \tag{7-46}$$

The exchange current density takes into account the actual concentration of both oxygen and proton species at the catalyst surface:

$$i_{0,c}^+ = i_{0,c}^{ref} \left(\frac{c_{O_2}}{c_{O_2}^{ref}} \right)^{\gamma_{O_2}} \left(\frac{c_{H^+}}{c_{H^+}^{ref}} \right)^{\gamma_{H^+}} \tag{7-47}$$

The movement of electrons in the solid portion of the catalyst layer is governed by ohm's law:

$$i_s = -\kappa_s^{eff} \frac{d\phi_s}{dz} \tag{7-48}$$

Electroneutrality in the catalyst layer must be preserved, thus:

$$\frac{di_s}{dz} + \frac{di}{dz} = 0 \tag{7-49}$$

In the gas diffusion layer, the transport of gaseous species is described by the Stefan–Maxwell equation (7-20) for three species, namely oxygen, nitrogen, and water vapor. Although in the catalyst layer oxygen is considered to be dissolved in water present in the electrolyte phase, the amount of oxygen and nitrogen dissolved in liquid water traveling through the gas diffusion layer is negligible compared with the amount of oxygen and nitrogen in the gas phase. The equation of conservation of species then takes the form:

$$\frac{dN_{O_2,g}}{dz} = 0 \tag{7-50}$$

Because nitrogen does not participate in the reaction, there is no net movement of nitrogen in the gas diffusion layer, thus:

$$N_{N_2,g} = 0 \tag{7-51}$$

Furthermore, water vapor in the pores of the gas diffusion layer is considered to be equilibrated with liquid water present in the pores, that is, $y_{H2O} = y_{H2O}^{sat}$, thus:

$$\frac{dy_{H_2O}}{dz} = 0 \tag{7-52}$$

where y_{H2O} is the molar fraction of water vapor in the gas diffusion layer, that is:

$$y_{O_2} + y_{N_2} + y_{H_2O} = 1 \qquad (7\text{-}53)$$

Incorporation of Equations (7-50) through (7-53) in the Stefan–Maxwell set of equations (7-20) yields:

$$\frac{P}{RT} \frac{D^{eff}_{H_2O-N_2}}{y_{N_2}} \frac{dy_{N_2}}{dz} = \frac{I}{4F} \frac{D^{eff}_{H_2O-N_2}}{D^{eff}_{N_2-O_2}} + N_{H_2O} \qquad (7\text{-}54)$$

where:

$$N_{H_2O} = \frac{I}{4F} y^{sat}_{H_2O} \left[1 - y^{sat}_{H_2O} - y_{N_2} + y_{N_2} \frac{D^{eff}_{N_2-O_2}}{D^{eff}_{H_2O-N_2}} \right]^{-1} \qquad (7\text{-}55)$$

Because there are no charged species traveling through the gas diffusion layer, water movement is solely due to pressure gradient, and Schlögl's equation (7-37) reduces to:

$$u = -\frac{K^{eff}_p}{\mu} \frac{dP}{dz} \qquad (7\text{-}56)$$

where K^{eff}_p accounts for the porous nature of the gas diffusion layer, including the wet-proofing.

Continuity for water species in the gas diffusion layer means:

$$\rho \frac{du}{dz} = -\frac{dN_{H_2O}}{dz} \qquad (7\text{-}57)$$

Unlike the general model presented in Section 7.1, "Theory and Governing Equations," the Bernardi–Verbrugge model needs the boundary conditions at the interfaces, namely the membrane/catalyst layer and the catalyst layer/ gas diffusion layer interfaces. The general model treats all three regions as a single domain and uses the same governing equations in all the regions; only the source terms are different. As a result of such an approach, no interfacial conditions are required to be specified at internal boundaries between the regions.

At the membrane, $z = 0$, the boundary conditions are:

$$\phi_m = 0 \qquad P = P_0 \qquad c_{O_2} = 0 \qquad (7\text{-}58)$$

At the membrane/catalyst layer interface, $z = \delta_m$, flux of liquid water, flux of dissolved oxygen, and current in the electrolyte phase are continuous, and therefore the following boundary conditions apply:

$$\kappa \left.\frac{d\phi_m}{dz}\right|_m = \kappa^{eff} \left.\frac{d\phi_m}{dz}\right|_{cl} \tag{7-59}$$

and

$$u|_m = \varepsilon_m\, u|_{cl}$$

where ε_m is volume fraction of polymer in the catalyst layer, and

$$D_{O_2} \left.\frac{dc_{O_2}}{dz}\right|_m = D_{O_2}^{eff} \left.\frac{dc_{O_2}}{dz}\right|_{cl} \tag{7-60}$$

Furthermore, at the membrane/catalyst layer interface, $z = \delta_m$, current in the solid phase is equal to zero:

$$\left.\frac{d\phi_s}{dz}\right|_{cl} = 0 \tag{7-61}$$

At the catalyst layer/gas diffusion layer interface, $z = \delta_m + \delta_{cl}$, the current in the solid phase and total flux of water are continuous:

$$\left.\begin{array}{l} \kappa_s^{eff} \left.\dfrac{d\phi_s}{dz}\right|_{cl} = \kappa_s^{eff} \left.\dfrac{d\phi_s}{dz}\right|_{gdl} \\[2ex] \rho\varepsilon_m\varepsilon_w u|_{cl} = \rho u|_{cl} + N_{H_2O} \end{array}\right\} \tag{7-62}$$

The dissolved oxygen concentration in the membrane phase of the catalyst layer is related to the oxygen gas-phase concentration in the gas diffuser by a Henry's law constant, K_{O2}, defined by:

$$c_{O_2}^{sat} = \left(1 - y_{N_2} - y_{H_2O}^{sat}\right) \frac{P_{in}}{K_{O_2}} \tag{7-63}$$

The ionic current at the catalyst layer/gas diffusion layer interface is equal to zero:

$$\left.\frac{d\phi_m}{dz}\right|_{gdl} = 0 \tag{7-64}$$

At the face of the gas diffuser, $z = \delta_m + \delta_{cl} + \delta_{gdl}$, which is in contact with the gas chamber:

$$p = p_{in}$$

$$y_{N_2} = y_{N_2}^{in} \tag{7-65}$$

Both p_{in} and y_{N2}^{in} are known.

Once the set of Equations (7-37 through 7-57) with boundary conditions (Equations 7-58 through 7-65) are simultaneously solved, this model provides useful information about the spatial variation of reaction rate, current density, and oxygen concentration within the modeling domain (Figure 7-6).

The current density is constant throughout the membrane region. In the catalyst layer, the electrolyte phase gradually transfers its current to the solid, electrically conductive phase, and the membrane phase current density decreases to zero at the catalyst layer/gas diffusion layer interface. For very low current density, the reaction rate is constant throughout the catalyst layer. At higher current densities, almost all the activity happens in a portion of the catalyst layer next to the gas diffusion layer, where oxygen is rapidly depleted.

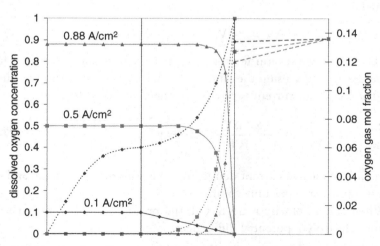

Figure 7-6 Sample of results generated by the Bernardi–Verbrugge model [5]—spatial distribution of current density (solid lines) and oxygen concentration (dotted line) in the electrolyte phase and oxygen mol fraction in the gas diffusion layer (dashed line).

7.3.2 One-Dimensional Catalyst Layer Model (You–Liu [12])

In this model, similar to Weisbrod et al. [13], the catalytic layer is described as a macro-pseudo-homogeneous film with the following assumptions:

- The potential drop in the solid matrix is negligible compared with the potential drop in the electrolyte phase
- The catalyst Pt particle is uniformly distributed throughout the layer
- The water content in the ionomer is constant
- The oxygen diffusion coefficient is constant
- Isothermal conditions

The model is derived from a basic mass/current balance by the control volume approach. A relaxation method is used to solve the two-point boundary problem. The effects of different parameters on the catalyst layer are studied in detail. The discussions in this chapter may help in the optimization design of the cathode catalyst layer in PEM fuel cells.

The composite catalyst layer is considered an effective, homogeneous medium of uniform thickness. The decrease of the oxygen molar flux in a control volume is balanced by the increase of the proton current density:

$$\frac{di_z}{dz} = -4F \frac{dN_{O_2}}{dz} \tag{7-66}$$

When the cathode electrochemical reaction, $O_2 + 4H^+ + 4e^- \rightarrow 2H_2O$, takes place, the electrode potential deviates from equilibrium potential, and it is defined as:

$$\Delta V_{act} = V_{ca} - V_r = \phi_s - \phi_m \tag{7-67}$$

The kinetic expression for the oxygen reduction rate per unit volume can be described by using the Butler–Volmer expression with the assumption that reduction current is positive for negative overpotential:

$$\frac{di_z}{dz} = A_c i_{0,c}^+ \left\{ \exp\left[\frac{-\alpha_c F \Delta V_{act,c}}{RT}\right] - \exp\left[\frac{(1-\alpha_c)F\Delta V_{act,c}}{RT}\right] \right\} \tag{7-68}$$

where i_0^+ is exchange current density at the equilibrium potential, and A_c is catalyst surface area per unit volume.

When water velocity is neglected, the proton current is related to the polymer electrolyte potential difference:

$$\frac{d\phi}{dz} = \frac{i_z}{\kappa^{eff}} \tag{7-69}$$

From Equations (7-66 and 7-69), and assuming the matrix of solid phase to be equipotential (as stated in assumption 1), changes in overpotential can be expressed as:

$$\frac{d(-\Delta V_{act,c})}{dz} = \frac{i_z}{\kappa^{eff}} \tag{7-70}$$

The oxygen flux is related to oxygen concentration according to Fick's diffusion law:

$$N_{O_2} = -D_{O_2}{}^{eff} \frac{d(C_{O_2})}{dz} \tag{7-71}$$

where D_{O2}^{eff} is the effective oxygen diffusion coefficient as per Equation (7-19).

Thus, four unknowns, i_z, nV_{act}, N_{O2}, and C_{O2} in the catalyst layer, are described by Equations (7-66), (7-68), (7-70), and (7-71). The appropriate boundary conditions are listed next.

At $z = 0$ (the gas diffuser/catalyst layer interface):

$$\left. \begin{array}{c} i_z = 0 \\[2mm] N_{O_2} = \dfrac{i_\delta}{nF} \\[2mm] C_{O_2} = C_{O_2}^{z=0} \end{array} \right\} \tag{7-72}$$

At $y = \delta$ (the catalyst layer/membrane interface):

$$\frac{d(-\Delta V_{act,c})}{dz} = \frac{i_\delta}{\kappa^{eff}} \tag{7-73}$$

The governing equations with a set of boundary conditions have been solved with a relaxation approach, starting with assumed overpotential at the catalyst layer/membrane interface. The model has been verified by comparing the results with experimental data. The results—spatial variation of current density, oxygen concentration, and overpotential across the catalyst layer—are shown in Figure 7-7. It should be noted that the results in Figure 7-7 are expressed as dimensionless current density, I^*, and dimensionless oxygen concentration, χ, defined as $I^* = I/I_{lim}$ and $\chi = C_{O2}/C_{O2}^{z=0}$, respectively. The results are similar to those from the Bernardi–Verbrugge model [5], indicating that at higher overpotentials, that is, higher overall current densities, the reaction happens in a portion of the catalyst

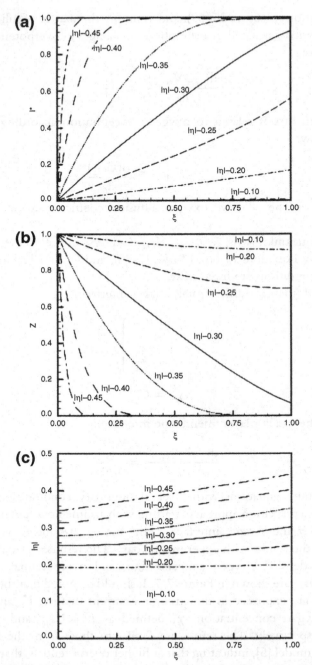

Figure 7-7 Sample of results generated with You–Liu model [12]: (a) variation of current density with overpotential, (b) variation of oxygen concentration with overpotential, (c) variation of overpotential across catalyst layer ($|\eta|$ in figures corresponds to nV_{act} in equations, and $\xi = z/\delta_c$). *(Reprinted by permission from International Association for Hydrogen Energy.)*

layer next to the gas diffusion layer. The model is useful to study the effect of catalyst layer parameters such as proton conductivity, effective porosity, and catalyst surface area.

7.3.3 Two-Dimensional Above-the-Channel Model (Jeng et al. [14])

Because the electrode portion covered by the ribs is not directly exposed to the channels, it suffers from a slow reactant gas mass transfer. The ribs can be regarded as barriers to mass transfer; however, they are necessary for electric current conduction. A complete understanding of the mass transfer phenomena within the gas diffusion layer (GDL), under the influence of current collector ribs, will facilitate a proper PEM fuel cell design. West and Fuller [15] studied the effects of rib sizing and the GDL thickness on the current and water distributions within a PEM fuel cell. They found that dimensions of the ribs only slightly altered the cathode potential for a given current density but had a significant influence on water management. Hental et al. [16] experimentally investigated the effects of both rib and channel widths on the performance of single PEM fuel cells. Naseri–Neshat et al. [17] studied the effect of gas flow channel spacing on current density through a 3-D model. Yan et al. [18] developed a two-dimensional mass transport model to investigate the anode gas flow channel cross-section and GDL porosity effects. They found that an increase in either the GDL porosity, channel width fraction, or the number of channels could lead to better cell performance. Nguyen and He, [19] based on the modeling results, concluded that more channels and shorter rib widths are preferred for an interdigitated flow field configuration.

The two-dimensional model for the oxygen transport through the GDL, such as the one proposed by Jeng et al., [14] takes the current collector ribs into account, as shown by the enlarged portion of a single cell appearing in Figure 7-8. L_1 and L_2 denote the midline of the GDL portion in contact with the rib and the channel, respectively. A GDL element bounded by L_1, L_2 has been selected for analysis in this two-dimensional model.

Jeng et al. [14] describe the two-dimensional oxygen mass transport within the GDL as:

$$N_{O_2} = -\left(\varepsilon^{3/2} D_{O_2}\right)\nabla C \qquad (7\text{-}74)$$

where N_{O2} is the oxygen molar flux (a vector quantity) in the GDL. By taking divergence of both sides of Equation (7-74) and by applying the

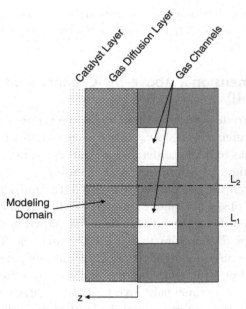

Figure 7-8 Modeling domain definition for the Jeng et al. model [14].

species conservation for oxygen, we obtain Laplace's equation, which governs the oxygen concentration distribution within the GDL:

$$\nabla^2 C = \frac{\partial^2 C}{\partial y^2} + \frac{\partial^2 C}{\partial z^2} \tag{7-75}$$

For this purpose, the catalyst layer can be regarded as an infinitely thin film and located on the left boundary of the GDL (Figure 7-8). The rate of the electrochemical reaction within the catalyst layer can be described using this thin-film model, yielding a Butler–Volmer rate expression. It is then simplified to give a Tafel-type equation in terms of the oxygen concentration at the catalyst layer as:

$$i = A_c i_0 \delta_c \frac{C(y,z)}{C_{ref}} \exp\left(\frac{\alpha_c F \Delta V_{act,c}}{RT}\right) \tag{7-76}$$

where:
  i = local current density, A cm^{-2}
  A_c = the specific area of the active surface, cm^2 cm^{-3}

i_0 = reference exchange current density, A cm^{-2}

δ_c = thickness of the catalyst layer, cm

$C(y,z)$ = the oxygen concentration at the catalyst layer, mol cm^{-3}

C_{ref} = the reference oxygen concentration associated with i_0, mol cm^{-3}

α_c = cathode transfer coefficient

$\Delta V_{act,c}$ = cathode overpotential, V

In this two-dimensional model, both the local current density and the oxygen concentration at the catalyst layer vary with y. Furthermore, under the steady-state condition and the zero reactant crossover assumption, the current is determined by the oxygen diffusion rate at the GDL/catalyst layer interface. This indicates that:

$$\frac{i}{4F} = N_{O_2}|_{z=\delta_C} = -\varepsilon^{3/2}D_{O_2}\frac{\partial C}{\partial z}\bigg|_{z=\delta_C} \qquad (7\text{-}77)$$

By combining Equations (7-76) and (7-77), the following boundary condition for the GDL/catalyst layer interface is obtained:

$$\frac{\partial C}{\partial z}\bigg|_{z=\delta_C} = -\frac{A_c i_0 \delta_c}{4F\varepsilon^{3/2}D_{O_2}}\frac{C(y,\delta_c)}{C_{ref}}\exp\left(\frac{\alpha_c F\Delta V_c}{RT}\right) \qquad (7\text{-}78)$$

where the expression is constant (K) for the given cathode overpotential nV and the physical parameter and property values A_c, i_0, δ_c, α_c, T, ε, D_{O2}, and C_{ref}.

Concentration of oxygen on the boundary facing the channel is assumed to be known, $C = C_0$. It is also assumed that the cathode overpotential, $nV_{act,c}$, is constant along the catalyst layer/GDL boundary.

Other boundary conditions follow from the domain geometry. Symmetrical boundary conditions on both the upper and lower boundaries of the GDL element apply; hence, the condition $\partial C/\partial y = 0$ is imposed on both L_1 and L_2. The molar flow rate across the GDL/rib interface is zero, so the $\partial C/\partial z = 0$ condition applies to that part of the boundary.

The two-dimensional Laplace's equation (7-75) associated with the corresponding boundary conditions has been discretized using the finite-difference approach and solved using an alternating-direction explicit (ADE) method. [20] From the solution, Jeng et al. obtained the oxygen concentration distribution in the GDL, that is, $C(y,z)$. The oxygen molar flow rate at the GDL/catalyst layer interface, $N_{O2}|_{z=dc}$, and the local current density, i, are then evaluated using Equation (7-77).

The average current density is simply:

$$i_{ave} = \frac{2}{w_1 + w_c} \int\limits_{0}^{(w_1+w_c)/2} i(y)dy \qquad (7\text{-}79)$$

The resulting oxygen concentrations inside the gas diffusion layer and current density at the GDL/catalyst layer interface are shown in Figure 7-9 for three different average current densities (0.1, 0.5, and 1.0 A cm^{-2}). Oxygen concentration is shown in a dimensionless form as C/C_0. At low average current density, oxygen concentration is high throughout the gas diffusion layer ($C > 0.78\ C_0$), resulting in relatively uniform current density distribution. At high current density, oxygen is almost depleted in the region of the gas diffusion layer above the rib, resulting in most of the reaction occurring directly above the channel and very little activity (0.1 i_{ave}) above the rib.

7.3.4 Two-Dimensional Along-the-Channel Model (Gurau et al. [21,22])

All of the previous one- and two-dimensional models assumed that oxygen concentration in the gas channel, or, more specifically, at the gas diffusion layer/gas channel interface, is known. However, oxygen concentration along the channel changes as oxygen is being depleted in the electrochemical reaction. At the same time, water content along the channel increases.

Yi and Nguyen [23] developed basically a one-dimensional along-the-channel model considering mass, species, and energy balances (not the momentum balance). The fluxes in z-direction (through the membrane electrode assembly) are taken into account as they affect the concentrations of reactants and products in the gas channel, but the changes of concentrations of reactants and products in z-direction are not considered, with the exception of water flux through the membrane. An interesting aspect of this model is that the fuel cell channels are modeled as the heat exchangers, allowing temperature variation along the channel, as a function of co- or counterflow of hydrogen, air, and coolant.

Furthermore, all the previous models consider oxygen transport in the gas diffusion layer solely as a result of concentration gradient, and they do not consider an interaction of the species in the channel with those in the gas diffuser due to velocity vector. Gurau et al. [21,22] presented the first unified approach by coupling the flow and transport governing equations in the

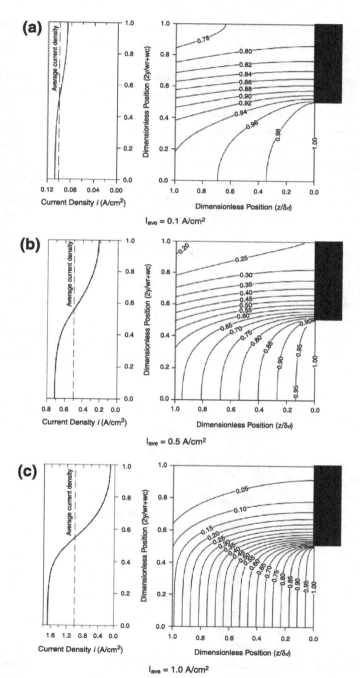

Figure 7-9 Oxygen concentration in gas diffusion layer and current density distribution at GDL–catalyst layer interface resulting from the Jeng et al. above-the-channel model [14]. *(Reprinted by Permission of Elsevier.)*

flow channel and gas diffuser with no boundary conditions at the interface. The modeling domain of this model is shown in Figure 7-10, and it is different for different species, namely:

- Oxygen domain includes cathode channel, cathode gas diffusion layer, and cathode catalyst layer, with no boundary conditions required at the interfaces between these layers.
- Hydrogen domain includes anode channel, anode gas diffusion layer, and anode catalyst layer, with no boundary conditions required at the interfaces between these layers.
- Liquid water domain includes the membrane and both catalyst layers. This implies that the volume occupied by liquid water in the gas channels coming from the gas diffusers is negligible (i.e., in gas channels only gas mixtures are present).

The model is based on the following set of assumptions:

- Only the steady state case is considered.
- The gas mixtures are considered perfect gases.
- The flow is laminar everywhere.
- The gas mixture flows are incompressible.
- The volume occupied by liquid water in the gas channels coming from the gas diffusers is negligible (in gas channels, only gas mixtures are present).

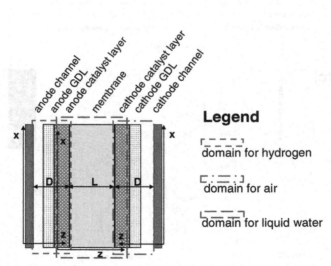

Figure 7-10 Domains for species in a PEM fuel cell used in Gurau et al. model [21].

- The dispersion of the fluids in the porous media is also disregarded. Its effects can be easily accounted for in the general diffusion coefficients.
- The gas diffusers, catalyst layers, and PEM are each considered as isotropic porous media.
- Dilation or contraction of the porous media is neglected.
- The contact electrical losses at the interfaces between different fuel cell elements are neglected.
- The catalyst layers are considered to have vanishing small thickness when the transport equations are solved, but the real values are taken into account when the membrane phase potential and current density are calculated.
- The heat generated under reversible conditions is neglected.

The distribution of the reactant concentrations along the electrodes is needed to calculate the transfer currents in electrochemical cells. The concentrations along the gas channels/gas diffusers/catalyst layers vary because of diffusion–convection transport and electrokinetics in the catalyst layers. These distributions depend therefore on the gas and medium properties, as well as on the reaction rates. These latter ones are in turn functions of the reactant concentrations, and an iterative procedure is required to predict them. A proper description of the species concentration distribution involves the use of the 2-D momentum equations in the coupled gas channel/gas diffuser/catalyst layer domain, coupled with the continuity and species concentration equations (such as the Stefan–Maxwell set of equations) and equations describing the electrochemical reactions (the Butler–Volmer equation). These latter ones are coupled with the transport equations in membrane via electroosmotic terms. The independent variables for constant geometry and material properties are the mass-flow rates, temperatures, humidities, and the pressures of the gas mixtures at the gas channel inlets, the external circuit resistance, and the temperature of the heating (cooling) medium. These are actually the only parameters that can be controlled in laboratory experiments or real-life fuel cell applications.

As already mentioned, boundary conditions for the dependent variables of the transport equations at the interfaces between different layers of the same domain are not required. At the gas channel entries, conditions of the first kind have to be prescribed for the gas mixture velocities, pressures, temperatures, and component concentrations. At the interfaces between the gas channels and the plate collectors, boundary conditions of the first kind have to be prescribed for the gas mixture velocity components (no slip condition) and for the temperature if the temperature of the cooling (heating) agent is known, or boundary conditions of the second kind are prescribed if the wall is adiabatic.

At the interfaces between the gas diffusers and the gas channels, the following boundary conditions are assumed for the liquid water velocity components:

$$\frac{\partial u}{\partial x} = \frac{\partial v}{\partial x} = 0 \qquad (7\text{-}80)$$

where u and v are velocities in x- and z-directions, respectively.

Similar boundary conditions are prescribed at the lower and upper boundaries that delimit the membrane:

$$\frac{\partial u}{\partial z} = \frac{\partial v}{\partial z} = 0 \qquad (7\text{-}81)$$

In the gas diffusers, water vapor is considered to be at equilibrium with the liquid water in the pores. Here also the liquid water temperature is set equal to the temperature of the gas mixtures.

For the membrane phase potential equation, the boundary conditions are:
- At the upper and lower limits, $\frac{\partial \phi}{\partial z} = 0$, that is, no protonic current leaves the domain
- At the interface between the catalyst layer and the gas diffuser, $\frac{\partial \phi}{\partial x} = 0$, that is, no protonic current leaves the domain

Solving the transport equations over domains covering more than one fuel cell element has to be done carefully. In the gas channels and in the gas diffusers, the objects of analysis are gas mixtures, whereas in the membrane, catalyst layers, and partially in the gas diffuser, it is the liquid water. The energy equation may be solved over the entire domain covering all the fuel cell elements; the continuity, momentum, and species conservation equations must be solved for each domain. It should be noted that the gas diffusers and catalyst layers each belong to two different domains.

The velocity and pressure fields for the gas mixtures are solved first in the coupled gas channel/gas diffuser domains, disregarding the changes in composition of the gas mixtures. This enables one to solve the flow and pressure fields for the gas mixtures first, and once these fields are found the equations for the other dependent variables may be solved. The gas species concentrations are dependent on the transfer current densities; therefore, the transport equations for the gas components are solved iteratively, together with the Butler–Volmer equations for anode and cathode catalyst layers.

After convergence is achieved, one proceeds to solve for the transport equations related to the liquid water flow, the membrane phase potential, and current densities. Because the source terms of the energy equations are functions of the current density, a new level of iterations is needed, except for the velocity and pressure fields of the gas mixtures.

There are generally two ways of solving for the electrochemical equations: Either the operating current density is given and different potential losses are calculated, or the so-called potentiostatic approach is used, when the cell potential is set and the current density is calculated. For a 2-D model, the operating current density is expressed as the average of the current density along the interface between the membrane and the catalyst layer.

Some of the results generated by the Gurau et al. model [21,22] are shown in Figures 7-11 and 7-12. As oxygen from air is consumed by electrochemical reaction at the catalyst layer, the oxygen mole fraction decreases along the flow direction. Because oxygen is only consumed by the catalyst layer, the model predicts a spanwise gradient of oxygen (Figure 7-11). The porous diffusion layer slows down oxygen transport to the catalyst layer while water vapor generated at the catalyst layer further inhibits oxygen from reaching the catalyst layer. Water vapor molar fraction increases along the length of the channel as water is being produced (Figure 7-12). Both oxygen and water vapor mole fraction distribution are a strong function of current, that is, reaction rate.

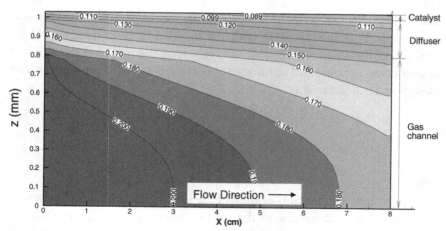

Figure 7-11 Oxygen mole fraction contours along cathode side gas channel *(results from Gurau et al. 2-D along-the-channel model [21,22]).*

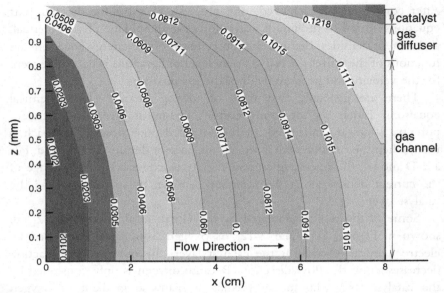

Figure 7-12 Water vapor mole fraction contours along cathode side gas channel *(results from Gurau et al. 2-D along-the-channel model [21,22]).*

Note that depending on operating conditions, air in the channel may be saturated already at the inlet, and it stays saturated along the channel, so any water added to the channel, such as product water, must be in liquid form. In that case a two-phase flow must be considered. The model by Gurau et al. does account for liquid water but only inside the membrane—liquid water presence in the gas diffusion layer and in the channels is ignored. You and Liu [24] built on the Gurau et al. model [21] by accounting for liquid water in all the regions of the model domain, albeit in isothermal conditions. Liquid water presence is included through a phase saturation term defined as:

$$s_{(l)} = \frac{\rho g_{H_2O} - \rho_g g_{H_2O(g)}}{\rho_l - \rho_g g_{H_2O(g)}} \tag{7-82}$$

$$s_{(g)} = 1 - s_{(l)} \tag{7-83}$$

where:

$s_{(l)}$ = volumetric fraction of the void space occupied by liquid water
$s_{(g)}$ = volumetric fraction of the void space occupied by gas mixture
g = species mass fraction
ρ = density (kgm^{-3})

The two-phase mixture density is:

$$\rho = \rho_g s_{(g)} + \rho_1 s_{(1)} \tag{7-84}$$

7.3.5 Three-Dimensional Models

Higher-dimension fuel cell models can be considered to study PEM fuel cell performance. As shown previously, two-dimensional models can be used to characterize fuel cell performance along the channel length (x-z plane) or across channels (y-z plane). In the y-z plane, the model can focus on the membrane electrolyte assemble or include the effect of gas transport in the porous diffusion layer and gas channel. In the y-z plane, several channels can be considered in order to investigate the effects such as gas mixing between the channels occurring in the diffusion layer. A three-dimensional model can be used to consider all the aforementioned phenomena, but computational limitations often limit the model's fidelity. The equations are the same as those in 1-D and 2-D models, but all the differential equations are applied in all three directions. Consequently, the boundary conditions must be applied for all three dimensions.

A three-dimensional distribution of oxygen mole fraction on the cathode side, resulting from modeling effort by Liu et al. [25,26], is shown in Figure 7-13. The modeling domain includes an entire length of the two halves of the adjacent gas channels separated by a "rib" of the bipolar plate (shown as an empty space in Figure 7-13) and corresponding gas diffusion layer above the channels and the rib. The channels show a typical downstream depletion of oxygen concentration in air due to consumption by electrochemical reactions at the catalyst layer. Oxygen mole fraction is low in the porous diffusion layer because of the low diffusion rate of oxygen into the porous structure, water generation as a byproduct of the electrochemistry occurring at the catalyst layer, and a majority of the diffusion layer being obstructed from the gas stream by the collector plate. This three-dimensional distribution of oxygen mole fraction shows that the most active catalyst layer is located directly above the gas channel. This trend is exhibited throughout the length of the channel. The trend is further validated by examining the current density distribution in the catalyst layer (Figure 7-14). The regions of high current density can be correlated to locations of high oxygen concentration on the catalyst layer. The location of highest current density is at the gas channel entrance,

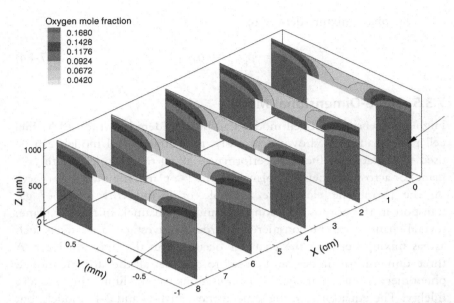

Figure 7-13 Three-dimensional distribution of oxygen mole fraction along cathode side gas channels *(Liu et al. [25])*.

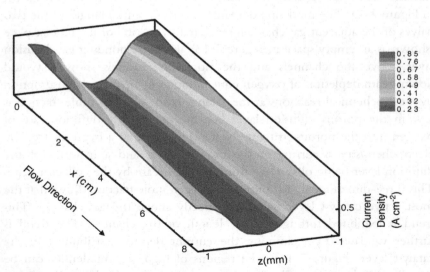

Figure 7-14 Current density distribution across the catalyst layer in the cathode *(results of 3-D model by Liu et al. [25])*.

where the highest oxygen concentration in air provides for the largest electrochemical reaction rates.

The same model may be applied to a slightly different geometry, such as an interdigitated flow field, where the flow is forced through the gas diffusion layer. In this case the oxygen mole fraction distribution in the channel and in the gas diffusion layer catalyst layers looks dramatically different from that in a straight or serpentine channel (Figure 7-15). Oxygen mole fraction is uniform throughout the inlet channel, and it is exhibiting significantly higher gradients in the gas diffusion layer, resulting in both higher limiting currents and more uniform distribution of current density at the gas diffusion layer/catalyst layer interface. In addition to Liu and coworkers at University of Miami [25–28], interdigitated flow field has been extensively studied by one of its first proponents, Nguyen and coworkers at the University of Kansas [19,29–33].

7.3.6 Through Gas Diffusion Layer Model/Pore-Network Model

In the cathode gas diffusion layer, oxygen and water flow in opposite directions. Liquid water may be present in the gas diffusion layer and it may block oxygen transport. Theoretical models of liquid water transport in PEMFC usually follow the macroscopic approach based on the two-phase Darcy's law. These models cannot incorporate the GDL morphology and require material-specific capillary pressure/liquid saturation and relative permeability/liquid saturation relationships, which may not be readily available in literature. Therefore, pore-scale models taking GDL microstructure and wetting characteristics into account are required. These models can be broadly divided into pore-network (PN) models, lattice-gas (LG) models, lattice-Boltzmann (LB) models, and molecular dynamic (MD) models. The pore-network modeling approach, being less computationally intensive while accounting for all the relevant two-phase flow physics, seems ideally suited to investigating liquid water transport in a GDL. In PN modeling, a porous medium is represented at the microscopic scale by a lattice of wide pores connected by narrower constrictions, called *throats*. Two-phase flow in such a porous structure is governed by capillary and viscous forces, and their relative magnitude governs the two-phase distribution and flow regimes. For a typical fuel cell application, where the viscosity ratio, defined as:

$$M = \frac{\mu_{nw}}{\mu_{wet}} \tag{7-85}$$

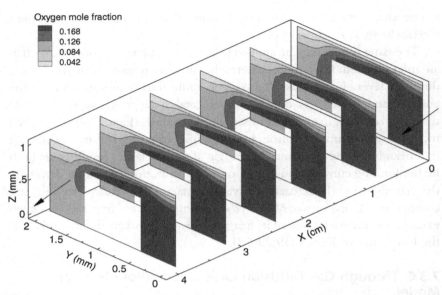

Figure 7-15 Three-dimensional distribution of oxygen mole fraction along cathode-side gas channels for interdigitated flow field *(Liu et al. [25])*.

where subscripts *nw* and *wet* stand for the nonwetting (water) and wetting (air) phases, respectively, is 17.5, and capillary number, defined as:

$$Ca = \frac{u\mu_{nw}}{\sigma} \qquad (7\text{-}86)$$

where u is the velocity of the nonwetting phase and σ is the surface tension, is of the order of 10^{-8}, two-phase flow in a GDL falls in the regime of capillary fingering.

Simha and Wang [34] employed randomly generated pore-network structures, which are created to match average properties of GDL materials such as porosity and permeability. However, these average GDL properties are insufficient to define a unique pore structure, because different porous media with totally different microstructures can have the same porosity and permeability. Luo et al. [35] introduced a topologically equivalent pore network (TEPN) modeling approach in which the pore network is extracted directly from high-resolution, three-dimensional microstructures of GDL materials, thus resulting in more realistic representation of various GDL materials and their relevant properties.

Modeling Assumptions

The models of two-phase flow in a pore network usually are based on a series of assumptions, such as:

- Wetting properties are assumed to be constant in the network.
- While the radius of a throat serves to define its hydraulic conductance, the volume contributed by all throats is assumed to be negligible.
- Fluid pressures are only defined in pores.
- A pore can only have single occupancy by liquid or gas. Physically, the nonwetting fluid is the bulk fluid and the wetting fluid stays as wetting films in corners, but for simplicity only bulk fluid is considered.
- The flow is laminar everywhere and governed by the Hagen–Poiseuille law.
- The fluids are Newtonian, incompressible, and immiscible.
- The injecting fluid is the nonwetting fluid.
- The resistance offered by a pore to flow is assumed negligible.
- Water enters GDL and flows through GDL in the liquid phase without phase change (evaporation and condensation).

Governing Equations (per Simha and Wang [34])

For liquid water to invade a throat, the pressure difference across a meniscus must exceed the throat entry capillary pressure, given by the Young–Laplace equation:

$$P_w - P_{air} > P_c, \quad P_c = \frac{2\sigma\cos\theta}{r_{th}} \tag{7-87}$$

where P_w and P_{air} represent liquid water and air pressure, respectively, P_c the capillary pressure, θ the contact angle between wetting phase and solid matrix, and r_{th} is the throat radius. As is clear from Equation (7-87), once a throat is invaded the connecting pore will be automatically invaded by liquid water owing to its larger size. Each phase, then, must obey volume conservation within each pore body:

$$V_i \frac{\partial S_i^\alpha}{\partial t} + \sum_{j \in N_i} Q_{ij}^\alpha = 0, \quad i = 1, 2 \ldots N \tag{7-88}$$

where S_i^α is the local saturation of the phase α in pore i, N the total number of pores, V_i the volume of pore i, N_i the number of pores connected to pore i, and Q_{ij}^α is the flow rate of phase between pore i and pore j. The flow rate of liquid water through a throat depends on the fluid configuration in the pores connected to that throat. For a liquid water-filled throat, connecting pores containing liquid water and air, the volumetric flow rate is given by:

$$Q_{ij}^w = \begin{cases} g_w(P_i - P_j - P_c) & \text{if } (P_i - P_j - P_c) > 0 \\ 0 & \text{if } (P_i - P_j - P_c) \leq 0 \end{cases} \quad (7\text{-}89)$$

Physically, Equation (7-89) means that flow through a throat is possible only if the pressure difference across the throat is higher than the throat entry capillary pressure, P_c. If pressure difference across a throat is less than the entry capillary pressure of the throat, it is called *capillary blocked*, and flow cannot occur through that throat. Whereas for a liquid water-filled throat connecting two liquid water-filled pores, the volumetric flow rate is given by:

$$Q_{ij}^w = g_w(P_i - P_j) \quad (7\text{-}90)$$

where g_w is water throat conductance ($m^4 \, s \, kg^{-1}$), defined by Poiseuille's law as:

$$g_w = \frac{2r_{th(i,j)}^4}{\pi \mu_w l_{(i,j)}} \quad (7\text{-}91)$$

where $l_{(i,j)}$ is the length of throat connecting pores i and j.

The constant injection rate at the inlet face is imposed by the following equation:

$$\sum_{\text{inlet pores}} [g_w(P_i - P_{i+1} - P_c)] = Q_{in} \quad (7\text{-}92)$$

The pressure field at every time step can be found by solving Equations (7-89) through (7-92). Once the pressure field is obtained, time steps are chosen so that only one pore reaches 100% liquid water saturation during any time step. Using the value of the current time step, liquid water saturation is updated in all the pores. The pores filled with liquid water are then tested for instability of the liquid water/air interface using the most recent pressure field. If the pressure drop across an interface is larger than the corresponding throat entry capillary pressure, the interface is called *unstable*; liquid water occupies the throat and the throat conductivity is updated accordingly. At any time step, volume averaged liquid water saturation, S^w, can be calculated as:

$$S^w = \frac{\sum_{i \in RV} V_i S_i^w}{\sum_{i \in RV} V_i} \qquad (7\text{-}93)$$

where RV denotes the representative volume over which averaging is done.

Initial and Boundary Conditions

The GDL is initially saturated with air. The inlet surface of the GDL is assumed to be in contact with a reservoir of the nonwetting fluid (i.e., liquid water), whereas the outlet surface is connected to a reservoir of the wetting fluid (air). A constant injection rate of liquid water is imposed on the inlet face under the assumption that all the water produced is in liquid form. A constant pressure boundary condition is imposed on the outlet face, whereas all other faces are subjected to the no-flow boundary condition.

Results

Water invades the GDL pores with the irregular fractal patterns typical of invasion percolation. Unlike viscous fingering, capillary fingering is characterized by no directional dependence. This provides significant in-plane spreading of liquid clusters, especially near the inlet face of GDL. Substantial in-plane spreading near the bottom of the GDL could explain the experimental observation that GDL flooding results in mass transport losses by blocking reactant transport through it. As liquid water invades the GDL, the liquid waterfront encounters multiple dead ends. The pressure difference across a gas/liquid interface must be larger than the capillary pressure at the interface for liquid water to invade further. Dead ends to front propagation appear when a liquid waterfront reaches a very narrow region with a very large entry capillary pressure. In such a case, the liquid water pressure increase at the inlet face makes the waterfronts unstable at several other locations, and liquid water invades further into the GDL there in order to maintain a constant flow rate.

Liquid water droplets reaching the hydrophobic GDL/channel interface grow to a critical size and then are removed from the GDL surface, either by drag force exerted by the gas flow or by capillary wicking onto the neighboring hydrophilic channel walls. Zhang et al. [36] predicted the droplet detachment diameter as a function of air velocity in the gas channel

based on simple force balance on a droplet. The breakout of liquid water
from preferential locations is consistent with the experimental observations,
where water droplets rather randomly pop out at the surface of the GDL
facing the flow channel. Figure 7–16 shows open liquid clusters at the steady

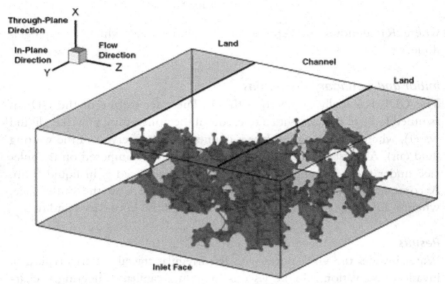

Figure 7-16 Open liquid clusters at the steady state during liquid water transport in
GDL *(Simha and Wang [34])*.

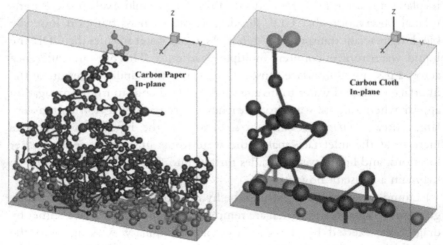

Figure 7-17 Liquid-saturated pores at the first breakthrough for two typical GDL
structures, namely carbon fiber paper (left) and carbon cloth (right) *(Luo et al. [35])*.

state during liquid water transport in GDL, obtained by Simha and Wang [34] using randomly generated pore-network structures. Figure 7-17 shows liquid-saturated pores at the first breakthrough for two typical GDL structures, namely carbon fiber paper and carbon cloth obtained by Luo et al. [35] using the topologically equivalent pore-network (TEPN) modeling approach, clearly showing different water behavior in these two structurally different gas diffusion media.

7.4 CONCLUSIONS

Modeling has emerged as a powerful and important tool for the fuel cell community. Several CFD software companies have developed fuel cell modules that can be used in conjunction with the original CFD codes (such as FLUENT and ANSYS CFX). Furthermore, open source codes are available on the Internet (e.g., OpenFOAM). Researchers, designers, and developers alike have used modeling to explore fuel cells, ranging from elucidating new fundamentals to characterizing fuel cell system performance. The user needs to select the appropriate model that balances model capabilities, robustness, and accuracy. Model accuracy is strongly dependent on the model's validity. The fuel cell model must be chosen such that it properly represents the problem being modeled, and it must be validated comprehensively to bring out its true predictive power.

This chapter is intended to provide an introduction to fuel cell modeling. The reader is encouraged to seek additional resources in this evolving field.

PROBLEMS

1. For a 2-D modeling domain representing the gas diffusion layer above two halves of the channels and an entire rib between them (as shown in the figure below) representing a portion of an interdigitated flow field, write the governing equations and corresponding boundary conditions. Assume isothermal conditions. Also assume that the catalyst layer has no thickness, that is, the reaction happens at the boundary between the gas diffusion layer and the catalyst layer. List the inputs (independent variables).

2. Write the governing equations and corresponding boundary conditions for electric current flux through the same domain defined in Problem 1.

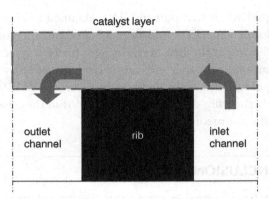

Figure with problems 1 and 2

3. Write the governing equations and corresponding boundary conditions for a 1-D model of a single fuel cell cathode channel (1-D means that only changes along the length of the channel should be taken into account). Assume isothermal conditions and that the reaction happens at the upper channel boundary. Also list other assumptions needed to develop this model.

 QUIZ

1. The relationship between electrical current and consumption of reactants in an electrochemical reaction is described by:
   a. Butler–Volmer equation
   b. Faraday's law
   c. Fick's law
2. Which of the following is *not* used in fuel cell modeling:
   a. Conservation of momentum
   b. Conservation of mass
   c. Conservation of temperature
3. Conservation of mass applies to:
   a. Any element of a fuel cell
   b. Any element of a fuel cell except the catalyst layer, where, because of the electrochemical reactions, there may be a change of mass
   c. Any element of a fuel cell except those where either evaporation or condensation takes place

4. The Stefan–Maxwell equation is used to model:
   a. Diffusion of a gas mixture through porous media
   b. Diffusion of a gas through another gas
   c. Diffusion of a gas through liquid

5. An assumption of isothermal conditions eliminates from modeling the:
   a. Energy conservation equation
   b. Momentum conservation equation
   c. Temperature conservation equation

6. The cathode overpotential is:
   a. Positive
   b. Negative
   c. It may be either positive or negative, depending on the sign of the anode overpotential

7. What is the source term in the energy conservation equation in the membrane?
   a. Heat of reaction in the cathode
   b. Ionic resistance
   c. Electronic resistance

8. In any volume of the anode, ionic current generated is always:
   a. Equal to the electronic current generated
   b. Greater than electronic current generated
   c. The same as ionic current generated in the adjacent anode volume

9. Total ionic current in the cathode is:
   a. Greater than the total ionic current in the anode
   b. Smaller than the total ionic current in the anode
   c. The same as the total ionic current in the anode

10. Symmetry boundary conditions are often used in fuel cell modeling to:
    a. Eliminate the need to model the anode (if geometry of the anode is symmetrical to the cathode)
    b. Reduce the number of equations used in the model and speed up the computation
    c. Simplify the modeling domain and speed up the computation

REFERENCES

[1] Barbir F. PEM Fuel Cells Design, Engineering, Modeling and Diagnostic Issues, lecture at the NSF Workshop on Engineering Fundamentals of Low Temperature PEM Fuel Cells (Arlington, VA, November 2001; http://electrochem.cwru.edu/NSF/presentations/NSF-Barbir_Diagnostics.pdf).

[2] Dagan G. Flow and Transport in Porous Formations. Berlin: Springer-Verlag; 1989.

[3] Bosnjakovic F. Technical Thermodynamics. New York: Holt Rinehart and Winston; 1965.

[4] Mazumder S, Cole JV. Rigorous 3-D Mathematical Modeling of PEM Fuel Cells. Journal of the Electrochemical Society 2003;Vol. 150(No. 11):A1503–9.

[5] Bernardi DM, Verbrugge MW. Mathematical Model of a Gas Diffusion Electrode Bonded to a Polymer Electrolyte. AIChE Journal 1991;Vol. 37(No. 8).

[6] Bird RB, Stewart WE, Lightfoot EN. Transport Phenomena. Second ed. New York: John Wiley & Sons; 2002.

[7] Gasteiger HA, Gu W, Makharia R, Matthias MF. Catalyst Utilization and Mass Transfer Limitations in the Polymer Electrolyte Fuel Cells, Tutorial. Orlando, FL: Electrochemical Society Meeting; 2003.

[8] Shimpalee S, Greenway S, Spuckler D, Van Zee JW. Predicting Water and Current Distributions in a Commercial-Size PEMFC. Journal of Power Sources 2004;Vol. 135:79–87.

[9] Oran ES, Boris JP. Numerical Simulation of Reactive Flow. Second ed. Cambridge, UK: Cambridge University Press; 2001.

[10] Kee RJ, Coltrin ME, Glarborg P. Chemically Reacting Flow: Theory and Practice. New York: John Wiley & Sons; 2003.

[11] Patankar SV. Numerical Heat Transfer and Fluid Flow. New York: Hemisphere; 1980.

[12] You L, Liu H. A Parametric Study of the Cathode Catalyst Layer of PEM Fuel Cells Using a Pseudo-Homogeneous Model. International Journal of Hydrogen Energy 2001;Vol. 26:991–9.

[13] Weisbrod KR, Grot SA, Vanderborgh N. Through-the-Electrode Model of a Proton Exchange Membrane Fuel Cell. In: Gottesfeld S, editor. Proc. 1st Intl. Symp. Proton Conducting Membrane Fuel Cells, Vols. 95–23. Pennington, NJ: Electrochemical Society; 1995. p. 152–66.

[14] Jeng KT, Lee SF, Tsai GF, Wang CH. Oxygen Mass Transfer in PEM Fuel Cell Gas Diffusion Layers. Journal of Power Sources 2004;Vol. 138:41–50.

[15] West AC, Fuller TF. Influence of Rib Spacing in Proton-Exchange Membrane Electrode Assemblies. Journal of Applied Electrochemistry 1996;Vol. 26:557–65.

[16] Hental PL, Lakeman JB, Mepsted GO, Adcock PL. New Materials for Polymer Electrolyte Membrane Fuel Cell Current Collectors. Journal of Power Sources 1999;Vol. 80:235.

[17] Naseri-Neshat H, Shimpalee S, Dutta S, Lee WK, Van Zee JW. Predicting the Effect of Gas-Flow Channel Spacing on Current Density in PEM Fuel Cells. Advanced Energy Systems 1999;Vol. 39:337–50. ASME.

[18] Yan WM, Soong CY, Chen FL, Chu HS. Effects of Flow Distributor Geometry and Diffusion Layer Porosity on Reactant Gas Transport and Performance of Proton Exchange Membrane Fuel Cells. Journal of Power Sources 2004;Vol. 125:27.

[19] Nguyen TV, He W. Interdigitated Flow Field Design. In: Vielstich W, Lamm A, Gasteiger H, editors. Handbook of Fuel Cell Technology: Fundamentals, Technology and Applications, Vol. 3. (Part 1). New York: John Wiley & Sons; 2003. p. 325–36.

[20] Anderson DA, Tannehill JC, Pletcher RH. Computational Fluid Mechanics and Heat Transfer. New York: McGraw-Hill; 1984.

[21] Gurau V, Liu H, Kakac S. Two-Dimensional Model for Proton Exchange Membrane Fuel Cells. AIChE Journal 1998;Vol. 44(No. 11):2410–22.

[22] Gurau V, Barbir F, Liu H. Two-Dimensional Model for the Entire PEM Fuel Cell Sandwich. In: Gottesfeld S, Fuller TF, editors. Proc. Proton Conduction Membrane Fuel Cells II, Vols. 98–27. Pennington, NJ: The Electrochemical Society; 1999. p. 479–503.

[23] Yi JS, Nguyen TV. An Along-the-Channel Model for Proton Exchange Fuel Cells. Journal of the Electrochemical Society 1998;Vol. 145:1149–59.

[24] You L, Liu H. A Two-Phase Flow and Transport Model for the Cathode of PEM Fuel Cells. International Journal of Heat and Mass Transfer 2002;Vol. 45:2277–87.

[25] Liu H, Zhou T, You L. Development of Unified PEM Fuel Cell Models. In: DOE Fuel Cells for Transportation Program. Oak Ridge National Laboratory: Annual National Laboratory R&D Meeting; June 6–8, 2001.

[26] Zhou T, Liu H. A General Three-Dimensional Model for Proton Exchange Membrane Fuel Cells. Journal of Transport Phenomena 2001;Vol. 3(No. 3):117–98.

[27] Kazim A, Liu H, Forges P. Modelling of Performance of PEM Fuel Cells with Conventional and Interdigitated Flow Fields. Journal of Applied Electrochemistry 1999;Vol. 29:1409–16.

[28] Wang L, Liu H. Performance Studies of PEM Fuel Cells with Interdigitated Flow Fields. Journal of Power Sources 2004;Vol. 134:185–96.

[29] Nguyen TV. Gas Distributor Design for Proton-Exchange-Membrane Fuel Cells. Journal of the Electrochemical Society 1996;Vol. 143(No. 5):L103–5.

[30] Wood III DL, Yi JS, Nguyen TV. Effect of Direct Liquid Water Injection and Interdigitated Flow Field on the Performance of Proton Exchange Membrane Fuel Cells. Electrochimica Acta 1998;Vol. 43:3795–809.

[31] Yi JS, Nguyen TV. Multicomponent Transport in Porous Electrodes of Proton Exchange Membrane Fuel Cells Using the Interdigitated Gas Distributors. Journal of the Electrochemical Society 1999;Vol. 146:38.

[32] He W, Yi JS, Nguyen TV. Two-Phase Flow Model of the Cathode of PEM Fuel Cells Using Interdigitated Flow Fields. AIChE Journal 2000;Vol. 46:2053–64.

[33] Natarajan D, Nguyen TV. Three-Dimensional Effects of Liquid Water Flooding in the Cathode of a PEM Fuel Cell. Journal of Power Sources 2003;Vol. 115:66–80.

[34] Sinha PK, Wang C-Y. Pore-Network Modeling of Liquid Water Transport in Gas Diffusion Layer of a Polymer Electrolyte Fuel Cell. Electrochimica Acta 2007;52:7936–45.

[35] Luo G, Ji Y, Wang C-Y, Sinha PK. Modeling Liquid Water Transport in Gas Diffusion Layers by Topologically Equivalent Pore Network. Electrochimica Acta 2010;55:5332–41.

[36] Zhang FY, Yang XG, Wang C-Y. Liquid Water Removal from a Polymer Electrolyte Fuel Cell. Journal of the Electrochemical Society 2006;153:A225–32.

CHAPTER EIGHT

Fuel Cell Diagnostics

Haijiang Wang

Institute for Fuel Cell Innovation, National Research Council Canada, Vancouver, BC, Canada V6T 1W5

Fuel cell science and technology cut across multiple disciplines, including materials science, interfacial science, transport phenomena, electrochemistry, and catalysis. It is always a major challenge to fully understand the thermodynamics, fluid mechanics, fuel cell dynamics, and electrochemical processes within a fuel cell. In addition to modeling, which is described in Chapter 7, diagnostics is another tool that can help elucidate the complex processes taking place in an operational fuel cell. On one hand, diagnostics can help distinguish the structure–property–performance relationships between a fuel cell and its components. On the other hand, results obtained from experimental diagnostics also provide benchmark-quality data for fundamental models, which further benefit in the prediction, control, and optimization of various transport and electrochemical processes occurring within fuel cells, as shown in Figure 8-1.

Various diagnostic tools for the accurate analysis of PEM fuel cells and stacks have been developed. At present, researchers who are conducting characterization of PEM fuel cells are mainly concentrated on the following issues [1,2]: (1) mass distribution, especially water distribution over the active electrode, including detection of flooding that leads to low catalyst utilization, (2) resistance diagnosis and detection of membrane drying, which closely relates to membrane conductivity, (3) optimization of electrode structures and components, fuel cell design, and operating conditions, (4) current density distribution in dimensionally large-scale fuel cells, (5) temperature variation resulting from a non-uniform electrochemical reaction and contact resistance in a single cell, and different interconnection resistances for a stack, and (6) flow visualization for direct observation of what is occurring within the fuel cell. Due to the complexity of the heat and mass transport processes occurring in fuel cells, there are typically a multitude of parameters to be determined. For all the previous reasons, it is important to examine the operation of PEM fuel cells or stacks with suitable techniques, which allow for evaluation of these parameters separately and can determine the influence of each on the global fuel cell performance.

PEM Fuel Cells
ISBN 978-0-12-387710-9

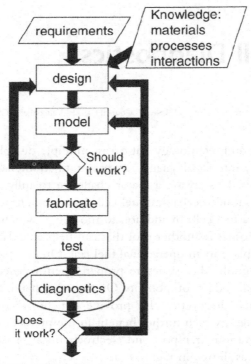

Figure 8-1 Role of diagnostics in the fuel cell design process.

The objective of this chapter is to provide a brief survey of diagnostic tools currently used in PEM fuel cell testing and research. For clarity, in this section the various testing and diagnostic tools are divided into two categories: electrochemical techniques and chemical/physical methods. Furthermore, for PEM fuel cell diagnosis by means of chemical and physical methods, various techniques for measuring species and for temperature and current distribution mapping are reviewed in detail.

8.1 ELECTROCHEMICAL TECHNIQUES

Electrochemical techniques, such as the polarization curve, current interruption, and electrochemical impedance spectroscopy, have been popularly employed in the diagnosis of fuel cells. Recent advances in the application of these electrochemical approaches to PEM fuel cells are described herein, and some novel methods are covered.

8.1.1 Polarization Curve

As discussed in Chapter 3, a plot of cell potential against current density under a set of constant operating conditions, known as a *polarization curve*, is the standard electrochemical technique for characterizing the performance of fuel cells (both single cells and stacks). It yields information on the performance losses in the cell or stack under operating conditions. The ideal polarization curve for a single hydrogen/air fuel cell has three major regions, which are shown in Figure 3-8. At low current densities (the region of activation polarization), the cell potential drops exponentially; the majority of these losses are due to the sluggish kinetics of the oxygen reduction reaction (ORR). At intermediate current densities (the region of ohmic polarization), the voltage loss caused by ohmic resistance becomes significant and results mainly from resistance to the flow of ions in the electrolyte and resistance to the flow of electrons through the electrode. In this region, the cell potential decreases nearly linearly with current density, while the activation overpotential reaches a relatively constant value. At high current densities (the region of concentration polarization), mass transport effects dominate due to the transport limit of the reactant gas through the pore structure of the gas diffusion layers (GDLs) and electrocatalyst layers (CLs), and cell performance drops drastically. Figure 3-8 also shows the difference between the theoretical cell potential (1.23 V) and the thermoneutral voltage (1.4 V), which represents energy loss under the reversible condition (the reversible loss). Very often, polarization curves are converted to power density versus current density plots by multiplying the potential by the current density at each point of the curve. The plot of power density versus current density can directly show the nominal power and maximum power of the cell.

A steady-state polarization curve can be obtained by recording the current as a function of cell potential or recording the cell potential as the cell current changes. A nonsteady state polarization curve can be obtained using a rapid current sweep [3]. The flow of the reactant gases may be kept constant at a sufficiently high rate to allow operation at the highest current density, or the flow rate may change in proportion to current density in prescribed stoichiometry. The former is suitable for quick sweeps, whereas the latter is only suitable for slow sweeps, since the flow rate adjustments may take some time. Changing the current ahead of the flow rate may result in starvation of the reactant gases and unwanted cell potential drop.

If the polarization curve is recorded in both directions, increasing and decreasing current, it may show hysteresis, that is, the two curves might not be on top of each other. This typically points to either flooding or membrane drying in the fuel cell. For example, if the cell is flooding on the cathode side, then operation at a higher current density would only make the situation worse because additional water would be produced. The polarization curve recorded with decreasing current would show lower voltage at higher currents than the previously taken polarization curve with increasing current. Conversely, if the cell is drying on the cathode side, additional water produced at high current densities would be beneficial, resulting in higher cell potential in a backward polarization curve (Figure 8-2). Similar behavior may result in a fuel cell that is either drying or flooding on the anode side. By measuring polarization curves, certain parameters such as the effects of the composition, flow rate, temperature, and relative humidity of the reactant gases on cell performance can be characterized and compared systematically [4].

Up to now, several modeling studies have been carried out to elucidate the electrochemical behavior of PEM fuel cells, and for this purpose, several empirical equations were introduced to mimic the polarization curves. By fitting the experimental results to one of the equations describing the polarization curve, insightful information may be gained about the parameters of the polarization curve, such as reversible cell potential, V_i; apparent exchange current density, i_o; Tafel slope, b; cell resistance, R_i; or limiting current i_L (as demonstrated by Ticianelli et al. [5]).

Polarization curves provide information on the performance of the cell or stack as a whole. Although they are useful indicators of overall performance

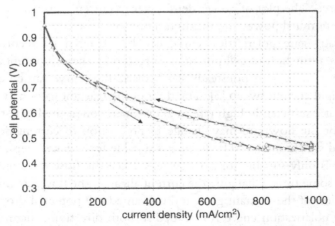

Figure 8-2 Fuel cell polarization curve hysteresis-cathode drying. (Cell temperature: 80°C; H_2/Air humidification: 80/60°C; H_2/Air stoichiometry: 1.5/5.0.)

under specific operating conditions, they fail to produce much information about the performance of individual components within the cell. They cannot be performed during normal operation of a fuel cell, and they take significant time to finish. In addition, they fail to differentiate between different mechanisms; for example, flooding and drying inside a fuel cell cannot both be distinguished in a single polarization curve. They are also incapable of resolving time–dependent processes occurring in the fuel cell and the stack. For the latter purpose, current interruption, electrochemical impedance spectroscopy measurements, and other electrochemical approaches are preferred. These techniques will be introduced in the following sections.

8.1.2 Current Interruption

In general, the current interruption method is used for measurement of the ohmic loss (i.e., cell resistance) in a PEM fuel cell. The principle of the technique is that the ohmic loss vanishes much more quickly than the electrochemical overpotential when the current is interrupted [6]. As shown schematically in Figure 8-3, a typical current interruption result is presented by recording the transient voltage when the interruption of the current

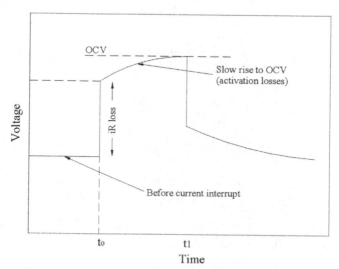

Figure 8-3 Cell voltage behavior after current interrupt. (The cell is operated at a fixed current. At $t = t0$, the current is interrupted and the ohmic losses vanish almost immediately. After the current interruption, overpotentials start to decay and the voltage rises exponentially towards the open-circuit voltage. At $t = t1$, the current is again switched on.)

occurs after the fuel cell has been operated at a constant current. The ohmic loss disappears almost immediately and the electrochemical (or activation) overpotential declines at a considerably slower rate. In this technique, rapid acquisition of the transient voltage data is of vital importance for adequate separation of the ohmic and activation loss. Figure 8-4 shows the typical equivalent circuit for a fuel cell, which consists of two resistors and a capacitor. The first resistor, R_R, represents the ohmic loss, and the second resistor, R_{act}, represents the activation loss. An example of the circuit used to perform a current interruption procedure can also be observed in Figure 8-4. When the switch is off the current is interrupted and no current flows through the first resistor in the fuel cell circuit. This makes the voltage increase instantaneously at first, but it then increases very slowly due to the discharging of the capacitor. The system reaches the open circuit voltage (OCV) once the capacitor is completely discharged. An electronic switch is needed in this technique to conduct the current interruption, and an oscilloscope is usually used to record the voltage signal.

The crucial issue for obtaining *in situ* ohmic loss measurements by the current interruption method is to separate the previously described two processes. Büchi et al.'s [7] experiment showed that the time scope for

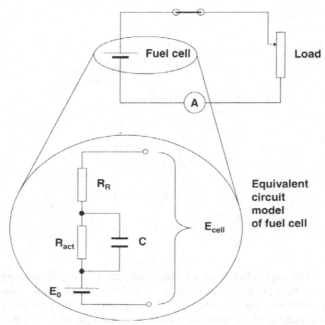

Figure 8-4 Equivalent circuit representing a fuel cell.

accurate current interruption measurements must be controlled between 0.5 ns and 10 ns. With the current interruption technique, Jaouen et al. [8] obtained specified values for various parameters such as the exchange current density, Tafel slope, oxygen solubility, and double-layer capacitance. Abe et al. [9] studied the effect of gas humidification temperature on the ohmic resistance by the current interruption technique. They found that the ohmic resistance increased by 3.5 mΩ when the cathode gas humidification temperature decreased from 80°C to 35°C. Using the current interruption method, the effect that operating conditions in a natural-breathing PEM fuel cell had on ohmic resistances was studied by Noponen et al. [10] Further, the current interruption method was employed by Mennola et al. [11] to determine the ohmic resistances of individual cells in a PEM fuel cell stack. In their experiments, a digital oscilloscope with a multichannel was connected in parallel to the individual cells to monitor the transient voltages. Their results showed good agreement between the ohmic loss in the entire stack and the sum of the ohmic losses of each individual cell.

Compared with other methods such as impedance spectroscopy, the current interruption method has the advantage of relatively straightforward data analysis. However, one of the weaknesses of this method is that the information obtained for a single cell or stack is limited. The second issue with this method is the difficulty in determining the exact point at which the voltage jumps instantaneously; thus, a fast oscilloscope should be used to record the voltage changes. Another difficulty is the so-called ringing effect caused by cable inductance [12], particularly for short delay times (*delay time* is the time between current interruption and voltage measurement).

8.1.3 Electrochemical Impedance Spectroscopy

In contrast to linear sweep and potential step methods, where the system is perturbed far from equilibrium, *electrochemical impedance spectroscopy* (EIS) applies a small AC voltage or current perturbation/signal (of known amplitude and frequency) to the cell, and the amplitude and phase of the resulting signal are measured as a function of frequency. This may be repeated through a wide range of frequencies (i.e., a large frequency spectrum). Basically, impedance is a measure of the ability of a system to impede the flow of electrical current. Thus, EIS is a powerful technique that can resolve various sources of polarization loss in a short time and has been widely applied in PEM fuel cells in a number of recent studies. Figure 8-5 shows the typical circuit used for an EIS test. A common use of EIS analysis in PEM fuel cells is to study the ORR, to characterize transport (diffusion)

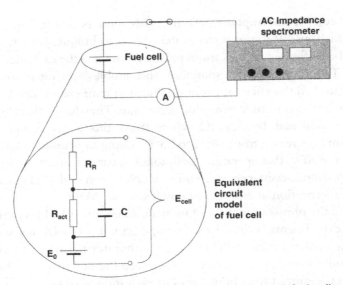

Figure 8-5 Electrochemical impedance spectroscopy setup and fuel cell equivalent circuit.

loss, to evaluate ohmic resistance and electrode properties such as charge transfer resistance and double layer capacitance, and to evaluate and optimize the membrane electrode assembly (MEA).

Impedance spectra are conventionally plotted in both Bode and Nyquist forms. In a *Bode plot*, the amplitude and phase of the impedance are plotted as a function of frequency; in a *Nyquist plot,* the imaginary part of the impedance is plotted against the real part at each frequency. Figure 8-6

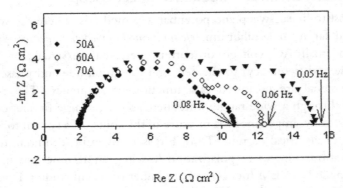

Figure 8-6 Typical impedance spectra of a PEM fuel cell. The spectra were obtained at 30°C using a Ballard Mark V six-cell stack with an active area of 280 cm^2 [13].

shows typical impedance spectra for PEM fuel cells in Nyquist form with two arcs, where the frequency increases from the right to the left. The high-frequency arc reflects the combination of the double-layer capacitance of the catalyst layers, the effective charge transfer resistance, and the ohmic resistance, in which the latter can be directly compared to the data obtained from current interruption measurements. The low-frequency arc always reflects the impedance due to mass transport limitations [13].

8.1.3.1 Cathode Behavior

Due to the fast hydrogen reduction reaction, the impedance spectrum of the fuel cell nearly equals the cathode impedance. As a result, the impedance of a fuel cell (H_2/O_2) is mainly used to study the cathode behavior.

High-Frequency Arc

Many researchers have studied the charge transfer of the cathode CL, which is represented by the high-frequency arc. The ohmic resistance, including the membrane resistance together with the GDL, bipolar plate, and contact resistances, is given by the intercept of the high-frequency arc with the real axis in the Nyquist plot. It is assumed that any change in this value during operation is caused by a change in membrane ionic resistivity due to membrane hydration. Therefore, the high-frequency resistance is always a measure of membrane water content.

A patent by General Motors disclosed a correlation between the degree of humidification and the high-frequency resistance of the membrane in a fuel cell stack [14]. General Motors was able to optimize the humidity level by monitoring the high-frequency resistance because it found that increased resistance signified cell drying, whereas a decreased resistance in conjunction with low performance signified cell flooding. By monitoring both cell resistance and pressure drop in an operational fuel cell stack, Barbir et al. [15] were able to diagnose either flooding or drying conditions inside the stack. Their results showed that drying typically caused a monotonous voltage decay, whereas flooding caused a rather erratic cell voltage behavior (i.e., sudden voltage fluctuation due to liquid water accumulation and expulsion inside the cell passages). Recently Oszcipok et al. [16] used EIS techniques to study the influence of cold start behavior on the membrane resistance and performance degradation of a PEM fuel cell.

While the high-frequency intercept corresponds to the ohmic resistance, the diameter of the arc relates to the charge transfer resistance of the catalyst layers at high frequencies. As discussed in Chapter 7, many models have been developed to explore the mechanism of the cathode catalyst layers under stationary conditions, such as the simple pore model, agglomerate model, macrohomogeneous model, and flooded–agglomerate model. Several equivalent circuit models have also been developed to mimic the spectra, such as the nonlinear least-squares procedure (NLSQ) [17] and transmission line models [18]. Consequently, the effects due to charge transfer, air diffusion through the CL pores, and diffusion in the ionomer layer surrounding the catalyst particles can be separated.

Based on a macrohomogeneous model of the gas diffusion electrode, Springer et al. [19] demonstrated that simultaneous fitting of impedance spectra obtained at different potentials could be used to evaluate the sources of PEM fuel cell performance losses. Three different types of loss, caused by interfacial kinetics, catalyst layer proton conductivity, and membrane conductivity, were resolved in their impedance spectra. Eikerling et al. [18] employed the macrohomogeneous model to describe the impedance responses of PEM fuel cells and further used the transmission–line model to simulate the impedance spectra. They also studied the relationship between the structure of the CL and the impedance spectra. The transmission–line model was also employed by Makharia et al. [20] to mimic the catalyst layers in H_2/N_2 and H_2/O_2 atmospheres. Parameters such as cell ohmic resistance, CL electrolyte resistance, and double-layer capacitance were extracted. More recently, Lefebvre et al. [21] and Jia et al. [22] used the transmission–line model to simulate the impedance behavior of PEM fuel cell catalyst layers under an N_2 atmosphere at the cathode side and, ultimately, the ionomer loading in the CL was optimized. Furthermore, they used EIS to study the capacitance and ion transport properties of catalyst layers under H_2/O_2 conditions [23]. Figure 8-7 illustrates the equivalent circuit of the transmission–line model, which consists of two parallel resistive elements, one for electron transport through the conducting carbon particles (R_{el}) and the other for proton transport in the CL (R_p). The resistive elements are connected by double-layer capacitances (C_{dl}) in parallel with the charge transfer resistance (R_{ct}).

Low-Frequency Arc

The appearance of the low–frequency arc can be attributed to the limitation of oxygen diffusion through the pores of the porous electrode, as illustrated in Figure 8-6. Evidence to support this conclusion comes from the absence of the low-frequency arc for operation with pure oxygen and an increase in the arc radius when operated with air or with increasing GDL thickness [24].

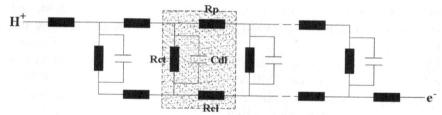

Figure 8-7 Transmission-line equivalent circuit describing the impedance behavior of the CL [18].

Cha et al. [25] employed impedance spectroscopy in their investigation of the microscale transport phenomena in flow channels for fuel cells based on the low-frequency feature. Their results showed that during flow channel scaling, the performance of the fuel cell was maximized at a certain flow channel size but declined as the channel size decreased, despite improved mass transport resulting from the velocity increase. Their explanation was that flooding blocked the flow channels and inhibited oxygen access. A more comprehensive study of the impedance response of PEM fuel cells has been conducted to investigate the effects of membrane thickness, cell temperature, and humidification conditions on fuel cell performance [17]. The low-frequency loop that appeared at high current density and low air flow rate, resulting in water accumulation in the GDL, was attributed to the mass transport limitation.

8.1.3.2 Anode Behavior

Carbon monoxide (CO) poisoning is a significant issue for PEM fuel cells, lowering performance due to deactivation of the Pt anode catalyst. Many efforts have been made to understand the mechanism of CO poisoning. EIS technique has been utilized by some researchers to study CO poisoning of the anode. However, two major problems remain. One complex issue with using EIS measurements to investigate anode poisoning is the separation of the anode and cathode impedances. Another problem is that the poisoning causes a change in the state of the fuel cell, which is reflected in the recorded impedance spectra.

Wagner et al. [26] observed that in carrying out the measurements in galvanostatic rather than potentiostatic mode, the change in fuel cell impedance was mainly due to that of the anode. Based on this measurement mode, the EIS technique was used to investigate the influence of CO poisoning on the Pt anode [27]. A combination of three mathematical procedures—real-time drift compensation, time-course interpolation, and the Z-HIT refinement—has been developed to validate and evaluate the EIS data of PEM fuel cell systems' change of state over time [28]. A similar EIS study of CO tolerance for different Pt-alloy anode catalysts in PEM fuel cells was presented as well [29]. More recently, Rubio et al. used EIS to diagnose performance degradation phenomena in PEM fuel cells due to anode catalyst poisoning by CO [30].

8.1.3.3 Fuel Cell Stack

Evaluation of stack performance is of importance for PEM fuel cell applications. For large fuel cell stacks, EIS application is limited because most commercial load banks operating at higher currents do not have good frequency responses. Yuan et al. successfully measured the EIS of a 500 W PEM fuel cell stack with the combination of a FuelCon test station, a TDI loadbank, and a Solartron 1260 frequency analyzer [13]. The effects of temperature, flow rate, and reactant humidity on the stack performance and impedance spectra were investigated. A rotary switch, developed in-house, was used in order for the individual cell impedances to be measured [31]. For simultaneous measurement of individual cell impedances, Hakenjos et al. [32] set up a measurement system containing a multichannel frequency response analyzer. As a proof of reliability, their results showed good agreement between the sum of the impedance of the single cells and the measured impedance of the whole stack. The absolute deviation was less than 2.5%.

8.1.3.4 Summary

EIS is an effective technique for fuel cell studies. This dynamic method can provide more information than steady-state experiments and can provide diagnostic criteria for evaluating PEM fuel cell performance. The main advantage of EIS as a diagnostic tool for evaluating fuel cell behavior is its ability to resolve, in the frequency domain, the individual contributions of the various factors determining the overall PEM fuel cell performance losses: ohmic, kinetic, and mass transport. Such a separation provides useful information for both optimization of the fuel cell design and selection of the most appropriate operating conditions.

However, the interpretation of impedance spectra hitherto has been difficult due to the complexity of the porous electrode and still remains a debated issue. Experimental results from Makharia et al. [20] showed that ohmic resistance estimated using EIS at 1 kHz also included a contribution of approximately 20 to 40 $m\Omega \cdot cm^2$, due to the electrolyte resistance in the CL. The low-frequency impedance, typically ascribed solely to limitations of oxygen transport to the cathode active sites through the flooded porous cathode, was also influenced at high current densities by dehydration of the membrane close to the anode [33]. The experiment of Andreaus et al. [34] demonstrated that this effect was more pronounced with a thicker membrane, where back diffusion of water from the cathode was less efficient. To summarize, there are still unresolved issues regarding the explanation of impedance spectra. For example, it is difficult to distinguish the individual contributions from the anode and cathode sides, although it is generally considered that the rapid kinetics and mass transport of the hydrogen oxidation reaction result in negligible impedance contribution from the anode CL. And as previously discussed, interpretation of the low-frequency feature can be very sophisticated. More information and details on EIS for fuel cells and electrochemical power sources in general can be found in the book by Barsoukov et al. [35]

8.1.4 Other Electrochemical Methods

8.1.4.1 Cyclic Voltammetry

Cyclic voltammetry (CV) is a commonly used *in situ* approach to fuel cell research, especially to characterize fuel cell catalyst activities. The *in situ* CV technique has proven to be quite effective for measuring the electrochemical surface area (ECA) of a gas diffusion electrode. In this technique the potential of a cell is swept back and forth between two voltage limits while the current is recorded. The voltage sweep is normally linear with time and the plot of the current versus voltage is called a *cyclic voltammogram* [36].

When someone carries out a CV measurement of a fuel cell, H_2 is fed to one electrode that acts as both the counter electrode and the reference electrode, functioning as a dynamic hydrogen electrode (DHE). The other electrode is flushed with inert gas (N_2 or Ar) and is taken as the working electrode. Voltammetric measurements are performed using a potentiostat/galvanostat and the cyclic voltammograms can be recorded at different voltage sweep rates. Low sweep rate is often used (e.g., 10 mV/s) to carry out the measurements at steady-state condition. A typical fuel cell cyclic voltammogram is shown in Figure 8-8. The two small redox peaks to the left, identified as hydrogen adsorption and desorption peaks, correspond to the hydrogen adsorption and desorption reaction on two types of crystal surface of the platinum catalyst. The two irreversible peaks to the right correspond to the surface oxide formation and reduction of platinum

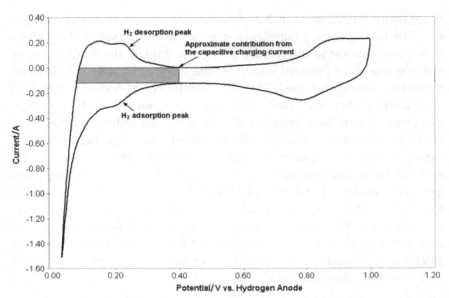

Figure 8-8 Schematic of a fuel cell CV. The two peaks identified represent the hydrogen adsorption and desorption peaks on the platinum fuel cell catalyst surface. The forward and reverse double-layer charging current density, $i_{dl\ charging}$, and the crossover current density, $i_{crossover}$, are also shown.

catalyst. The forward and reverse double-layer current density is also shown in this figure.

The ECA of the working electrode can be obtained based on the hydrogen adsorption reaction or desorption reaction. The adsorption charge or desorption charge can be calculated from the areas underneath the adsorption peaks or desorption peaks in the voltammogram. To be accurate, the double-layer charging current has to be subtracted. The H_2 adsorption charge on a smooth Pt electrode has been measured at $210\ mC/cm^2$. The ECA of an electrode is then calculated using the following equation [37]:

$$ECA(cm^2 Pt/gPt) = \frac{\text{Charge}(\mu C/cm^2)}{210(\mu C/cm^2\ \text{Pt}) \times \text{Catalyst Loading}(g\ Pt/cm^2)},$$

(8-1)

The disadvantage of this technique for measuring supported electrocatalysts is that the carbon support can alter the H_2 adsorption and desorption characteristics by increases in the double-layer current and the redox reactions of surface active groups on carbon. To avoid carbon oxidation, the voltage of the working electrode is always set below 1.0 V (versus DHE).

8.1.4.2 CO Stripping Voltammetry

The difficulty of using hydrogen adsorption/desorption on a platinum catalyst surface to measure ECA is that there is no well-defined baseline for the hydrogen redox reaction. Subtraction of background current is quite subjective. Oxidation of CO adsorbed on the platinum catalyst surface shows a well-behaved peak with a well-defined baseline. Therefore CO stripping voltammetry is a more accurate technique for measuring the ECA of fuel cell electrodes. The principle of this approach is the same as for CV [38], the only difference being that the working electrode of the fuel cell has to be covered with a monolayer of CO before the CV measurement. This is done through the following procedures: A gas mixture of CO and inert gas (Ar or N_2 with about 1% CO) is first flowed through the working electrode for several minutes. Pure inert gas is then used to remove the CO in the gas phase by flowing pure inert gas through the working electrode for several more minutes. An example of a CO stripping voltammogram is shown in Figure 8-9. When calculating ECA based on CO stripping voltammetry, Equation 8-1 can still be used by replacing 210 $\mu C/cm^2$ with the value of 484 $\mu C/cm^2$ for polycrystalline Pt. One study shows that ECA calculated by means of CO adsorption seems not to be dependent on platinum loading and is generally lower than that obtained by hydrogen adsorption [39].

The CO stripping peak potential can provide information on the active surface sites of the catalyst layer. Experimental results have also demonstrated

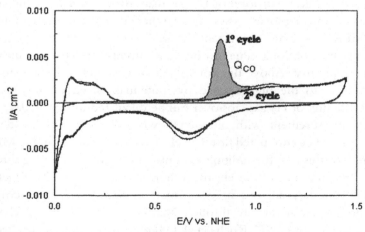

Figure 8-9 Cyclic voltammograms (10mV/s) at 25°C in the potential range 0–1400 mV vs. NHE on Pt/C (E-TEK) with (first cycle) and without (second cycle) a CO adsorbed adlayer. The filled area represents the charge related to the CO oxidation reaction [39].

that the CO stripping peak potential can provide information on the composition of an unsupported metal alloy surface and is useful for exploring the reaction mechanism of a metal alloy with enhanced CO tolerance [40]. Song et al. [41] used this technique to investigate the effect of different electrode fabrication procedures on the structural properties of MEAs. It has also been found that exposing CO to platinum, and the subsequent removal of that CO by electrochemical stripping, is an excellent method of cleaning and activating Pt [42]. The fuel cell achieved its maximum performance after several CO adsorption/CO_2 desorption cycles.

8.1.4.3 Linear Sweep Voltammetry

Crossover of hydrogen and oxygen through the membrane, which lowers the fuel cell performance and fuel efficiency, is considered one of the most important phenomena in PEM fuel cells. *Linear sweep voltammetry* (LSV) is a convenient technique to measure the rate of gas crossover. The technique is also handy to check for electrical shorts of the MEA. The experimental procedure is similar to the CV technique but without the backward voltage scan. When conducting LSV, humidified H_2 and N_2 are supplied to the anode and cathode sides of the fuel cell, respectively. A linear voltage that usually ranges from 0 V to 0.8 V is applied across the cathode and anode, and the current is recorded and plotted against the voltage. Higher voltages are avoided to prevent Pt oxidation. In LSV, since there are no electroactive species introduced into the cathode side, the electric current is attributable to the electrochemical oxidation of H_2 gas that crossed over from the anode side through the membrane, assuming that there is no electrical short current in the MEA. The crossover current typically increases with the voltage and rapidly reaches a limiting value (the limiting current) when the potential gets to around 300 mV. Above this voltage, all crossover H_2 is instantaneously oxidized due to the high applied overpotential. Based on the limiting current, the flux of H_2 gas crossover can be calculated using Faraday's law. An LSV measurement with nitrogen flowing through both anode and cathode should be conducted first if an electrical short current in the MEA is suspected. In this case the limiting current with hydrogen flowing through the anode is the sum of the electrical short current and the hydrogen gas crossover current. Using this diagnostic method, Song et al. [43] determined the hydrogen crossover rate through Nafion 112 membrane at elevated temperatures up to 120°C. Kocha et al. [44] recently examined the effects of various operating temperatures, gas pressures, and relative humidity on hydrogen crossover.

To quantitatively measure the flux of hydrogen crossover, *chrono-coulometry* (CC) needs to be used. A fixed voltage is applied across the cathode and anode instead of a voltage scan. The voltage applied has to be at the limiting current range (e.g., 0.5 V). The current is recorded as a function of time and can be easily transformed to charge as a function of time. The charge, which corresponds to the oxidation of the H_2 crossed over from the anode to the cathode, can be readily converted to the amount of H_2 crossed over. The rate of hydrogen crossover is thus measured.

Crossover current, measured before and after durability tests, is a very good indicator of membrane degradation over time. In Liu et al.'s [45] experiment, crossover current measurement was used to study membrane durability. They found that the reinforced Gore composite membranes exhibited an order of magnitude longer lifetime than the Nafion membrane of comparable thickness. This *in situ* method was further employed by Yu et al. [46] to investigate degradation mechanisms under low humidification of the feed stream. More recently, Wu et al. [47] employed this method to investigate the degradation mechanisms of a six-cell PEM fuel cell stack under close to open-circuit conditions. As shown in Figure 8-10, the

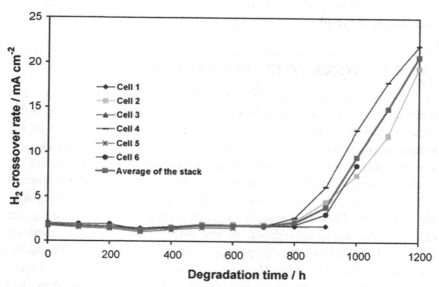

Figure 8-10 The change of hydrogen crossover rate through the membrane of each cell with time of 1200 h (cell temperature: 70°C; humidifier temperature of nitrogen/hydrogen: 70/70°C; flow rate of nitrogen/hydrogen: 3.3/3.0 SLPM) [47].

average crossover current after an 800-hour durability test was much higher than it was before 800 hours. This dramatic increase in the crossover current after 800 hours is a clear indication of membrane failure [47].

8.1.4.4 Cathode Discharge

Stumper et al. [48] developed an *in situ* cathode discharge method to determine the MEA resistance and electrode diffusivity (MRED) of a fuel cell. This method was based on the galvanostatic discharge of a fuel cell with an interrupted reactant supply. During a cathode discharge experiment, the cathode compartment was separated from the gas supply by closing both inlet and outlet valves, whereas the anode side continued to be supplied with H_2. Then the load was switched on with a constant current, and the cell voltage was recorded during the discharge of the fuel cell. The pure ohmic resistance of the fuel cell could be determined by fitting the equation developed by Srinivasan et al. [49] to the transient polarization curves, which were obtained from a series of cell discharge voltage profiles measured at different current densities. The mass transport coefficient of the electrode could also be determined at the same time. The MRED method can provide valuable experimental data for the investigation of the structure-performance relationships for fuel cell electrodes, but its data processing includes some assumptions and empirical models, which could compromise the accuracy of the results.

8.2 PHYSICAL AND CHEMICAL METHODS

Due to the complexity of the electrochemical, fluid dynamics, and thermodynamic processes within the fuel cell and the confounding influences of different factors, electrochemical techniques alone cannot satisfy the various needs of PEM fuel cell diagnosis. In addition, most of the electrochemical techniques measure the average properties of the whole electrode, so local effects are hard to determine using these techniques. For industrial-sized PEM fuel cells, local effects are vitally important to both performance and durability. Therefore, mapping techniques become an important group of diagnostic tools, in addition to the physical and chemical methods developed to measure specific properties of fuel cell materials or specific processes within fuel cells. Here we present the physical and chemical methods according to three diagnostic categories: species, temperature, and current distribution mapping.

8.2.1 Species Distribution Mapping

8.2.1.1 Pressure-Drop Measurements

In general, the performance of a PEM fuel cell is stable within a relatively narrow operational conditions window, which is mainly related to water management issues inside the cell. To take the proper corrective action(s), it is necessary to have a simple, quick, reliable monitoring and diagnostic tool. Pressure drop can be an important design parameter and diagnostic tool, especially at the cathode, where the product water is produced. The cause of pressure drop is the friction between the reactant gases and the flow field passages and/or through the GDL (especially for interdigitated flow fields). Liquid water may be present inside the channel, either in the form of little droplets or as a film. Formation of larger water droplets effectively reduces the channel cross-sectional area and/or diverts the flow through other channels, in either case causing an increase in pressure drop. A pump or blower must provide any increase in pressure, resulting in an increase in parasitic power (load) loss. However, a higher pressure drop also results in more effective removal of excess liquid from the fuel cell. Therefore, the pressure drop must be carefully considered. In flooding conditions, too much liquid water in the flow channels increases gas flow resistance, which will impede reactant gas transport and lead to performance losses related to mass transport. Pressure drop on the cathode side increases with cell flooding, while it remains unchanged with cell drying, thus clearly distinguishing between the two phenomena. An increase in pressure drop, particularly on the cathode side of the PEM fuel cell, is a reliable indicator of PEM fuel cell flooding.

Pressure drops within fuel cells have been monitored to determine various parameters, such as flow field design (both anode and cathode), flooding, etc. For example, Barbir et al. [50] and He et al. [51] used pressure drop as a diagnostic tool for the detection of flooding in the fuel cell because of the strong dependence that the gas permeability of the porous electrodes has on liquid water content. In the above-mentioned studies the pressure drop was observed in a fuel cell with interdigitated flow fields under a number of operating conditions, which caused either flooding or drying inside the fuel cell. Another example of using pressure drop as a diagnostic tool is the method and apparatus designed by General Motors for monitoring H_2/O_2 fuel cells to detect and correct flooding [52]. In this method, pressure drop across a flow field (anode or cathode) is monitored and compared to predetermined thresholds of acceptability. If the pressure drop

exceeds the determined threshold, corrective measures, such as dehumidi-
fying the gases, increasing the gas mass flow rate, and reducing gas pressure
and/or reducing current drain, are automatically initiated.

Rodatz et al. [53] studied a large PEM fuel cell stack used in automotive
applications under different operating conditions. One of the main
parameters studied was the pressure drop within the stack and its relationship
to the flow field (bends in the channels) and the single- and two-phase flows.
It was observed that once the current in the fuel cell stack was reduced (i.e.,
by applying dry conditions) the pressure drop decreased slowly until it
reached a new value. This was attributed to the fact that the current
reduction reduces the flux of product water in the flow channels and thus
reduces the total mass flow in the flow channels. A transparent PEM fuel cell
with a single straight channel was designed by Ma et al. [54] to study liquid
water transport in the cathode channel. The pressure drop between the
channel inlet and outlet on the cathode side was used as a diagnostic signal to
monitor liquid water accumulation and removal. The proper gas velocities
for different currents were determined according to the pressure drop
curves. Pei et al. [55] studied the hydrogen pressure drop characteristics in
a PEM fuel cell to use the pressure drop as a diagnostic tool for prediction of
liquid water flooding in fuel cell stacks before flow channels have been
blocked.

8.2.1.2 Gas Composition Analysis

The species distribution within a PEM fuel cell is critical to fully characterize
the local performance and accurately quantify the various modes of water
transport. The most commonly used analytical technique for measuring the
gas composition within a fuel cell is *gas chromatography* (GC). Mench et al.
[56] demonstrated the measurement of water vapor, hydrogen, and oxygen
concentration distributions at steady state. A micro gas chromatograph was
utilized to measure the samples, which were extracted from eight different
sampling ports at various locations along the anode and cathode flow paths
of a specially designed fuel cell. While GC provides molar percent-level
accuracy for the species, it still requires on the order of 5 min per data point
and is consequently limited to analyzing steady-state species distribution.
This is why Dong et al. [57] used an Agilent Technologies real-time gas
analyzer that enabled the measurement of various species in near-real time
(about 1 s per data point). Another way to improve the accuracy and
repeatability of the technique shown by Mench et al. [56] was presented by
Yang et al. [58], who utilized two sets of multiposition microactuators and

micro GCs to measure gas concentration distributions on the anode and cathodes sides simultaneously. However, in general the weakness of utilizing GC to measure species distribution is that only discrete data at desired positions can be determined.

GC has also been employed in several studies to measure the crossover rate of hydrogen and oxygen through the membrane in PEM fuel cells. H_2 or O_2 is supplied in one side of the fuel cell and N_2 or He is fed to the opposite side, functioning as a carrier gas. The carrier gas transports the H_2 or O_2, permeating the membrane to a GC detector. Using this method, Broka et al. [59] investigated the crossover rates of hydrogen and oxygen through the Nafion 117 membrane and recast Nafion film at different temperatures and gas relative humidities. Recently, Liu et al. [60] measured the crossover rates of oxygen through Nafion/PTFE composite membranes by means of GC.

Mass spectrometry is another analytical technique capable of determining the gas composition in an operating fuel cell. Partridge et al. [61] utilized spatially resolved capillary inlet mass spectrometry to successfully conduct measurements (with a temporal resolution of 104 ms) at realistic humidity levels, despite the concern that liquid water could, throughout the active area of the fuel cell, block the capillaries used for gas sampling. They analyzed the effect of load switching on the species concentrations and observed concentration gradients and nonuniformities.

8.2.1.3 Neutron Imaging

To date, neutron imaging has been successfully used to visualize water dynamics in the flow channels of operating PEM fuel cells [62]. Because the neutron incoherent scattering length for hydrogen is nearly two orders of magnitude larger than the length for almost all other elements, neutrons are ideal for studying hydrogen-containing compounds such as water. Equipped with good-quality neutrons to probe the cell and a high-sensitivity scintillator/charge coupled device (CCD) camera as the detector system to record the images, the neutron imaging technique shows the potential for discerning two-phase flows within the fuel cell in real time or steady state.

Bellows et al. [63] first used the neutron imaging technique to measure water gradient profiles within the Nafion membrane of an operating PEM fuel cell. Preliminary neutron intensity gradients showed qualitative agreement with the expected response of the membrane water content to changes in feed gas humidification and fuel cell current. Geiger et al. [64] for the first time reported the gas/liquid two-phase flow patterns in the flow fields using neutron imaging. However, in the presented radiographs, the inadequate spatial resolution (500 μm/pixel) and low image acquisition rates (5 s/image), which included the exposure and read-out time for the CCD camera, limited the investigation of the two-phase flow to real time. In Satija et al.'s experimental design [65], spatial and temporal resolutions for CCD images were improved to 160 μm/pixel and 2 s/image, respectively. The water content in a PEM fuel cell as a function of time over 2000 s was presented, masking techniques were used to differentiate the location of the water in the cell, and tomography was performed to create a digital three-dimensional representation of the fuel cell stack. Figure 8-11 shows a typical setup for neutron imaging of a fuel cell [65], in which neutrons are converted to light using a scintillator screen and the light is focused on the CCD chip.

Recently, a neutron imaging technique developed at the Penn State Breazeale Nuclear Reactor significantly improved the spatial resolution to 129 μm/pixel and the temporal resolution to 30 images/s [66]. Results showed that liquid water had a tendency to accumulate at specific locations, depending on operating conditions. Kramer et al. [67] employed the neutron imaging technique to reveal the relationship between water content in a PEM fuel cell and operating current density, with a spatial resolution of 115 μm/pixel and exposure times ranging from 0.5 s to 15.0 s, and for the first time quantified liquid water in the GDL was reported with proper algorithms. More recently, rather than a CCD camera, a flat-panel amorphous silicon detector was utilized by Hickner et al. [68] to measure water content in a PEM fuel cell under varying current densities and operating temperatures. The images obtain by this detector could reach 127 μm/pixel in spatial resolution and 3 images/s in temporal resolution.

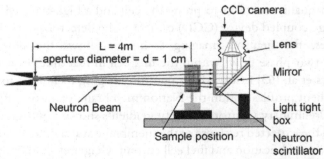

Figure 8-11 Typical experimental setup for neutron imaging of a fuel cell [65].

With ongoing improvements in spatial and temporal resolution, neutron imaging can be expected to play a greater role in any fuel cell development related to water transport. However, the requirement of a neutron source with a high fluence rate limits the wide application of neutron imaging. In addition, cell rotation is imperative for present tomographic imaging methods to gain three-dimensional information on an operating PEM fuel cell.

8.2.1.4 Magnetic Resonance Imaging

Magnetic resonance imaging (MRI), based on the nuclear magnetic resonance (NMR) phenomenon, is an imaging technique using gradient radio frequency (RF) pulses in a strong magnetic field. The basic principle of MRI is that certain atomic nuclei within an object, if placed in a magnetic field, can be stimulated by the correct RF pulses. After this stimulation the nuclei relax while energy is induced into a receiver antenna to further obtain a viewable image. The nondestructive nature of MRI enables one to obtain unique *in situ* information from a multitude of systems. As for a PEM fuel cell, the hydrogen atom is always observed in MRI experiments due to the excellent correlation between its signal intensity and water content in the membrane. According to Tsushima and his coworkers' MRI experiments, the effects of operating condition, membrane thickness, and liquid water supply on water distribution within polymer electrolyte membranes in fuel cells were demonstrated [69,70].

A few studies have used unique approaches to obtain increasing amounts of information. ^2H has been used as a labeling species in MRI, which is of particular interest because it does not produce a signal. Feindel et al. [71] switched between ^2H_2O and H_2O humidification and between an H_2 and $^2H_2$ fuel supply to observe the interactions and transport throughout the PEM (based on the decrease or increase in signal strength) at resolutions of 234 μm and 128 s/image. Dunbar and Masel [72] obtained three-dimensional images in miniature PEM fuel cells with MRI. Each voxel measured 138 μm × 138 μm × 200 μm and the acquisition time for each sequence was 259 s. Of interest is their observation that liquid water accumulated in stationary waves with occasional slipping, rather than the typical slug-type flow that was expected. However, their unique channel shape (1 mm wide × 3 mm deep) may have been the cause of this phenomenon. Zhang et al. [73] obtained high-resolution MRI images by using gold–plated, printed circuit board (PCB) flow fields as current collectors and as the resonating plates of the probe. This approach allowed visualization of the through-plane water distribution in the membrane with a resolution of 6 μm and an acquisition time of 360 s/image.

The weakness of the MRI technique is mainly in the requirement that the materials have to be nonmagnetic. For this reason the water content in the catalyst layer and GDL, made from either nonwoven carbon paper or woven carbon cloth, will be difficult to visualize with MRI.

8.2.1.5 X-Ray Imaging

Due to the inadequate spatial and temporal resolutions of the previously mentioned neutron imaging and MRI techniques, it is imperative to develop a high-resolution, *in situ* technique for imaging pore-scale flow and multiphase transport in individual components of PEM fuel cells. As a consequence, X-ray imaging has recently gained significant attention in academia and industry for detecting and quantifying water distribution within a fuel cell. The basic principle of X-ray imaging is that the intensity of an X-ray beam is attenuated as it traverses through a material. The transmitted radiation, received by an array of detectors, produces a two-dimensional or three-dimensional map based on the variation in X-ray adsorption throughout the sample.

The X-ray microtomography technique was pioneered by Sinha et al. [74] to obtain high-resolution, three-dimensional images of liquid water distribution in a GDL during gas purging. A comparatively high spatial resolution of 10 μm and temporal resolution of 0.07 s for a single image was achieved in their study. Synchrotron X-ray radiography was employed by Hartnig et al. [75] to investigate water evolution and liquid water transport from the CL through the GDL to the gas channels in an operating fuel cell at a microscopic level of 3 μm and a time resolution of 5 s. By means of the X-ray radiography technique, Lee et al. [76] quantitatively visualized the water distribution in the region between the flow plate and the GDL in a PEM fuel cell. A high spatial resolution of 9 μm was achieved in their study. Albertini et al. [77] also recently reported the hydration profile of a Nafion membrane in an operational fuel cell using X-ray radiography. The spatial resolution was 10 μm but the time resolution was relatively low (75 s).

X-ray imaging is a promising technique for the quantification of liquid water in different components of PEM fuel cells due to its resolution of ~10 μm. Though high spatial and temporal resolutions can be realized, X-ray imaging still needs much more development and refinement for fuel cell research. For example, further improvement in the spatial resolution down to 1–5 μm is preferable for the quantitative study of liquid water distribution in thin membranes, the CL, or the MPL.

8.2.1.6 Optically Transparent Fuel Cells

To delineate the origin and development of flooding with high spatial and temporal resolution in PEM fuel cells, recent research has resorted to transparent cell design based on optical diagnostics.

Weng et al. [78] designed a transparent PEM fuel cell to visualize and study the distribution of water and water flooding inside the cathode gas channels (serpentine channels) and to explain the phenomenon of membrane dehydration. The cell consisted of two transparent acrylic cover plates and two flow field plates (anode and cathode) made out of brass, as shown in Figure 8-12 [78]. Liu et al. [79] developed a number of transparent fuel cells, each with different flow fields, to study water flooding, two–phase flow of the reactant and products, and pressure drop in the cathode flow channels. Ge et al. [80] used a transparent cell to visualize liquid water and ice formation during startup of a fuel cell at subzero temperatures. More recently, Spernjak et al. [81] characterized *in situ* water dynamics using

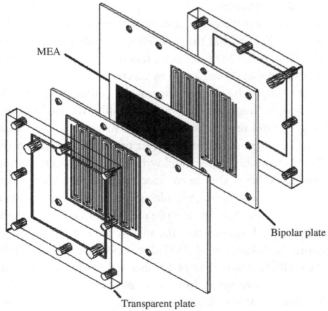

Figure 8-12 Schematic of a transparent PEM fuel cell [78].

optical imaging in parallel, single-serpentine, and interdigitated flow fields in PEM fuel cells. Chen [82] investigated the air/water two-phase flow behavior in parallel channels with porous media inserts by using a self-designed transparent assembly. The findings from his experiment suggested that flow channels with porous media inserts might be an effective design in handing the flow-distribution problem and optimizing water management in PEM fuel cells.

The effects of species other than water have also been investigated through the use of transparent cells. Inukai et al. [83] conducted novel work with an oxygen-sensitized dye complex dispersed in an oxygen-permeable polymer matrix. By irradiating the dye with a laser and recording the image with a CCD camera, they were able to visualize oxygen depletion from inlet to outlet. With the use of other dye materials, the cell could also be employed to visualize temperature, CO_2 concentration, and water vapor concentration, among others. Murahashi et al. [84] used a transparent cell in combination with EIS to study the effects of CO poisoning on an operating fuel cell. In addition to a near five-fold increase in impedance after 300 h at 100 ppm CO, the transition point to a two-phase flow moved toward the cathode exit for low humidity and high CO concentration conditions.

8.2.1.7 Embedded Sensors

Another method for mapping the distribution of species within PEM fuel cells is to embed sensors within the fuel cell to directly measure the presence of the species of interest. One promising technology is the use of fiber optic sensors, with a signal that is unaffected by electrical fields and a structure that should minimally affect the performance of the cell components. As part of the DOE Hydrogen Program, McIntyre et al. [85] developed two distinct fiber optic temperature probes (free space and monolithic) and a gas species sensor based on capillary mass spectrometer probe measurements (for H_2O, H_2, and O_2 concentrations) for deployment in fuel cells. Various other sensor technologies have also been successfully embedded in operating PEM fuel cells. Takaichi et al. [86] placed platinum wires between eight 25-μm-thick sheets of Nafion membrane to measure, throughout the resulting 200-μm-thick membrane, the mixed potential caused by reactant gas permeation. Nishikawa et al. [87] developed a cell with six humidity sensors in the cathode plates. The plates also included three current sensors, and they were able to correlate humidity and current distribution changes with the current density as well as the humidity and oxygen utilization. The results showed that high utilization levels shifted the highest performance towards the cathode inlet, while low humidity shifted the highest performance towards the outlet. Büchi and Reum [88] placed 10-μm-thick gold wires in between the CL and GDL. The wires, spaced 0.2 mm apart, acted as potential sensors to calculate local current density and membrane resistance (influenced by water content) at subchannel resolutions.

8.2.2 Temperature Distribution

Heat generation always accompanies the operation of a fuel cell due to inefficiencies in the basic fuel cell electrochemical reaction, fuel crossover (residual diffusion through the solid-electrolyte membrane), and electrical heat from interconnection resistances. Spatial temperature variation can occur if any of these heat-generating processes occur preferentially in different parts of the fuel cell stack.

8.2.2.1 IR Transparent Fuel Cells

In recent years, a number of studies on temperature distribution along the active area of a fuel cell have used infrared (IR) cameras. Hakenjos et al. [89] designed a cell for the combined measurement of current and temperature distribution; for the latter, an IR transparent window made out of zinc selenide was located at the cathode side of the fuel cell. Sailler et al. [90] conducted combined measurements of current density and temperature distribution. The current density was measured with magnetic sensors, whereas the temperature distribution was measured with an array of nine thermocouples and an IR camera. The current density and temperature distributions highlighted heterogeneous distribution influenced by cell geometry and collector locations.

8.2.2.2 Embedded Sensors

Temperature distribution mapping can also be obtained by inserting commercially available or specially fabricated fine temperature sensors into different positions in PEM fuel cells, including thermocouples, thermistors, in-fiber Bragg grating sensors, and bandgap temperature sensors. Wilkinson et al. [91] developed a simple, *in situ*, noninvasive method of measuring the temperature distribution of a fuel cell with micro-thermocouples. In this study, the thermocouples were located in the landing area of the flow field plates (in contact with the GDL of the MEA). The temperature data taken at different locations along the flow channel was then used to find each temperature slope, which in turn was related through mathematical equations to the local current density of each location. Thus, the current density

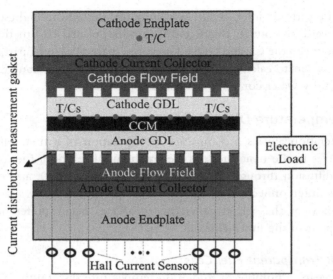

Figure 8-13 Schematic diagram of an experimental PEM fuel cell for simultaneous measurement of current and temperature distributions [92].

distribution in the fuel cell was determined via simple temperature measurements. More recently, Zhang et al. [92] measured simultaneously the in-plane current and temperature distributions in a PEM fuel cell by using a specially designed current distribution measurement gasket in the anode and 10 fine thermocouples between the catalyst layer and the GDL in the cathode. A single cell with an active area of 4 cm × 4 cm was used, and the experimental setup is shown schematically in Figure 8-13 [92].

Fiber optic sensors are an alternative to thermocouples as embedded temperature distribution mapping sensors. As described in Section 8.2.1.7, McIntyre et al. [85] recently developed two distinct fiber optic temperature probe technologies for fuel cell applications (free space probes and optical fiber probes). Both sensor technologies showed similar trends in fuel cell temperature and were also used to study transient conditions.

8.2.3 Current Distribution Mapping

Interesting and useful information about the inner workings of a fuel cell may be obtained by current density mapping. In PEM fuel cells, uniformity of the current density across the entire active area is critical for optimizing fuel cell performance. A nonuniform current density in the fuel cell can drastically affect different parameters, such as reduced reactant and catalyst utilization along the active area, decrease in total efficiency and lifetime, and durability failure modes. Thus, determination of the current density distribution information is vitally useful during the cell design process, particularly for verification and calibration of the numerical models. A number of methods for measuring current distribution in PEM fuel cells have been demonstrated, and the following sections discuss some of these in further detail.

8.2.3.1 Partial MEA

The partial MEA approach involves the use of several MEAs, each with different sections of the catalyzed active area covered, thus reducing the total active area of each MEA [93]. Appointed sectional performance can be achieved by subtracting one steady-state polarization curve from another. The main advantage of this method is that it is relatively simple to implement with respect to other techniques. In addition, this technology allows cell performance to be analyzed in a steady state, albeit with low spatial resolution. Although the resolution can be improved by increasing the number of segments or portions to be tested, significant errors arise due to inherent variations in electrical, transport, and kinetic properties between different MEAs.

8.2.3.2 Segmented Cells

A number of research groups have presented segmented cell approaches and combined them with electrochemical methods, for example, EIS. These diagnostic approaches provide direct information not just on the current distribution of the cell but also on other phenomena that are occurring inside the cell under various operating conditions. To date, a parallel effort has been made to achieve sufficient spatial and temporal resolution by designing segmented flow-field plates. The basic concept of the segmented cell (or segmented flow-field plate) approach is to divide the anode and/or cathode plate into conductive segments that are electrically isolated from each other.

Subcells

Stumper et al. [93] presented the subcell approach to measure localized currents and localized electrochemical activity in a fuel cell. In this method a number of subcells were situated in different locations along the cell's active area, and each subcell was electrically insulated from the others and from the main cell. Separate load banks controlled each subcell. The current–voltage characteristics for the subcells, compared to those of the main cell, are indicative of local fuel cell performance [93]. Although this method provides a good understanding of the current distribution along a flow field, the manufacturing of the modified flow-field plates and MEA makes it very complex and difficult. Recently, Mench et al. [94], Yang et al. [95], and Rajalakshmi et al. [96] also used the subcell approach to determine the current distributions under various operating conditions in PEM fuel cells with a single serpentine flow field.

Segmented Plates

In order to have spatially resolved performance data, Stumper et al. [93] used a passive resistor network made from resin-isolated graphite blocks located between the flow-field plate and the current collector. The potential drops across these blocks were monitored using Ohm's law to establish the current flowing through them. By scanning the entirety of the graphite blocks, they mapped the electrode's current distribution. One of the main advantages to this approach is that time-dependent phenomena can be monitored in real time, enabling researchers to observe sudden changes after certain parameters have been modified. But a key issue about this configuration arises from the fact that the graphite blocks are not part of the flow-field plate; thus, with these passive resistors there is the possibility of low spatial resolution due to the lateral in-plane current through the flow-field plate. The setup of these resistors also makes this approach quite complex and tedious. In another study done by Wieser et al. [97], a magnetic loop array was employed to determine the currents generated by segmented fuel cells. A flow-field plate divided into 40 electrically isolated segments was used to avoid lateral currents between adjacent segments. The current sensors, which consisted of Hall sensors fixed in the air gaps of annular soft magnetic ferrites, were placed around a gudgeon on each flow-field segment. The advantage of this method is that a combination of high spatial and time resolution can be achieved and, in principle, integration into fuel cell stacks is possible. However, installation of the Hall sensors into the flow-field plate makes maintenance difficult and the investigation of different flow-field structures expensive.

Printed Circuit Board

Cleghorn et al. [98] performed pioneering work using printed circuit board (PCB) technology to create a segmented anode current collector and anode flow field to measure current distribution in PEM fuel cells. In this study, the gas diffusion backing and catalyst layer in the anode side were also segmented and each segment had an active area of 4.4 cm^2. Figure 8-14

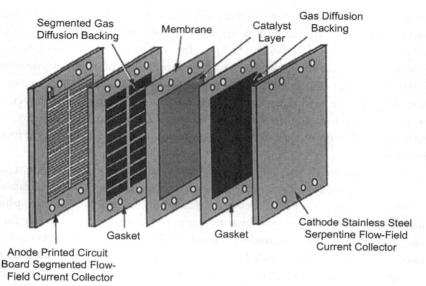

Figure 8-14 Schematic of the segmented fuel cell design using printed circuit board (PCB) technology [98].

shows this segmented fuel cell designed using PCB technology. Through the use of current distribution they were able to investigate different flow-field designs and optimize utilization of the active electrode area with the best humidification conditions and reactant stoichiometries. It is important to note that some of the mentioned segmented cell approaches have also been used to develop other diagnostic tools. For example, Brett et al. [99] used a segmented cell to demonstrate localized EIS response over a frequency range of 0.1 Hz to 10 kHz as a function of position in a PEM fuel cell. Their results proved that integral EIS measurement alone was not sufficient to properly characterize the operation of a fuel cell.

8.3 CONCLUSIONS

This chapter has reviewed various diagnostic techniques currently employed in PEM fuel cell research and combines the work of many researchers. All the diagnostic approaches explained here were divided into two main categories: electrochemical techniques and chemical and physical methods. Due to the complexity of the electrodes' behavior within the fuel cell and the confounding influences of different factors, a variety of techniques are needed for different purposes. The polarization curve is, nevertheless, the simplest and easiest way of characterizing the fuel cell. EIS is considered an effective technique for investigating electrochemical systems. This method allows the separate examination of different processes in the PEM fuel cell, such as anode kinetics, anode mass transport, cathode kinetics, cathode mass transport, and membrane conductivity. Moreover, it is an effective tool for determining electrode parameters as functions of structure and composition, which is a prerequisite for optimizing electrode structure.

As for chemical and physical methods, the level of development of relevant equipment and materials plays an important role in the application of the techniques to PEM fuel cells. For example, neutron radiographic imaging has already proven to be a unique experimental method to visualize and quantify liquid water inside an operating PEM fuel cell. However, it is currently limited in both spatial and temporal resolution, making it difficult to capture transient two-phase flow phenomena. Further, it is difficult to differentiate between liquid transport inside the anode and cathode and their individual components, and the technique is limited to locations with applicable neutron sources. Other visualization approaches that help to understand phenomena within a fuel cell have been discussed (IR tests for thermal imaging and transparent cells for liquid water visualization). However, these methods also fail to give a better understanding of the transport mechanisms of liquid water in microstructures (i.e., inside and through the GDL).

Current density distribution along the active area in a fuel cell is a very critical parameter that needs to be understood completely to improve the design of each cell component. Hitherto, one limitation prevailing in much of the published work on segmented cells is the inability to produce a similar level of current density as is possible with nonsegmented cells. Difficulties in the design, fabrication, and assembly of the segmented cells may result in higher and nonuniform contact resistance between components of the fuel cell, which leads to lower performance. Another limitation is that the test conditions of many studies are far different from a typical automotive drive cycle. Only when these limitations are conquered can fuel cell redesign approaches have the potential to become standard diagnostic tools for fuel cell research and development.

As shown in this chapter, test equipment integrated with several diagnostic techniques is preferable for a deeper insight into the mechanisms that cause performance losses and spatially nonuniform distribution. As a consequence, more information, obtained simultaneously with these various diagnostic tools, will strongly support the development of empirical models or validate theoretical models predicting performance as a function of operating conditions and fuel cell characteristic properties.

PROBLEMS

1. Assume the polarization data in the following table fits the equation:

$$V_{cell} = V_0 - b \ln(i/i_0) - Ri$$

Apply the least-square method to estimate the values of the parameters in the preceding equation, namely V_0, b, i_0, and R.

i (mA cm^{-2})	5	10	20	50	100	200	400	600	800
V (V)		0.933	0.912	0.891	0.859	0.830	0.791	0.734	0.683	0.635

2. Based on Equation (8-1), derive the equation for calculating the electrochemical surface area by means of CO stripping voltammetry.
3. Compare the advantages and disadvantages of each of the current distribution mapping techniques.

QUIZ

1. Hysteresis of the fuel cell polarization curve is caused by:
   a. Either cell drying or flooding
   b. Cell flooding only
   c. Cell resistance
2. The electrochemical surface area of the catalyst layer can be measured by:
   a. Cyclic voltammetry
   b. CO stripping voltammetry
   c. Both (a) and (b)

3. In the current interruption measurement, voltage drop due to cell resistance is:
   a. The remaining cell voltage immediately after current interruption
   b. The difference between the cell voltage immediately after and before current interruption
   c. The difference between the cell voltage immediately after current interruption and the open circuit voltage
4. A fuel cell equivalent circuit consists of:
   a. At least one resistor and one capacitor
   b. At least two resistors and one capacitor
   c. At least two capacitors and one resistor
5. A Nyquist diagram plots:
   a. Frequency vs. phase angle
   b. Cell impedance vs. cell resistance
   c. Imaginary vs. real portion of the impedance
6. Cell resistance may be measured:
   a. With low-frequency AC impedance
   b. With high-frequency AC impedance
   c. Only with the entire frequency spectrum
7. Which of the following methods can be used to check the cell resistance in a nonoperational fuel cell?
   a. Current interruption
   b. Magnetic resonance imaging
   c. Electrochemical impedance spectroscopy
8. In a Nyquist diagram, the cell resistance is represented by:
   a. The radius of the resulting semicircle
   b. The distance between the origin and the semicircle's center
   c. The distance between the origin and the closest point on the semicircle
9. Cell flooding is likely to result in:
   a. Increase in pressure drop through the cathode and no changes in cell resistance
   b. No changes in pressure drop and decrease in cell resistance
   c. Increase in both pressure drop and cell resistance
10. Which of the following methods cannot be used to visualize water dynamics in an operating PEM fuel cell?
    a. Neutron imaging
    b. Magnetic resonance imaging
    c. Pressure drop measurement

REFERENCES

[1] Wang CY. Fundamental Models for Fuel Cell Engineering. Chemical Reviews 2004;104(10):4727–66.

[2] Hinds G. Performance And Durability of PEM Fuel Cells, 2004 NPL Report DEPC-MPE 002. Teddington, UK: National Physical Laboratory; 2004.

[3] Lim CY, Haas HR. A Diagnostic Method for an Electrochemical Fuel Cell and Fuel Cell Components. WO Patent March 2006. 2,006,029,254.

[4] Wu JF, Yuan XZ, Wang HJ, Blanco M, Martin JJ, Zhang JJ. International Journal of Hydrogen Energy 2008;33(6):1735–46.

[5] Ticianelli EA, Derouin CR, Redondo A, Srinivasan S. Methods to Advance Technology of Proton Exchange Membrane Fuel Cells. Journal of the Electrochemical Society 1998;135(9):2209–14.

[6] Wruck WJ, Machado RM, Chapman TW. Current Interruption-Instrumentation and Applications. Journal of the Electrochemical Society 1987;134:539–46.

[7] Büchi FN, Marek A, Scherer GG. Situ Membrane Resistance Measurements in Polymer Electrolyte Fuel Cells by Fast Auxiliary Current Pulses. Journal of the Electrochemical Society 1995;142:1895–901.

[8] Jaouen F, Lindbergh G, Wiezell K. Transient Techniques for Investigating Mass-transport Limitations in Gas Diffusion Electrodes. Journal of the Electrochemical Society 2003;150:A1711–7.

[9] Abe T, Shima H, Watanabe K, Ito Y. Study of PEFCs by AC Impedance, Current Interrupt, and Dew Point Measurements. Journal of the Electrochemical Society 2004;151:A101–5.

[10] Noponen M, Hottinen T, Mennola T, Mikkola M, Lund P. Determination of Mass Diffusion Overpotential Distribution with Flow Pulse Method from Current Distribution Measurements in a PEMFC. Journal of Applied Electrochemistry 2002;32(10):1081–9.

[11] Mennola T, Mikkola M, Noponen M. Measurement of Ohmic Voltage Losses in Individual Cells of a PEMFC Stack. Journal of Power Sources 2002;112(1):261–72.

[12] Smith M, Johnson D, Scribner L. Electrical Test Method for Evaluating Fuel Cell MEA Resistance. Fuel Cell Magazine March 2004:15.

[13] Yuan XZ, Sun JC, Blanco M, Wang HJ, Zhang JJ, Wilkinson DP. AC Impedance Diagnosis of a 500 W PEM Fuel Cell Stack: Part I: Stack Impedance. Journal of Power Sources 2006;161(2):920–8.

[14] Mathias MF, Grot SA. System and Method for Controlling the Humidity Level of a Fuel Cell. US Patent April 2002. 6,376,111.

[15] Barbir F, Gorgun H, Wang X. Relationship Between Pressure Drop and Cell Resistance as a Diagnostic Tool for PEM Fuel Cells. Journal of Power Sources 2005;141(1):96–101.

[16] Oszcipok M, Riemann D, Kronenwett U, Kreideweis M, Zedda M. Statistic Analysis of Operational Influences on the Cold Start Behaviour of PEM Fuel Cells. Journal of Power Sources 2005;145(2):407–15.

[17] Ciureanu M, Roberge R. Electrochemical Impedance Study of PEM Fuel Cells. Experimental Diagnostics and Modeling of Air Cathodes. Journal of Physical Chemistry B 2001;105(17):3531–9.

[18] Eikerling M, Kornyshev AA. Electrochemical Impedance of the Cathode Catalyst Layer in Polymer Electrolyte Fuel Cells. Journal of Electroanalytical Chemistry 1999;475(2):107–23.

[19] Springer TE, Zawodzinski TA, Wilson MS, Gottesfeld S. Characterization of Polymer Electrolyte Fuel Cells Using AC Impedance Spectroscopy. Journal of the Electrochemical Society 1996;143(2):587–99.

[20] Makharia R, Mathias MF, Baker DR. Measurement of Catalyst Layer Electrolyte Resistance in PEFCs Using Electrochemical Impedance Spectroscopy. Journal of the Electrochemical Society 2005;152(5):A970–7.

[21] Lefebvre MC, Martin RB, Pickup PG. Characterization of Ionic Conductivity Profiles within Proton Exchange Membrane Fuel Cell Gas Diffusion Electrodes by Impedance Spectroscopy. Electrochemical and Solid-State Letters 1999;2: 259–61.

[22] Jia N, Martin RB, Qi Z, Lefebvre MC, Pickup PG. Modification of Carbon Supported Catalysts to Improve Performance in Gas Diffusion Electrodes. Electrochimica Acta 2001;46(18):2863–9.

[23] Easton EB, Pickup PG. An Electrochemical Impedance Spectroscopy Study of Fuel Cell Electrodes. Electrochimica Acta 2005;50(12):2469–74.

[24] Freire TJP, Gonzalez ER. Effect of Membrane Characteristics and Humidification Conditions on the Impedance Response of Polymer Electrolyte Fuel Cells. Journal of Electroanalytical Chemistry 2001;503(1–2):57–68.

[25] Cha SW, O'Hayre R, Prinz FB. AC Impedance Investigation of Transport Phenomena in Micro Flow Channels in Fuel Cells. Proceedings of Second International Conference on Fuel Cell Science, Engineering and Technology. Rochester, NY: American Society of Mechanical Engineers; June 14–16, 2004.

[26] Wagner N, Gülzow E. Change of Electrochemical Impedance Spectra (EIS) with Time during CO-poisoning of the Pt-anode in a Membrane Fuel Cell. Journal of Power Sources 2004;127(1–2):341–7.

[27] Schiller CA, Richter F, Gülzow E, Wagner N. Relaxation Impedance as a Model for the Deactivation Mechanism of Fuel Cells due to Carbon Monoxide Poisoning. Physical Chemistry Chemical Physics 2001;3(11):2113–6.

[28] Schiller CA, Richter F, Gülzow E, Wagner N. Validation and Evaluation of Electrochemical Impedance Specta of Systems with States That Change with Time. Physical Chemistry Chemical Physics 2001;3(3):374–8.

[29] Leng YJ, Wang X, Hsing IM. Assessment of CO-tolerance for Different Pt-alloy Anode Catalysts in a Polymer Electrolyte Fuel Cell Using AC Impedance Spectroscopy. Journal of Electroanalytical Chemistry 2002;528(1–2):145–52.

[30] Rubio MA, Urquia A, Dormido S. Diagnosis of Performance Degradation Phenomena in PEM Fuel Cells. International Journal of Hydrogen Energy 2010;35(7):2586–90.

[31] Yuan XZ, Sun JC, Wang HJ, Zhang JJ. AC Impedance Diagnosis of a 500 W PEM Fuel Cell Stack: Part II: Individual Cell Impedance. Journal of Power Sources 2006;162(2):929–37.

[32] Hakenjos A, Zobel M, Clausnitzer J, Hebling C. Simultaneous Electrochemical Impedance Spectroscopy of Single Cells in a PEM Fuel Cell Stack. Journal of Power Sources 2006;154(2):360–3.

[33] Paganin VA, Oliveira CLF, Ticianelli EA, Springer TE, Gonzalez ER. Modelistic Interpretation of the Impedance Response of a Polymer Electrolyte Fuel Cell. Electrochimica Acta 1998;43(24):3761–6.

[34] Andreaus B, McEvoy AJ, Scherer GG. Analysis of Performance Losses in Polymer Electrolyte Fuel Cells at High Current Densities by Impedance Spectroscopy. Electrochimica Acta 2002;47(13-14):2223–9.

[35] Barsoukov E, Macdonald JR. Impedance Spectroscopy: Theory, Experiment and Applications. New York: John Wiley & Sons; 2005.

[36] O'Hayre R, Cha SW, Colella W, Brinz FB. Fuel Cell Fundamentals. New York: Wiley; 2006.

[37] Ralph TR, Hards GA, Keating JE, Campbell SA, Wilkinson DP, David M, et al. Low Cost Electrodes for Proton Exchange Membrane Fuel Cells. Journal of the Electrochemical Society 1997;144(11):3845–57.

[38] Brett DJL, Atkins S, Brandon NP, Vesovic V, Vasileiadis N, Kucernak AR. Investigation of Reactant Transport within a Polymer Electrolyte Fuel Cell using Localised CO Stripping Voltammetry and Adsorption Transients. Journal of Power Sources 2004;133(2):205–13.

[39] Pozio A, De Francesco M, Cemmi A, Cardellini F, Giorgi L. Comparison of High Surface Pt/C Catalysts by Cyclic Voltammetry. Journal of Power Sources 2002;105(1):13–9.

[40] Mukerjee S, Lee SJ, Ticianelli EA, McBreen J, Grgur BN, Markovic NM, et al. Investigation of Enhanced CO Tolerance in Proton Exchange Membrane Fuel Cells by Carbon Supported PtMo Alloy Catalyst. Electrochemical and Solid-State Letters 1999;2:12–5.

[41] Song SQ, Liang ZX, Zhou WJ, Sun GQ, Xin Q, Stergiopoulos V, et al. Direct Methanol Fuel Cells: The Effect of Electrode Fabrication Procedure on MEAs Structural Properties and Cell Performance. Journal of Power Sources 2005;145(2):495–501.

[42] Xu Z, Qi Z, Kaufman A. Activation of Proton-exchange Membrane Fuel Cells via CO Oxidative Stripping. Journal of Power Sources 2006;156(2):281–3.

[43] Song Y, Fenton JM, Kunz HR, Bonville LJ, Williams MV. High-performance PEMFCs at Elevated Temperatures using Nafion 112 Membranes. Journal of the Electrochemical Society 2005;152(3):A539–44.

[44] Kocha SS, Yang JD, Yi JS. Characterization of Gas Crossover and Its Implications in PEM Fuel Cells. AIChE Journal 2006;52(5):1916–25.

[45] Liu W, Singh A, Rusch G, Ruth K. The Membrane Durability in PEM Fuel Cells. Journal of New Materials for Electrochemical Systems 2001;4(4):227–31.

[46] Yu J, Matsuura T, Yoshikawa Y, Islam MN, Hori M. In Situ Analysis of Performance Degradation of a PEMFC under Nonsaturated Humidifications. Electrochemical and Solid-State Letters 2005;8:A156–8.

[47] Wu JF, Yuan XZ, Martin JJ, Wang HJ, Yang DJ, Qiao JL, et al. Proton Exchange Membrane Fuel Cell Degradation under Close to Open-circuit Condition: Part I: In Situ Diagnosis. Journal of Power Sources 2010;195(4):1171–6.

[48] Stumper J, Haas H, Granados A. In Situ Determination of MEA Resistance and Electrode Diffusivity of a Fuel Cell. Journal of the Electrochemical Society 2005;152(4):A837–44.

[49] Srinivasan S, Velev OA, Parthasarathy A, Manko DJ, Appleby AJ. High Energy Efficiency and High Power Density Proton Exchange Membrane Fuel Cells - Electrode Kinetics and Mass Transport. Journal of Power Sources 1991;36(3):299–320.

[50] Barbir F, Gorgun HH, Wang X. Relationship between Pressure Drop and Cell Resistance as a Diagnostic Tool for PEM Fuel Cells. Journal of Power Sources 2005;141(1):96–101.

[51] He W, Lin G, Nguyen TV. Diagnostic Tool to Detect Electrode Flooding in Proton-exchange-membrane Fuel Cells. AIChE Journal 2003;49(12):3221–8.

[52] Bosco AD, Fronk MH. Fuel Cell Flooding Detection and Correction. US Patent August 2000. 6,103,409.

[53] Rodatz P, Büchi F, Onder C, Guzzella L. Operational Aspects of a Large PEFC Stack under Practical Conditions. Journal of Power Sources 2004;128(2):208–17.

[54] Ma HP, Zhang HM, Hu J, Cai YH, Yi BL. Diagnostic Tool to Detect Liquid Water Removal in the Cathode Channels of Proton Exchange Membrane Fuel Cells. Journal of Power Sources 2006;162(1):469–73.

[55] Pei P, Ouyang M, Feng W, Lu L, Huang H, Zhang J. Hydrogen Pressure Drop Characteristics in a Fuel Cell Stack. International Journal of Hydrogen Energy 2006;31(3):371–7.

[56] Mench MM, Dong QL, Wang CY. *In Situ* Water Distribution Measurements in a Polymer Electrolyte Fuel Cell. Journal of Power Sources 124(1):90–98.

[57] Dong Q, Kull J, Mench MM. Real-time Water Distribution in a Polymer Electrolyte Fuel Cell. Journal of Power Sources 2005;139(1):106–14.

[58] Yang XG, Burke N, Wang CY, Tajiri K, Shinohara K. Simultaneous Measurements of Species and Current Distributions in a PEFC under Low-humidity Operation. Journal of the Electrochemical Society 2005;152(4):A759–66.

[59] Broka K, Ekdunge P. Oxygen and Hydrogen Permeation Properties and Water Uptake of Nafion® 117 Membrane and Recast Film for PEM Fuel Cell. Journal of Applied Electrochemistry 1997;27(2):117–23.

[60] Liu F, Yi B, Xing D, Yu J, Zhang H. Nafion/PTFE Composite Membranes for Fuel Cell Applications. Journal of Membrane Science 2003;212(1–2): 213–23.

[61] Partridge WP, Toops TJ, Green JB, Armstrong TR. Intra-Fuel Cell Stack Measurements of Transient Concentration Distributions. Journal of Power Sources 2006;160(1):454–61.

[62] Mukundan R, Borup RL. Visualising Liquid Water in PEM Fuel Cells Using Neutron Imaging. Fuel Cells 2009;9(5):499–505.

[63] Bellows RJ, Lin MY, Arif M, Thompson AK, Jacobson D. Neutron Imaging Technique for *In Situ* Measurement of Water Transport Gradients within Nafion in Polymer Electrolyte Fuel Cells. Journal of the Electrochemical Society 1999;146(3):1099–103.

[64] Geiger AB, Tsukada A, Lehmann E, Vontobel P, Wokaun A, Scherer GG. *In Situ* Investigation of Two-phase Flow Patterns in Flow Fields of PEFCs Using Neutron Radiography. Fuel Cells 2002;2(2):92–8.

[65] Satija R, Jacobson DL, Arif M, Werner SA. Situ Neutron Imaging Technique for Evaluation of Water Management Systems in Operating PEM Fuel Cells. Journal of Power Sources 2004;129(2):238–45.

[66] Kowal JJ, Turhan A, Heller K, Brenizer J, Mench MM. Liquid Water Storage, Distribution, and Removal from Diffusion Media in PEFCs. Journal of the Electrochemical Society 2006;153(10):A1971–8.

[67] Kramer D, Zhang J, Shimoi R, Lehmann E, Wokaun A, Shinohara K, et al. In Situ Diagnostic of Two-phase Flow Phenomena in Polymer Electrolyte Fuel Cells by Neutron Imaging: Part A. Experimental, Data Treatment, and Quantification. Electrochimica Acta 2005;50(13):2603–14.

[68] Hickner MA, Siegel NP, Chen KS, Mcbrayer DN, Hussey DS, Jacobson DL, et al. Real-time Imaging of Liquid Water in an Operating Proton Exchange Membrane Fuel Cell. Journal of the Electrochemical Society 2006; 153(5):A902–8.

[69] Teranishi K, Tsushima S, Hirai S. Analysis of Water Transport in PEFCs by Magnetic Resonance Imaging Measurement. Journal of the Electrochemical Society 2006;153(4):A664–8.

[70] Tsushima S, Teranishi K, Hirai S. Magnetic Resonance Imaging of the Water Distribution Within a Polymer Electrolyte Membrane in Fuel Cells. Electrochemical and Solid-State Letters 2004;7:A269–72.

[71] Feindel KW, Bergens SH, Wasylishen RE. Use of Hydrogen–Deuterium Exchange for Contrast in ^{1}H NMR Microscopy Investigations of an Operating PEM Fuel Cell. Journal of Power Sources 2007;173(1):86–95.

[72] Dunbar Z, Masel RI. Magnetic Resonance Imaging Investigation of Water Accumulation and Transport in Graphite Flow Fields in a Polymer Electrolyte Membrane Fuel Cell: Do Defects Control Transport? Journal of Power Sources 2008;182(1):76–82.

[73] Zhang Z, Martin J, Wu J, Wang H, Promislow K, Balcom BJ. Magnetic Resonance Imaging of Water Content Across the Nafion Membrane in an Operational PEM Fuel Cell. Journal of Magnetic Resonance 2008;193(2):259–66.

[74] Sinha PK, Halleck P, Wang CY. Quantification of Liquid Water Saturation in a PEM Fuel Cell Diffusion Medium Using X-ray Microtomography. Electrochemical and Solid-State Letters 2006;9(7):A344–8.

[75] Hartnig C, Manke I, Kuhn R, Kleinau S, Goebbels J, Banhart J. High-resolution In-plane Investigation of the Water Evolution and Transport in PEM Fuel Cells. Journal of Power Sources 2009;188(2):468–74.

[76] Lee SJ, Lim NY, Kim S, Park GG, Kim CS. X-ray Imaging of Water Distribution in a Polymer Electrolyte Fuel Cell. Journal of Power Sources 2008;185(2): 867–70.

[77] Albertini VR, Paci B, Nobili F, Marassi R, Michiel MD. Time/Space-Resolved Studies of the Nafion Membrane Hydration Profile in a Running Fuel Cell. Advanced Materials 2009;21(5):578–83.

[78] Weng FB, Su A, Hsu CY, Lee CY. Study of Water-flooding Behaviour in Cathode Channel of a Transparent Proton-exchange Membrane Fuel Cell. Journal of Power Sources 2006;157(2):674–80.

[79] Liu X, Guo H, Ma CF. Water Flooding and Two-phase Flow in Cathode Channels of Proton Exchange Membrane Fuel Cells. Journal of Power Sources 2006;156(2): 267–80.

[80] Ge S, Wang CY. Situ Imaging of Liquid Water and Ice Formation in an Operating PEFC during Cold Start. Electrochemical and Solid-State Letters 2006;9(11):A499–503.

[81] Spernjak D, Prasad AK, Advani SG. In Situ Comparison of Water Content and Dynamics in Parallel, Single-serpentine, and Interdigitated Flow Fields of Polymer Electrolyte Membrane Fuel Cell. Journal of Power Sources 2010;195(11): 3553–68.

[82] Chen J. Experimental Study on the Two-Phase Flow Behavior in PEM Fuel Cell Parallel Channels with Porous Media Inserts. Journal of Power Sources 2010;195(4):1122–9.

[83] Inukai J, Miyatake K, Takada K, Watanabe M, Hyakutake T, Nishide H, et al. Direct Visualization of Oxygen Distribution in Operating Fuel Cells. Angewandte Chemie International Edition 2008;47(15):2792–5.

[84] Murahashi T, Kobayashi H, Nishiyama E. Combined Measurement of PEMFC Performance Decay and Water Droplet Distribution under Low Humidity and High CO. Journal of Power Sources 2008;175(1):98–105.

[85] McIntyre TJ, Allison SW, Maxey LC, Partridge WP, Cates MR, Lenarduzzi R, et al. Fiber Optic Temperature Sensors for PEM Fuel Cells. DOE Hydrogen Program FY 2005 Progress Report 2005:989–99.

[86] Takaichi S, Uchida H, Watanabe M. Distribution Profile of Hydrogen and Oxygen Permeating in Polymer Electrolyte Membrane Measured by Mixed Potential. Electrochemistry Communications 2007;9(8):1975–9.

[87] Nishikawa H, Kurihara R, Sukemori S, Sugawara T, Kobayasi H, Abe S, et al. Measurements of Humidity and Current Distribution in a PEFC. Journal of Power Sources 2006;155(2):213–8.

[88] Büchi FN, Reum M. Measurement of the Local Membrane Resistance in Polymer Electrolyte Fuel Cells (PEFC) on the Sub-mm Scale. Measurement Science and Technology 2008;19:085702–7.

[89] Hakenjos A, Muenter H, Wittstadt U, Hebling C. A PEM Fuel Cell for Combined Measurement of Current and Temperature Distribution, and Flow Field Flooding. Journal of Power Sources 2004;131(1-2):213–6.

[90] Sailler S, Rosini S, Chaib MA, Voyant JY, Bultel Y, Druart F, et al. Electrical and Thermal Investigation of a Self-Breathing Fuel Cell. Journal of Applied Electrochemistry 2007;37(1):161–71.

[91] Wilkinson M, Blanco M, Gu E, Martin JJ, Wilkinson DP, Zhang JJ, et al. *In Situ* Experimental Technique for Measurement of Temperature and Current Distribution in Proton Exchange Membrane Fuel Cell. Electrochemical and Solid-State Letters 2006;9:A507–11.

[92] Zhang G, Guo L, Ma L, Liu H. Simultaneous Measurement of Current and Temperature Distributions in a Proton Exchange Membrane Fuel Cell. Journal of Power Sources 2010;195(11):3597–604.

[93] Stumper J, Campbell SA, Wilkinson DP, Johnson MC, Davis M. *In situ* Methods for the Determination of Current Distributions in PEM Fuel Cells. Electrochimica Acta 1998;43(24):3773–83.

[94] Mench MM, Wang CY. In Situ Current Distribution Measurements in Polymer Electrolyte Fuel Cells. Journal of the Electrochemical Society 2003;150(8):A1052–9.

[95] Yang XG, Burke N, Wang CY, Tajiri K, Shinohara K. Simultaneous Measurements of Species and Current Distributions in a PEFC under Low-Humidity Operation. Journal of the Electrochemical Society 2005;152(4):A759–66.

[96] Rajalakshmi N, Raja M, Dhathathreyan KS. Evaluation of Current Distribution in a Proton Exchange Membrane Fuel Cell by Segmented Cell Approach. Journal of Power Sources 2002;112(1):331–6.

[97] Wieser C, Helmbold A, Gülzow E. A New Technique for Two-dimensional Current Distribution Measurements in Electrochemical Cells. Journal of Applied Electrochemistry 2000;30(7):803–7.

[98] Cleghorn SJC, Derouin CR, Wilson MS, Gottesfeld S. A Printed Circuit Board Approach to Measuring Current Distribution in a Fuel Cell. Journal of Applied Electrochemistry 1998;28(7):663–72.

[99] Brett DJL, Atkins S, Brandon NP, Vesovic V, Vasileiadis N, Kucernak A. Localized Impedance Measurements along a Single Channel of a Solid Polymer Fuel Cell. Electrochemical and Solid-State Letters 2003;6:A63–6.

CHAPTER NINE

Fuel Cell System Design

A *system* is defined as a group of units, objects, or items so combined as to form a whole and to operate in unison. In the case of a fuel cell, the system includes all the components needed to operate a fuel cell stack and deliver electrical current. A fuel cell stack is obviously the heart of a fuel cell system; however, without the supporting equipment the stack itself would not be very useful. The fuel cell system typically involves the following subsystems:

- Oxidant supply (oxygen or air)
- Fuel supply (hydrogen or hydrogen rich gas)
- Heat management
- Water management
- Power conditioning
- Instrumentation and controls

Depending on the available or chosen fuel and oxidant, the fuel cell systems may be categorized as:

- Hydrogen/oxygen systems
- Hydrogen/air systems
- Reformate/air systems

9.1 HYDROGEN/OXYGEN SYSTEMS

Because of technical difficulties, the added size and weight of oxygen storage, and related safety concerns, pure oxygen systems are typically used only in applications where air is not available, such as in submarines and space applications.

9.1.1 Oxygen Supply

Stored oxygen is already under pressure, so the supply of oxygen to a fuel cell involves a pressure regulator to reduce the pressure to the fuel cell. Oxygen should be supplied in excess of that stoichiometrically required, typically with stoichiometric ratios of 1.2 to 1.3 (i.e., 20% to 30% excess). The reason for this is that the excess oxygen has to carry the product water from the cell. Excess oxygen may be vented at the fuel cell exit, but most practical systems operate in a closed loop configuration, where excess

PEM Fuel Cells
ISBN 978-0-12-387710-9

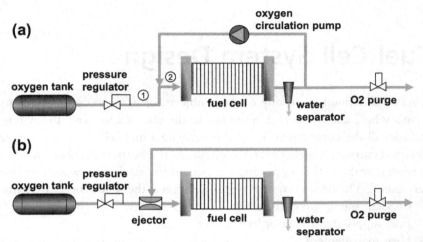

Figure 9-1 Closed loop oxygen supply system: (a) with circulation pump; (b) with ejector.

oxygen is returned back to the stack inlet. An active (pump) or a passive (ejector) device is needed to bring the gas from the low pressure at the stack exit to a higher pressure at the stack inlet. Liquid water at the exhaust may be separated in a simple water/gas separator, and the warm, saturated gas from the exit may be mixed with a dry gas from the tank to obtain desired humidity at the stack inlet. The resulting oxygen supply system is shown in Figure 9-1.

From the mass and energy balance it is possible to calculate the achievable oxygen inlet temperatures for various combinations of operating conditions, namely temperature, pressure, and stoichiometry.

Mass Balance

Because oxygen from the tank is completely dry, the amount of water at the stack inlet is equal to the amount of water vapor at the stack outlet. This also means that the amount of liquid water separated at the stack outlet is equal to water generated in the stack plus any net water transport from the anode through the membrane. The mass balance equation is:

$$\dot{m}_{H_2O,in(v)} + \dot{m}_{H_2O,in(l)} = \dot{m}_{H_2O,out(v)} \tag{9-1}$$

where:

Indexes *in* and *out* refer to in and out of the fuel cell stack, and
Indexes (*v*) and (*l*) refer to vapor and liquid form, respectively.

$\dot{m}_{H_2O,in(v)}$ = the mass flow of water vapor at stack inlet:

$$\dot{m}_{H_2O,in(v)} = \min\left[\frac{IN_{cell}}{4F}M_{H_2O}S_{O2}\frac{P_{sat}(t_{in})}{P_{out}-P_{sat}(t_{in})}, \dot{m}_{H_2O,out(v)}\right] \quad (9\text{-}2)$$

and

$\dot{m}_{H_2O,out(v)}$ = mass flow of water vapor at stack outlet:

$$\dot{m}_{H_2O,out(v)} = \frac{IN_{cell}}{4F}M_{H_2O}(S_{O2}-1)\frac{P_{sat}(t_{out})}{P_{out}-P_{sat}(t_{out})} \quad (9\text{-}3)$$

Depending on the pressure and temperature at the stack inlet, not all water may be in a vapor form. The amount of liquid water at the stack inlet, $\dot{m}_{H_2O,in(l)}$, may be calculated from Equation (9-1). However, temperature at the stack inlet, t_{in}, is unknown and must be found from the energy balance. The energy balance for mixing of two streams is (the numbers in subscript correspond to those in Figure 9-1):

$$H_1 + H_4 = H_2 \quad (9\text{-}4)$$

where:

H_1 = enthalpy of oxygen from the tank:

$$H_1 = \frac{IN_{cell}}{4F}M_{O2}c_{p,O2}t_{tank} \quad (9\text{-}5)$$

H_4 = enthalpy of oxygen and water vapor at the stack outlet:

$$H_4 = \frac{IN_{cell}}{4F}(S_{O2}-1)\left[M_{O2}c_{p,O2}t_{st,out} + M_{H_2O}\frac{P_{sat}(t_{st,out})}{P_{out}-P_{sat}(t_{st,out})}\right.$$
$$\left. \times \left(c_{p,H_2O(v)}t_{st,out} + h_{fg}^0\right)\right]$$
$$(9\text{-}6)$$

H_2 = enthalpy of the mix, that is, oxygen and water at the stack inlet:

$$H_2 = \frac{IN_{cell}}{4F}S_{O2}M_{O2}c_{p,O2}t_{st,in} + \dot{m}_{H_2O,in(v)}\left(c_{p,H_2O(v)}t_{st,in} + h_{fg}^0\right)$$
$$+ \dot{m}_{H_2O,in(l)}c_{p,H_2O(l)}t_{st,in} \quad (9\text{-}7)$$

The results are shown in Figure 9-2. Oxygen from the tank is assumed to be at 20°C and dry. It should be noted that for the points above the

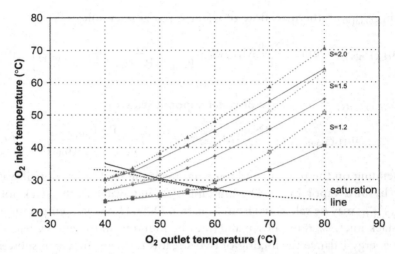

Figure 9-2 Achievable oxygen temperature at the stack inlet when the oxygen exhaust is recirculated back to the inlet, as shown in Figure 9-1. Dashed lines indicate atmospheric pressure, and solid lines are for 300 kPa.

saturation line, the oxygen at the stack inlet is oversaturated, that is, it contains liquid water, but below the saturation line the oxygen is undersaturated. In either case the resulting temperature at the oxygen inlet is below the stack operating temperature.

If this simple humidification scheme is not sufficient, active humidification of oxygen must be applied using the liquid water collected from the stack exhaust and the heat from the stack (Figure 9-3). This way it is possible to reach saturation at a temperature closer to the stack operating temperature. Sometimes humidification is accomplished in a separate section of the stack also using the stack heat, as shown in Chapter 6.

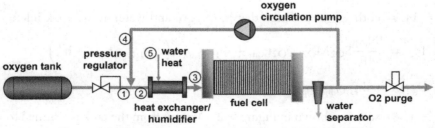

Figure 9-3 Closed loop oxygen supply system with humidifier.

The amount of water that needs to be added in the humidifier can be calculated from the mass balance:

$$
\dot{m}_{H_2O,in} = \frac{IN_{cell}}{4F} M_{H_2O} \left[S_{O2} \frac{\varphi P_{sat}(T_{in})}{P_{in} - \varphi P_{sat}(T_{in})} \right.
$$

$$
\left. - (S_{O2} - 1) \frac{P_{sat}(T_{out})}{P_{out} - P_{sat}(T_{out})} \right]
$$

(9-8)

The amount of heat needed to ensure that all the water at the fuel cell entrance is in vapor form can be calculated from the energy balance (the numbers in subscript correspond to those in Figure 9-3):

$$
H_{in} = H_3 - H_1 - H_4 - H_5
$$

(9-9)

where:

H_3 = enthalpy of oxygen and water vapor at the stack inlet(W):

$$
H_3 = \frac{IN_{cell}}{4F} S_{O2} \left[M_{O2} c_{p,O2} t_{st,in} + M_{H_2O} \frac{\varphi P_{sat}(t_{st,in})}{P_{in} - \varphi P_{sat}(t_{st,in})} \right.
$$

$$
\left. \times \left(c_{p,H_2O(v)} t_{st,in} + h_{fg}^0 \right) \right]
$$

(9-10)

H_1 = enthalpy of oxygen from the tank as defined previously (Equation 9-5)

H_4 = enthalpy of oxygen and water vapor at the stack outlet, also defined previously (Equation 9-6)

H_5 = enthalpy of liquid water added to the humidifier(W):

$$
h_5 = \dot{m}_{H_2O,in} c_{p,H_2O(l)} t_w
$$

(9-11)

9.1.2 Hydrogen Supply

The fuel for PEM fuel cells is hydrogen. Hydrogen is the lightest and most abundant element in the universe; however, on Earth it is not present in its molecular form but instead in many chemical compounds, such as water or hydrocarbons. Hydrogen is therefore not an energy source but a synthetic fuel that must be produced. For fuel cell systems, hydrogen may be produced elsewhere and then stored as a part of the system, or hydrogen generation may be a part of a fuel cell system. The systems with hydrogen storage are typically much simpler and more efficient, but hydrogen storage requires a lot of space, even when hydrogen is compressed to very high pressures or even liquefied. Table 9-1 shows some hydrogen properties.

The most common way of storing hydrogen is in high-pressure cylinders. Typical storage pressures are between 200 bars and 450 bars (3,000 or 6,600 psi) and even 690 bars (10,000 psi) have been reported [4]. Table 9-2 shows the required volume to store 1 kg of hydrogen at different pressures (1 kg of hydrogen happens to be the energy equivalent of 1 gallon of gasoline). Hydrogen storage in conventional steel cylinders (at 2,200 psi) is not practical for almost any application because they are too heavy (although

Table 9-1 Hydrogen Properties (Compiled from [1–3])

Property	Unit	Value
Molecular weight	$kg\,kmol^{-1}$ $(gmol^{-1})$	2.016
Density	$kg\,m^{-3}$ (gl^{-1})	0.0838
Higher heating value	$MJ\,kg^{-1}$	141.9
	$MJ\,m^{-3}$	11.89
Lower heating value	$MJ\,kg^{-1}$	119.9
	$MJ\,m^{-3}$	10.05
Boiling temperature	K	20.3
Density as liquid	$kg\,m^{-3}$	70.8
Critical temperature	K	32.94
Critical pressure	bar	12.84
Critical density	$kg\,m^{-3}$	31.40
Self ignition temperature	K	858
Ignition limits in air	(vol.%)	4—75
Stoichiometric mixture in air	(vol.%)	29.53
Flame temperature in air	K	2318
Diffusion coefficient	$m^2 s^{-1}$	0.61
Specific heat	$kJ\,kg^{-1}\,K^{-1}$ $(J\,g^{-1}\,K^{-1})$	14.89

Table 9-2 Required Volume to Store 1 kg of Hydrogen as Compressed Gas at 20°C

Pressure MPa	Volume Liters
0,1013	11,934.0
100	128.7
200	68.4
300	48.4
350	42.7
450	34.9
700	25.7

a fuel cell-powered submarine built by Perry Technologies in 1989 used them for storage of both hydrogen and oxygen [5]). Lightweight composite tanks constructed with an aluminum body wrapped by composite fiber and epoxy resin have been developed for storage of hydrogen in automotive applications, allowing storage density as high as 5% hydrogen (by weight). However, when the tank support, valves, and pressure regulators are taken into account, practical storage densities are between 3% and 4% hydrogen by weight. These tanks come in different sizes, usually 30–40 liters storing 1.3 to 1.5 kg of hydrogen at 350 bar [6]. For automotive applications two pressure standards have been adopted, namely 350 and 700 bar (5,000 and 10,000 psi).

Another option is to store hydrogen in a liquid form. Hydrogen is liquid at 20.3 K. This is a common way to store relatively large quantities of hydrogen. Smaller tanks for use in automobiles have been developed and demonstrated by BMW [7]. They can reach storage efficiency of 14.2% hydrogen by weight and require about 22 liters to store 1 kg of hydrogen. These tanks must be specially constructed and heavily insulated to minimize hydrogen boil-off. A relatively simple evaporator is sufficient to produce gaseous hydrogen needed for fuel cell applications.

Yet another way of storing hydrogen is in metal hydrides. Some metals (such as various alloys of magnesium, titanium, iron, manganese, nickel, chromium, and others) form metal hydrides when exposed to hydrogen. Hydrogen atoms are packed inside the metal lattice structure, and because of that, higher storage densities may be achieved than with compressed hydrogen (1 kg hydrogen can be stored in 35–50 liters). The problem with this storage is that the metals are intrinsically heavy; storage efficiency of 1.0% to 1.4% hydrogen by weight can be achieved. Higher storage efficiencies have been reported with some metal hydrides, but those are typically high-temperature metal hydrides (above 100°C) and thus not practical with low-temperature PEM fuel cells. To release hydrogen from metal hydrides, heat is required. Waste heat from the fuel cell, in both water-cooled and air-cooled systems, is sufficient to release hydrogen from low-temperature metal hydrides. Because hydrogen is stored in basically solid form, this is considered one of the safest hydrogen storage methods.

Several chemical ways of storing hydrogen have been proposed and some practically demonstrated, such as hydrazine, ammonia, methanol, ethanol, lithium hydride, sodium hydride, sodium borohydride, lithium borohydride, diborane, calcium hydride, and so forth. Although attractive because most of them are in liquid form and offer relatively high hydrogen

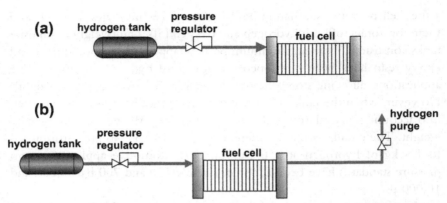

Figure 9-4 Hydrogen supply schemes: (a) dead-end; (b) dead-end with intermittent purging.

storage efficiencies (up to 21% by weight for diborane), they require some kind of a reactor to release hydrogen. In addition, some of them are toxic, and some can cause severe corrosion problems.

Once hydrogen is released from the storage tank, the simplest way to supply hydrogen to a fuel cell is in the dead–end mode (Figure 9-4a). Such a system would only require a preset pressure regulator to reduce the pressure from the stack to the fuel cell. Long–term operation in a dead–end mode may be possible only with extremely pure gases, both hydrogen and oxygen. Any impurities present in hydrogen will eventually accumulate in the fuel cell anode. This also includes water vapor that may remain (when the back diffusion is higher than the electroosmotic drag), which may be the case with very thin membranes and when operating at low current densities. In addition, inerts and impurities may diffuse from the air side until an equilibrium concentration is established. To eliminate this accumulation of inerts and impurities, purging of the hydrogen compartment may be required (Figure 9-4b). This may be programmed either as a function of cell voltage or as a function of time.

If purging of hydrogen is not possible or preferred for reasons of safety, mass balance, or system efficiency, excess hydrogen may be flown through the stack (S > 1) and unused hydrogen returned to the inlet, either by a passive (ejector) or an active (pump or compressor) device (Figure 9-5). In either case, it is preferred to separate and collect any liquid water that may be present at the anode outlet. The amount of liquid water to be collected depends on operating conditions and membrane properties. In thinner

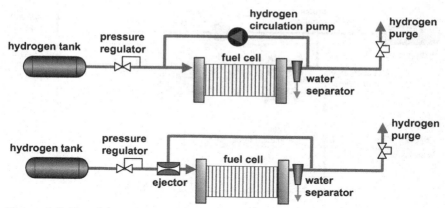

Figure 9-5 Closed loop hydrogen supply system with pump (above) and ejector (below).

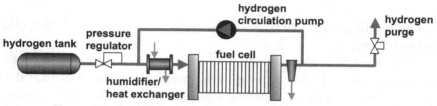

Figure 9-6 Closed loop hydrogen supply system with humidifier.

membranes, back diffusion may be higher than the electroosmotic drag, and some of the product water may exit the stack at the anode side.

Hydrogen typically must be humidified up to 100% relative humidity before entering the fuel cell stack to avoid drying the membrane due to electroosmotic drag. In that case a humidifier/heat exchanger is needed at the stack inlet (Figure 9-6). Hydrogen may be humidified by water injection and simultaneously or subsequently heated to facilitate evaporation of water, or by membrane humidification, which allows water and heat exchange.

9.1.3 Water and Heat Management: System Integration

In addition to supplying the reactants to the fuel cell stack, the fuel cell system also must take care of the fuel cell byproducts: water and heat. Water plays an important role in fuel cell operation. As discussed in Chapter 4, water is essential for proton transport across the polymer membrane. For that reason, in principle both cathode and anode reactants must be humidified

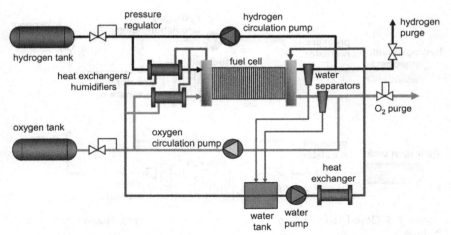

Figure 9-7 An example of closed loop hydrogen/oxygen fuel cell system.

before entering a fuel cell stack. Water must be collected at the fuel cell exhaust for reuse.

In a closed loop configuration, the system is always operating in water production mode. Both reactant gases are stored dry and humidified at the inlet, and unused gases are recirculated from the outlet back to the inlet after the liquid water has been removed and collected in a tank. The tank must be sized so that it can store the produced water. On a system level, including hydrogen and oxygen storage tanks, the mass of the system does not change, that is, hydrogen and oxygen are converted to water. The mass of water increases with time as the mass of hydrogen and oxygen in the tanks decreases.

The same water may be used for humidification and to remove the heat from the stack. Heat is discharged from the system through a radiator or a liquid/liquid heat exchanger (depending on the application). Figure 9-7 shows an example of a hydrogen/oxygen closed loop system [8].

9.2 HYDROGEN/AIR SYSTEMS

For most terrestrial systems it is more practical to use oxygen from air than to carry oxygen as a part of the fuel cell system. Oxygen content in air is 20.95% by volume. This dilution has a penalty on fuel cell voltage (about 50 mV), as shown in Chapter 3. Additional penalty in both power output and efficiency is because oxygen and almost four times that of nitrogen must be somehow pumped through the fuel cell.

9.2.1 Air Supply

In hydrogen/air systems, air is supplied by a fan or a blower (for low-pressure systems) or by an air compressor (for pressurized systems). In the former case, the exhaust from the fuel cell is opened directly into the environment (Figure 9-8a), whereas in a pressurized system, pressure is maintained by a preset pressure regulator (Figure 9-8b).

In any case, a fan, blower, or compressor is run by an electric motor that requires electrical power and thus represents power loss or parasitic load. Compression may be either isothermal or adiabatic. The former implies an infinitesimally slow process allowing temperature equilibration with the environment. The latter implies quite the opposite—a process so fast that no heat is exchanged with the environment during the compression. This is much closer to real life, where the speed of compression is such that it does not allow heat exchange with the environment.

The ideal power needed for adiabatic compression of air from pressure P_1 to pressure P_2 is:

$$W_{comp,ideal} = \dot{m}_{Airin} \cdot c_p \cdot T_1 \left[\left(\frac{P_2}{P_1} \right)^{\frac{k-1}{k}} - 1 \right] \qquad (9\text{-}12)$$

where:

$\dot{m}_{Airin}$ = air flow rate, $g\,s^{-1}$
c_p = specific heat, $J\,g^{-1}\,K^{-1}$
T_1 = temperature before compression, K
P_2 = pressure after compression, Pa
P_1 = pressure before compression, Pa
k = ratio of specific heats (for diatomic gases $k = 1.4$)

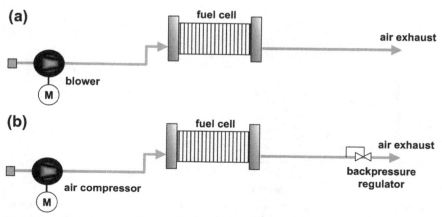

Figure 9-8 Air supply for a fuel cell system: (a) ambient pressure; (b) pressurized system.

However, no compression is ideal, that is, there are inefficiencies associated with the compression process that result in more power needed:

$$W_{comp} = \frac{\dot{m}_{Airin} \cdot c_p \cdot T_1}{\eta_{comp}} \left[\left(\frac{P_2}{P_1} \right)^{\frac{k-1}{k}} - 1 \right] \tag{9-13}$$

where the efficiency, η_{comp}, is defined as a ratio between ideal (adiabatic) and actual compression power.

From the energy balance it is possible to calculate the temperature at the end of compression:

$$T_2 = T_1 + \frac{T_1}{\eta_{comp}} \left[\left(\frac{P_2}{P_1} \right)^{\frac{k-1}{k}} - 1 \right] \tag{9-14}$$

The actual electric power draw for compression is even larger than W_{comp} from Equation (9-13) because of additional mechanical and electrical inefficiencies:

$$W_{EM} = \frac{W_{comp}}{\eta_{mech} \cdot \eta_{EM}} \tag{9-15}$$

At higher pressures, above 150 kPa, a compressor's electric motor may consume a significant portion of the fuel cell power output, depending not only on the compressor efficiency but also on the stoichiometric ratio and cell operating voltage, as shown in Figure 9-9.

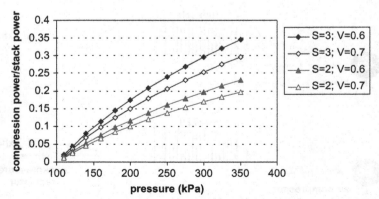

Figure 9-9 Compression power as a function of pressure (assuming inlet pressure is atmospheric, inlet temperature of 20°C, and compressor efficiency of 70%).

The net power output, W_{net}, is the fuel cell power, W_{FC}, less the power delivered to the auxiliary components, W_{aux}, which includes the compressor or the blower:

$$W_{net} = W_{FC} - W_{aux} \qquad (9\text{-}16)$$

The net system efficiency is then:

$$\eta_{sys} = \eta_{FC} \frac{W_{net}}{W_{FC}} = \eta_{FC}(1 - \xi_{aux}) \qquad (9\text{-}17)$$

where ξ_{aux} is the ratio between auxiliary power (also called parasitic losses) and fuel cell power output, W_{aux}/W_{FC}.

One of the main reasons for operating a fuel cell at an elevated pressure is to get more power out of it. However, when compression power is taken into account, operation at higher pressure may not result in more power, as is illustrated in the following example.

Example 9-1

Consider a fuel cell stack that can operate at either 300 kPa or 170 kPa (inlet pressure) with an oxygen stoichiometric ratio of 2. The stack performance (polarization curve) is shown in Figure 5.1. Calculate the net power output and the net efficiency at both pressures. Assume current density of 800 mA cm^{-2}.

Solution

Operation at higher pressure (300 kPa) would yield 0.66 V per cell (from the polarization curve).

Resulting power density is: $0.8 \times 0.66 = 0.528$ W cm^{-2}

Air flow rate is (Equation 5-42):

$$\dot{m}_{Airin} = \frac{S_{O2}}{r_{O2}} \frac{M_{Air}}{4F} i = \frac{2}{0.21} \frac{29}{4 \times 96485} 0.8 = 5.725 \times 10^{-4} \text{ g s}^{-1} \text{ cm}^{-2}$$

Compression power (from Equation 9-13 or from Figure 9-9, assuming the compressor total efficiency of 0.7):

$$W_{comp} = \frac{\dot{m}_{Airin} \cdot c_p \cdot T_1}{\eta_{comp}} \left[\left(\frac{P_2}{P_1} \right)^{\frac{k-1}{k}} - 1 \right]$$

$$= \frac{5.725 \times 10^{-4} \times 1 \times 293}{0.7} \left[\left(\frac{300}{101.3} \right)^{\frac{1.4-1}{1.4}} - 1 \right] = 0.088 \text{ W cm}^{-2}$$

Net power density (Equation 9-16): $0.528 - 0.088 = 0.440$ W cm^{-2}

System efficiency (Equation 9-17): $\dfrac{0.66}{1.482} \left(1 - \dfrac{0.088}{0.528} \right) = 0.37$

Similar calculation for low-pressure operation (170 kPa) would yield the following results:

Cell potential: 0.60 V per cell
Power density: $0.480 \, W \, cm^{-2}$
Compression power: $0.040 \, W \, cm^{-2}$
Net power density: $0.480 - 0.040 = 0.440 \, W \, cm^{-2}$
System efficiency: 0.37

Therefore, for this particular case there is no apparent advantage in operating this fuel cell at elevated pressure, that is, both systems result in same net power output and same system efficiency. If the voltage difference between these two pressures were less than 60 mV, low pressure would result in more power output and better system efficiency. However, there may be other determining factors, such as operating temperature, that may impact system complexity and performance at different pressures.

The desired air flow rate is maintained by regulating the speed of the compressor. For low-pressure systems with insignificant compressor power draw, it may be convenient to operate at a constant flow rate, that is, at the constant compressor speed. For high-pressure operating systems, operation at constant flow rate would have a detrimental effect on the system efficiency at partial loads, as shown in the following example.

Example 9-2

For the fuel cell from Example 9-1, calculate the ratio between auxiliary power and fuel cell power output, W_{aux}/W_{FC}, at nominal power, at 50% net power output, and at 25% net power output if the compressor operates at constant speed, that is, providing constant air flow rate.

Solution

At nominal power ($440 \, mW/cm^2$):

$W_{aux}/W_{FC} = 88/528 = 0.167$ at 300 kPa
$W_{aux}/W_{FC} = 40/480 = 0.083$ at 170 kPa

At 50% nominal power ($220 \, mW/cm^2$):

$W_{aux}/W_{FC} = 88/(220 + 88) = 0.286$ at 300 kPa
$W_{aux}/W_{FC} = 40/(220 + 40) = 0.145$ at 170 kPa

At 25% nominal power ($110 \, mW/cm^2$):

$W_{aux}/W_{FC} = 88/(110 + 88) = 0.444$ at 300 kPa
$W_{aux}/W_{FC} = 40/(110 + 40) = 0.267$ at 170 kPa

The system efficiency at 25% nominal power output at 300 kPa would drop to 0.3 because of high parasitic losses, as opposed to 0.485 if the flow rate were kept proportional to current.

From the previous example it follows that in pressurized systems it is important to regulate the air flow rate in proportion to current being generated.

In general, there are two types of compressors that may be used in pressurized fuel cell systems:

- Positive displacement compressors, such as piston, diaphragm, scroll, screw, or rotary vane compressors, or
- Centrifugal compressors, radial or axial.

Because of their pressure-flow characteristics, the flow regulation is different for these two types of compressors.

For a positive displacement compressor, the flow rate may be changed without the need to change the back pressure (Figure 9-10) by simply reducing the speed of the motor, N_p. The pressure is somewhat higher at higher flow rates as a function of pressure drop through the stack. Note that the compressor efficiency may not be the same along the operation line. The compressor efficiency at both partial load and nominal power should be taken into account when selecting the compressor for a fuel cell system.

However, the centrifugal compressors have different pressure-flow rate characteristics. The centrifugal compressors must not be operated in the low-flow region left of the surge line. In fuel cell systems, this means that the flow regulation must be accompanied by the pressure regulation as well, in order to keep the operating point to the right of the compressor surge line (Figure 9-11). This happens to be beneficial in terms of compressor efficiency. By changing both flow rate and pressure, it is possible to operate with high efficiencies through a wide range of flow rates.

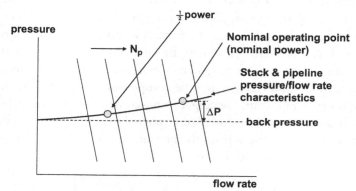

Figure 9-10 Reduction of flow rate does not need a change in back pressure for positive displacement compressors.

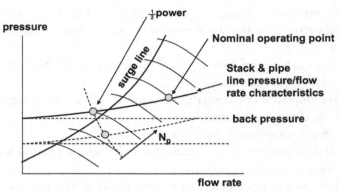

Figure 9-11 Reduction of flow rate requires reduction of pressure for centrifugal compressors.

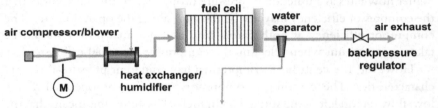

Figure 9-12 Schematic diagram of air supply for a fuel cell system.

Air typically has to be humidified before entering the fuel cell stack. Various humidification schemes are discussed in Section 9.2.4, "Humidification Schemes." At the stack outlet, typically there is some liquid water that may be easily separated from the exhaust air in a simple, off-the-shelf gas/liquid separator. Water collected at the exhaust may be stored and reused, either for cooling or for humidification (Figure 9-12).

In a pressurized system, air at the exhaust is warm and still pressurized at somewhat lower pressure than it is at the inlet. This energy may be utilized in an expander or a turbine to generate work that may offset some of the work needed to compress the air. The compressor and the turbine may be installed on the same shaft, creating a turbo-compressor (Figure 9-13).

The amount of work that may be extracted from the hot air at the fuel cell exhaust by expanding it from the pressure at the fuel cell exhaust, P_{out}, to the atmospheric pressure, P_0, is:

$$W_{exp} = \dot{m}_{Airout} \cdot c_p \cdot T_{out} \left[1 - \left(\frac{P_0}{P_{out}} \right)^{\frac{k-1}{k}} \right] \eta_{exp} \qquad (9\text{-}18)$$

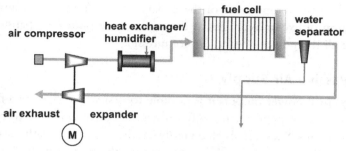

Figure 9-13 Schematic diagram of a fuel cell system with an expander at the exhaust.

The expander efficiency, η_{exp}, is a ratio between actual work and ideal isentropic work between these two pressures.

Because the air at the exhaust also carries some water vapor, the mass flow rate, specific heat, and the coefficient k (ratio of specific heats) have to be adjusted accordingly. This adjustment typically results in a power increase of less than 5%.

The temperature at the end of expansion is:

$$T_{end} = T_{out} - T_{out}\left[1 - \left(\frac{P_0}{P_{out}}\right)^{\frac{k-1}{k}}\right]\eta_{exp} \qquad (9\text{-}19)$$

Because of the inefficiencies of both compression and expansion processes and because of the pressure drop through the stack, an expander may recover only a portion of the compression work, as shown in Figure 9-14.

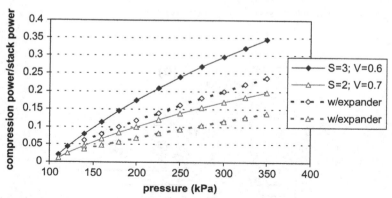

Figure 9-14 Compression power as a function of fuel cell operating pressure for a system with compressor/expander (assuming ambient pressure of 101.3 kPa and ambient temperature of 20°C, and both compressor and expander efficiency of 70%).

However, if the temperature of the exhaust is increased—for example, by combustion of excess hydrogen at the stack anode outlet—expansion may generate all the power needed for compression.

9.2.2 Passive Air Supply

For very small power outputs it is possible to design and operate a fuel cell with passive air supply relying only on natural convection due to concentration gradients. Such a fuel cell typically has the front of the cathode directly exposed to the atmosphere, therefore without the bipolar plates; or it is in bipolar configuration, where the cathode flow field has openings to the atmosphere on the side (Figure 9-15). In either case an oxygen concentration gradient is formed between the open atmosphere and the catalyst layer where oxygen is being consumed in the electrochemical reaction. The performance of such fuel cells is typically not limited only by the oxygen diffusion rate but also by water and heat removal, both dependent on the temperature gradient. Maximum current density that can be achieved with free-convection fuel cells is typically limited to about $0.1–0.15\,A\,cm^2$.

These ambient or free-convection fuel cells result in a very simple system, needing only hydrogen supply (Figure 9-16). Several cells may be connected sideways in series (the anode of one cell electrically connected to the cathode of the adjacent cell) to get the desired voltage output. Figure 9-16 shows such a multicell configuration where the free-convection cathodes are employed on both sides (and the hydrogen flow field is contained in the middle). It is interesting that this fuel cell can generate power at either 6 V or 12 V, depending on how the two sides are electrically connected (in parallel or in series, respectively).

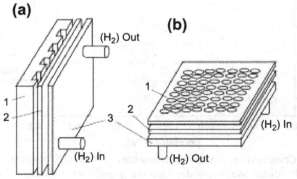

Figure 9-15 Examples of free-convection fuel cells [9].

Figure 9-16 Simple fuel cell system based on passive air supply [10].

9.2.3 Hydrogen Supply

Hydrogen supply for hydrogen/air systems does not differ from hydrogen supply for hydrogen/oxygen systems, described previously in the section "Hydrogen Supply." Design options include dead-end or flow-through with recirculation of unused hydrogen. However, operation in a closed loop system without intermittent purging would not be possible for an extended period because of accumulation of inerts, either those present in hydrogen or nitrogen that would over time permeate the fuel cell membrane because of concentration gradient.

One additional option in hydrogen/air systems is to use hydrogen from the fuel cell exhaust in a burner to generate more heat, which then may be converted into work in a turbine/expander (Figure 9-17). The turbine/expander mounted on the same shaft with the air compressor may minimize, or under some conditions eliminate, the parasitic losses associated with powering the air compressor.

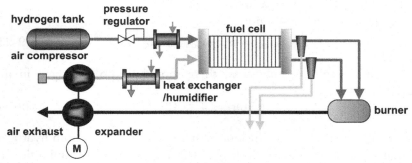

Figure 9-17 Use of hydrogen exhaust in a burner to enhance operation of the expander/turbine.

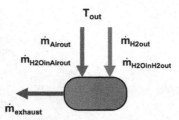

Figure 9-18 Mass balance of a tail gas burner.

The amount of work that can be extracted from the exhaust gas in the turbine is:

$$W_{exp} = \dot{m}_{exh} \cdot c_{pexh} \cdot T_B \left[1 - \left(\frac{P_0}{P_{out}} \right)^{\frac{k-1}{k}} \right] \eta_{exp} \qquad (9\text{-}20)$$

The composition of the exhaust gas, its specific heat, mass flow rate, and temperature, may be calculated from the burner mass and energy balance, as shown in Figure 9-18.

The mass flow rate out of the burner consists of unburned oxygen, nitrogen, and water vapor:

$$\dot{m}_{exp} = \dot{m}_{O2Bout} + \dot{m}_{N2Bout} + \dot{m}_{H_2OBout,V} \qquad (9\text{-}21)$$

Oxygen mass flow is equal to the mass flow of oxygen before the burner minus oxygen that was used for combustion of hydrogen (assuming complete combustion):

$$\dot{m}_{O2Bout} = \dot{m}_{O2FCout} - \frac{1}{2} \dot{m}_{H_2FCout} \frac{M_{O2}}{M_{H_2}} \qquad (9\text{-}22)$$

where:

M_{O2} = molecular weight of oxygen

M_{H_2} = molecular weight of hydrogen

Nitrogen mass flow should not change:

$$\dot{m}_{N2Bout} = \dot{m}_{N2FCout} = \dot{m}_{N2FCin} \qquad (9\text{-}23)$$

In case of complete combustion, there should be no hydrogen in the exhaust:

$$\dot{m}_{H_2Bout} = 0 \qquad (9\text{-}24)$$

Water vapor mass flow rate includes water vapor present in both hydrogen and air outlet from the fuel cell plus water generated by combustion of hydrogen in the burner:

$$\dot{m}_{H_2OBout,V} = \dot{m}_{H_2OinAirout,V} + \dot{m}_{H_2OinH_2out,V} + \dot{m}_{H_2FCout} \frac{M_{H_2O}}{M_{H_2}} \qquad (9\text{-}25)$$

The energy balance requires that the sum of the enthalpies of all the flows coming into the burner must be equal to the sum of enthalpies leaving the burner:

$$\sum H_{FCout} = \sum H_{Bout} \qquad (9\text{-}26)$$

The sum of enthalpies coming into the burner includes hydrogen and oxygen exhaust from the fuel cell together with water vapor present:

$$\sum H_{FCout} = t_{out} \Big[\dot{m}_{Airout} C_{pAir} + \dot{m}_{H_2out} C_{pN2} $$

$$+ \left(\dot{m}_{H_2OinAirout,v} + \dot{m}_{H_2OinH_2out,v} \right) C_{pH_2O,V} \Big] \qquad (9\text{-}27)$$

$$+ \left(\dot{m}_{H_2OinAirout,v} + \dot{m}_{H_2OinH_2out,v} \right) h_{fg}^0$$

The sum of enthalpies at the burner outlet is:

$$\sum H_{Bout} = t_B \left(\dot{m}_{O2Bout} C_{pO2} + \dot{m}_{N2FCout} C_{pN2} \right.$$

$$\left. + \dot{m}_{H_2OBout,v} C_{pH_2O,V} \right) + \dot{m}_{H_2OBout,v} h_{fg}^0 \qquad (9\text{-}28)$$

By combining Equations (9-21 through 9-28), one may obtain the resulting temperature of the exhaust gas.

Obviously, a higher hydrogen stoichiometry ratio would leave more hydrogen in the fuel cell exhaust and consequently would result in more work produced from the turbine, reducing the parasitic load and therefore increasing the fuel cell net power output. However, burning excess hydrogen results in lower fuel cell system efficiency (as shown in Figure 9-19).

9.2.4 Humidification Schemes

In principle, both air and hydrogen streams must be humidified at the fuel cell inlet. Hydrogen must be humidified to ensure that the electroosmotic drag does not dry out the anode side of the membrane. Air must be humidified in spite of water produced on the cathode side to ensure that the excess of dry gas, particularly in the entrance region, does not remove water at a rate higher than water is being generated in the electrochemical reaction.

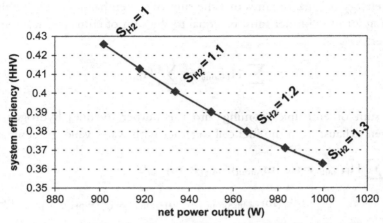

Figure 9-19 Net power output and system efficiency of a fuel cell system with a tail gas burner and an expander/turbine for various hydrogen utilization (stack power output 1 kW at 0.7 V/cell, 60°C, 300 kPa; compressor and expander efficiencies 0.70).

Several humidification methods and schemes can be applied for air humidification. Humidification of a gas can be achieved by:

- Bubbling the gas through water
- Direct water or steam injection
- Exchange of water (and heat) through a water-permeable medium
- Exchange of water (and heat) on an adsorbent surface (enthalpy wheel)

Bubbling is often used in laboratory settings for relatively low flow rates (corresponding to small, single cells) but rarely in practical systems. In this method, air is dispersed through a porous tube immersed in heated liquid water. In this way, bubbles of air in water form a relatively large contact area between gas and liquid water where the evaporation process can take place. The desired level of humidification is achieved by controlling the temperature of liquid water, in laboratory settings typically with electrical heaters. If the device is sized properly, the gas emerging at the liquid surface is saturated at a desired, preset temperature. The efficiency of humidification depends on the water level, so the required water level must be maintained. The device must be designed so that the outgoing humidified air does not carry water droplets. In practical systems this method is rarely used because of its size and cumbersome controls (temperature, water level) and the need to collect liquid water at the fuel cell outlet, which in some cases may require not only gas/liquid separators but a condensing heat exchanger as well. In addition, pushing air through a porous medium results in additional pressure

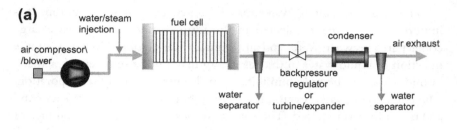

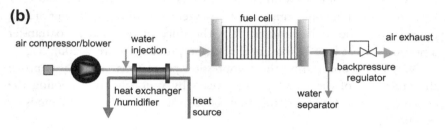

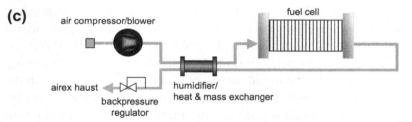

Figure 9-20 Practical air humidification schemes: (a) water/steam injection, (b) water injection with heat exchange, (c) exchange of water and heat with the cathode exhaust.

drop, which then must be overcome by the air supply device, resulting in higher pumping power, that is, higher parasitic power and lower system efficiency.

Direct water injection is a more elegant, more compact, and easier-to-control method (Figure 9-20a). The required amount of water to be injected can be easily calculated for any combination of operating conditions (temperature, pressure, gas flow rate, and desired relative humidity):

$$\dot{m}_{H_2O} = \dot{m}_{Air} \frac{M_{H_2O}}{M_{Air}} \left(\frac{\varphi P_{sat}(T)}{P - \varphi P_{sat}(T)} - \frac{\varphi_{amb} P_{sat}(T_{amb})}{P_{amb} - \varphi_{amb} P_{sat}(T_{amb})} \right) \quad (9\text{-}29)$$

where φ, T, P, and φ_{amb}, T_{amb}, AND P_{amb} are relative humidity, temperature, and pressure at the fuel cell inlet and of the ambient air, respectively.

The exact amount of water may be dosed by a metering pump. It is important that water is injected in the form of fine mist so that a large contact area between water and air facilitates evaporation. However, simple injection of liquid water in the gas stream may not be sufficient to actually humidify the gas, because humidification also requires heat for evaporation. The enthalpy of water, even if the water is hot, is usually not sufficient and additional heat is required. The sources of heat may be the air compressor (obviously applicable only in pressurized systems) and the fuel cell stack itself. In most of the operating conditions, there is a sufficient amount of heat generated in the fuel cell stack. It is the duty of the system to transfer a portion of that heat to the humidification process (Figure 9-20b).

Water injection during the compression process may actually improve the efficiency of the compression process by simultaneously cooling the compressed gas; however, this method is not applicable for all kinds of compressors.

Direct steam injection eliminates the need for additional heat exchange; however, steam must be generated somewhere in the system, which means that this method is applicable only for the systems where heat is generated at a temperature above water's boiling temperature (100°C at 101.3 kPa).

The *exchange of heat and water* from the cathode exhaust with incoming air is an elegant way of taking advantage of water and heat produced in the fuel cell stack (Figure 9-20c). This can be achieved through a water-permeable medium, such as porous plate (metal, ceramic, or graphite) or a water-permeable membrane (such as Nafion), or through an enthalpy wheel. These devices are essentially mass and heat exchangers, allowing both heat and water to be exchanged between warm and oversaturated fuel cell exhaust and dry incoming air. The heat and water fluxes in a humidifier at low fuel cell operating pressure are in the same direction (from fuel cell exhaust to inlet), whereas at elevated pressure due to compression, the inlet gas may be hot, and in that case the heat and water flow in opposite directions. In either case, it is not possible to humidify the incoming gas to 100% relative humidity at stack operating temperature with this method because of finite temperature and water concentration differences between the two sides of the humidifier needed for heat and mass transfer.

Instead of humidification of reactant gases, the membrane may be humidified by intermittently short-circuiting the stack for a very short period of time. This creates additional water in the electrochemical reaction on the cathode side, which then back-diffuses through the membrane.

9.2.5 Water and Heat Management: System Integration

Water and heat are the byproducts of the fuel cell operation, and the supporting system must include the means for their removal. Both water and heat from the fuel cell stack may be at least partially reused—for example, for humidification of the reactant gases or for facilitating hydrogen release from metal hydride storage tanks.

Water and heat handling may be integrated into a single subsystem if deionized water is used as a stack coolant, as shown in Figure 9-21. In that case water removes the heat from the stack and the same water and heat are used to humidify the reactant gases. The remaining heat has to be discarded to the surroundings through a heat exchanger; for hydrogen/air systems that is typically a radiator. The amount of heat to be discarded must be calculated from the stack and humidifier energy balances.

The flow rate of the coolant is:

$$\dot{m}_{coolant} = \frac{Q}{c_p \cdot \Delta T} \qquad (9\text{-}30)$$

ΔT is typically a design variable. Most typically, ΔT is below 5°C and rarely above 10°C. Smaller ΔT results in more uniform stack temperature distribution, but it requires larger coolant flow rate, which in turn increases parasitic losses. Sometimes larger stack temperature variations are needed to maintain the water in the desired state [11]; in that case, the coolant ΔT is dictated by the stack temperature requirements.

The size of the radiator heat exchanger depends on the temperature difference between the coolant and the ambient air. For that reason, it is preferred to operate the fuel cell system at a higher temperature for systems where the size of the components is critical. However, the operating

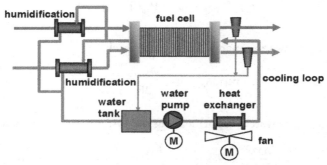

Figure 9-21 Water and heat management integrated in a single loop.

pressure and water balance must be taken into account in deciding on the operating temperature.

A hydrogen/air fuel cell system must be designed so that it does not need any backup water supply. Water is generated inside the stack, but water is needed for humidification of one or both reactant gases. Sometimes, depending on the intended operation of the fuel cell system and depending on the stack's capabilities to recirculate water internally (e.g., with the use of thinner membranes and with hydrogen and air in counterflow), humidification of only one of the reactants may be sufficient. Water for humidification must be collected at the stack exhaust, on the cathode side (although sometimes it is necessary on both cathode and anode exhausts). Depending on the fuel cell operating conditions (pressure, temperature, and flow rates), water at the exhaust may be in either liquid or gaseous form. Liquid water is relatively easily separated from the gaseous stream in liquid/gas separators. If liquid water collected at the stack exhaust is not sufficient for humidification, the exhaust must be cooled so that additional water may condense and be separated.

On the system level, water balance is very simple: The amount of water entering the system in ambient air plus water generated inside the fuel cell must be larger than or equal to the amount of water leaving the system with exhaust air and hydrogen. Hydrogen exhaust can be either continuous (flow-through mode) or periodic stream (dead-end or recirculation with periodic purging).

The system water balance is given with the following equation:

$$\dot{m}_{H_2O,Airin} + \dot{m}_{H_2O,gen} = \dot{m}_{H_2O,Airout} + \dot{m}_{H_2O,H_2out} \qquad (9\text{-}31)$$

where:

$$\dot{m}_{H_2O,Airin} = \frac{S_{O2}}{r_{O2}} \frac{\varphi_{amb}P_{vs}(T_{amb})}{P_{amb} - \varphi_{amb}P_{vs}(T_{amb})} \frac{M_{H_2O}}{4F} I \cdot n_{cell} \qquad (9\text{-}32)$$

$m_{H_2O,gen}$ = water generated inside the stack:

$$\dot{m}_{H_2O,gen} = \frac{M_{H_2O}}{2F} I \cdot n_{cell} \qquad (9\text{-}33)$$

$\dot{m}_{H_2O,Airout}$ = water (vapor) leaving the system with air exhaust:

$$\dot{m}_{H_2O,Airout} = \frac{S_{O2} - r_{O2in}}{r_{O2in}} \frac{P_{vs}(T_{st})}{P_{ca} - \Delta P_{ca} - P_{vs}(T_{st})} \frac{M_{H_2O}}{4F} I \cdot n_{cell} \qquad (9\text{-}34)$$

$\dot{m}_{H_2O,H_2out}$ = water leaving the system with hydrogen exhaust:

$$\dot{m}_{H_2O,H_2out} = (S_{H_2} - 1) \frac{P_{vs}(T_{st})}{P_{an} - \Delta P_{an} - P_{vs}(T_{st})} \frac{M_{H_2O}}{2F} I \cdot n_{cell} \quad (9\text{-}35)$$

Water balance therefore depends on:
- The oxygen (and hydrogen) flow rate, that is, the stoichiometric ratio
- Stack operating temperature, that is, temperature of the exhaust
- Stack operating pressure, or more precisely the stack outlet pressure, that is, the pressure at which the liquid water is separated from the exhaust gases
- Ambient conditions (pressure, temperature, and relative humidity)

It should be noted that water balance on the system level does not depend on current and number of cells (because $I \cdot n_{cell}$ product appears in each of the previous equations, it gets cancelled when Equations (9-32) through (9-35) are introduced in Equation (9-31)).

Figure 9-22 shows the required exhaust temperature that results in neutral water balance as a function of air flow rate and operating pressure (ambient conditions are assumed to be 20°C, 101.3 kPa, and 60% relative humidity). For the ambient pressure operation and oxygen stoichiometry of 2.0, the stack should not be operated above 60°C. If it is necessary to operate at a higher temperature, then a higher operating pressure should be selected or an additional heat exchanger may be needed to achieve a neutral water balance (which may defeat the purpose of operating at a higher temperature).

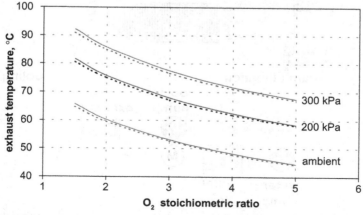

Figure 9-22 Required stack exhaust temperature to achieve neutral system water balance.

It should be noted that hydrogen does not carry much water out of the system. The dashed lines in Figure 9-22 take into account water that may be taken away from the system by hydrogen exhaust, assuming hydrogen stoichiometry of 1.2 at the inlet and saturated conditions at the outlet.

The presence of water in the system makes fuel cell systems susceptible to freezing if used outdoors in cold climates. In such a case, the coolant loop is separated from the water system, as shown in Figure 9-23. This allows for antifreeze coolants (such as aqueous solutions of ethylene-glycol or propylene-glycol) to be used instead of deionized water. Nevertheless, water cannot be completely eliminated from the system; after all, the PSA membrane contains up to 35% water. Operation, or actually survival and startup, of a fuel cell system in a cold environment is an issue that must be addressed by the system design. The system, when operational, can keep itself at operating temperature. Therefore, one way to keep the system warm is to operate it either periodically or constantly at some very low power level, which may significantly affect the overall system efficiency and may not be practical for long periods. Another way to avoid freezing is to drain the water from the system at shutdown. In systems that are used in backup power applications, small electric heaters may be used to keep the system above freezing and ready for quick startup.

Smaller stacks, less than 2-3 kW, may be air cooled. In that case a fan replaces the coolant pump (as shown in Figure 9-24). Reuse of waste heat

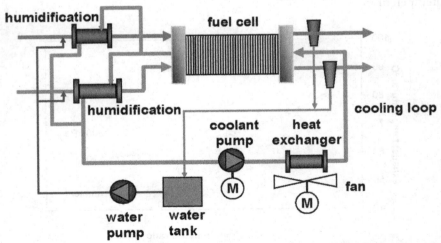

Figure 9-23 Fuel cell system with separate subsystems: water for humidification and antifreeze coolant for cooling.

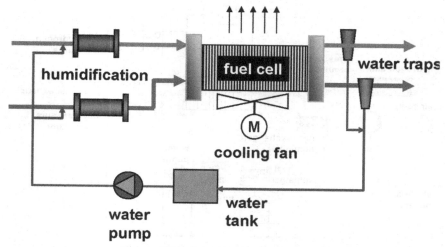

Figure 9-24 Air cooled fuel cell system.

collected by air as coolant is not practical, with an exception in systems where hydrogen is stored in metal hydride bottle(s). In that case heat is needed to release hydrogen from metal hydride, so the warm exhaust of the cooling air may be blown over the metal hydride tanks.

Even smaller stacks with relatively large outside surface area (relative to the active area inside) may be passively cooled by natural convection and radiation. If necessary, the surface area may be enlarged by fins.

The system configuration greatly depends on application. In some cases a very simple system, consisting of nothing more than a bottle of hydrogen and a fuel cell (such as the one shown in Figure 9-16), may be sufficient. In other cases the system needs most of the components and subsystems discussed earlier. Figure 9-25 shows a schematic diagram of an actual hydrogen/air fuel cell system employed in a fuel cell utility vehicle [12]. Figure 9-26 shows the actual layout in the vehicle.

9.3 FUEL CELL SYSTEMS WITH FUEL PROCESSORS

To bring fuel cells to the marketplace even before hydrogen becomes a readily available fuel, conventional fuels such as natural gas for stationary applications, gasoline for transportation, and methanol for portable power may be used if hydrogen generation from these fuels is made a part of the

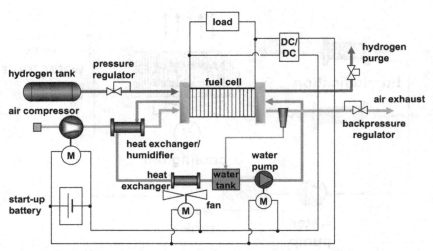

Figure 9-25 An example of a complete hydrogen/air fuel cell system.

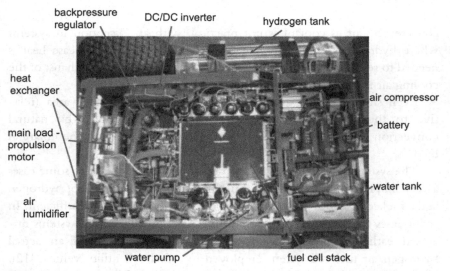

Figure 9-26 An actual fuel cell system installed in a John Deere Gator utility vehicle showing all the components from the schematic diagram in Figure 9-22.

fuel cell system. Hydrogen can be generated from hydrocarbon fuels such as natural gas, gasoline, or methanol by several processes, such as:

- Steam reformation
- Partial oxidation
- Autothermal reformation, which is essentially a combination of steam reformation and partial oxidation

In addition, several other processes must be employed to produce hydrogen pure enough to be used in PEM fuel cells, such as:

- Desulfurization—to remove sulfur compounds present in fuel
- Shift reaction—to reduce the content of CO in the gas produced by the fuel processor
- Gas cleanup, involving preferential oxidation, methanation, or membrane separation—to further minimize the CO content in the reformate gas

9.3.1 Basic Processes and Reactions

Table 9-3 summarizes the basic processes and reactions used in fuel processing. In addition to general equations, examples are given for methane, isooctane, and methanol. Methane may be used as a good representative of natural gas (natural gas contains up to 95% methane), and octane may be used as a representative of liquid hydrocarbon fuels. Gasoline is actually a blend of various hydrocarbons and cannot be represented with a single chemical formula.

Each reaction either needs energy (endothermic) or generates heat (exothermic). In the equations in Table 9-3, endothermic heat is shown on the left side of the equation, whereas exothermic heat is shown on the right side as a result of the reaction. The heats of the reactions were calculated from the heat of formation of participating species (shown in Table 9-4) and therefore refer to 25°C. Because the reactions take place at considerably higher temperatures, additional heat must be brought to the process to bring the reactants to the reaction temperature, and a significant amount of heat leaves the system with the products. For the reactions involving water, the form in which water participates may have a significant impact on the heat required or generated. In most of the processes steam is used, and therefore it would be appropriate to use the equations with water vapor. However, for system-level analyses such as those presented in this chapter, energy for steam generation must also be taken into account. In that case the equations with liquid water are used. This is also convenient when we're dealing with the efficiencies. As discussed in Chapter 2, the use of higher heating value for efficiency calculation is more appropriate, although in all combustion processes water leaves the system as vapor and has no chance to contribute to the process efficiency.

9.3.2 Steam Reforming

Steam reforming is an endothermic process, which means that heat must be brought to the reactor. This heat is typically generated by combustion of an

Table 9-3 Basic Processes and Reactions in Fuel Reforming

Combustion	$C_mH_nO_p + (m + n/4 - p/2)O_2 \rightarrow mCO_2 + (n/2)H_2O + heat$
Water vapor generated	$CH_4 + 2O_2 \rightarrow CO_2 + 2H_2O(g) + 802.5\,kJ$
	$C_8H_{18} + 12.5O_2 \rightarrow 8CO_2 + 9H_2O(g) + 5063.8\,kJ$
	$CH_3OH + 1.5O_2 \rightarrow CO_2 + 2H_2O(g) + 638.5\,kJ$
Liquid water generated	$CH_4 + 2O_2 \rightarrow CO_2 + 2H_2O(l) + 890.5\,kJ$
	$C_8H_{18} + 12.5O_2 \rightarrow 8CO_2 + 9H_2O(l) + 5359.8\,kJ$
	$CH_3OH + 1.5O_2 \rightarrow CO_2 + 2H_2O(l) + 726.5\,kJ$
Partial oxidation	$C_mH_nO_p + (m/2)O_2 \rightarrow (m - p)CO + pCO_2 + (n/2)H_2 + heat$
	$CH_4 + {}^1/_2O_2 \rightarrow CO + 2H_2 + 39.0\,kJ$
	$C_8H_{18} + 4O_2 \rightarrow 8CO + 9H_2 + 649.8\,kJ$
	$CH_3OH + {}^1/_2O_2 \rightarrow CO_2 + 2H_2 + 154.6\,kJ$
Steam reforming	$C_mH_nO_p + mH_2O + heat \rightarrow (m - p)CO + pCO_2 + (m + n/2)H_2$
	$CH_4 + H_2O(g) + 203.0\,kJ \rightarrow CO + 3H_2$
Water vapor used	$C_8H_{18} + 8H_2O(g) + 1286.1\,kJ \rightarrow 8CO + 17H_2$
	$CH_3OH + H_2O(g) + 87.4\,kJ \rightarrow CO_2 + 3H_2$
	$CH_4 + H_2O(l) + 247.0\,kJ \rightarrow CO + 3H_2$
Liquid water used	$C_8H_{18} + 8H_2O(l) + 1638.1\,kJ \rightarrow 8CO + 17H_2$
	$CH_3OH + H_2O(l) + 131.4\,kJ \rightarrow CO_2 + 3H_2$
Gas shift reaction	
Water vapor used	$CO + H_2O(g) \rightarrow CO_2 + H_2 + 37.5\,kJ$
Liquid water used	$CO + H_2O(l) + 6.5\,kJ \rightarrow CO_2 + H_2$
Preferential Oxidation	$CO + 0.5O_2 \rightarrow CO_2 + 279.5\,kJ$
Water evaporation	$H_2O(l) + 44.0\,kJ \rightarrow H_2O(g)$

additional amount of fuel. The two reactions are physically separated by a thermally conductive wall. Because of that, the resulting gas does not include any nitrogen. The steam reformation reaction is reversible, and the product gas is a mixture of hydrogen, carbon monoxide, carbon dioxide (some shift reaction also taking place), water vapor, and unconverted fuel. Figure 9-27 shows a schematic diagram of the steam-reforming process. The actual composition is a function of temperature, pressure, and composition of the feed gas. Figure 9-28 shows equilibrium concentrations of methane steam reformation as a function of temperature. Because the product gas still contains a significant amount of carbon monoxide, it is taken to the shift reactor, where CO reacts with additional steam and is converted to more hydrogen and CO_2. Depending on the desired CO content and applied

Table 9-4 Heat of Formation of Some Common Fuel Cell Gases and Liquids (from [13])

	Molecular Mass g/mol	Heat of Formation kJ/mol
Hydrogen, H_2	2.016	0
Oxygen, O_2	31.9988	0
Nitrogen, N_2	14.0067	0
Carbon monoxide, CO	28.0106	-113.8767
Carbon dioxide, CO_2	44.010	-393.4043
Water vapor, H_2O (g)	18.0153	-241.9803
Water liquid, H_2O (l)	18.0153	-286.0212
Methane, CH_4	16.043	-74.85998
Methanol, CH_3OH	32.0424	-238.8151
Octane, C_8H_{18}	114.230	-261.2312

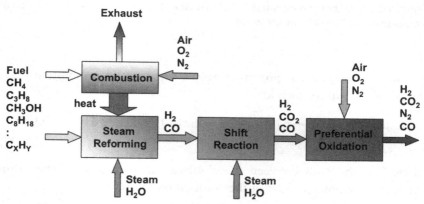

Figure 9-27 Schematic representation of a steam-reforming process.

catalyst, the shift reaction may be split into a high-temperature and a low-temperature shift reaction. The resulting gas still contains about 1% of CO, which would be detrimental to PEM fuel cell. The CO content is further reduced in the preferential oxidation process, where CO is catalytically oxidized with oxygen from air. Selection of the catalyst and control of operating conditions is critical in order to avoid, or at least minimize, combustion of hydrogen present in the gas.

The overall equation for steam reforming including the shift reaction is:

$$C_mH_nO_p + (2m - p)H_2O + Xheat \rightarrow mCO_2 + (2m + n/2 - p)H_2$$

$$(9\text{-}36)$$

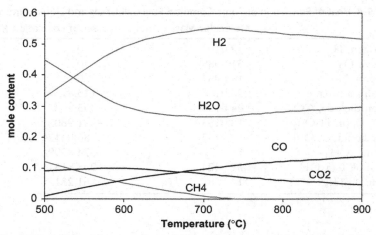

Figure 9-28 Equilibrium concentrations of methane steam reformation as a function of temperature (at atmospheric pressure).

The heat needed for the process, Xheat, is obtained by combustion of an additional amount of fuel, k:

$$kC_mH_nO_p + k(m + n/4 - p)O_2 \rightarrow kmCO_2 + k(n/2)H_2O + Xheat$$

$$(9\text{-}37)$$

where k is the ratio between the absolute values of heat of the steam reforming reaction and heat of the combustion reaction:

$$k = \frac{\Delta H_{SR} + (m - p)\Delta H_{shift}}{\Delta H_{comb}} \qquad (9\text{-}38)$$

Note that ΔH_{comb} is also the heating value of particular fuel used in the combustion process (Table 9-3).

The theoretical efficiency of the steam-reforming process is then a ratio between the heating value of hydrogen produced and the heating value of fuel consumed in both reactions:

$$\eta = \frac{\left(2m + \dfrac{n}{2} - p\right)}{(1 + k)} \frac{\Delta H_{H_2}}{\Delta H_{fuel}} \qquad (9\text{-}39)$$

It can be easily shown that the theoretical efficiency of the steam reforming process is 100%, if the heating values are replaced by the heats of formation of the constituents of the corresponding reactions (i.e., hydrogen combustion, fuel combustion, steam reforming, and shift reaction). The actual efficiency is always lower because of:

- Heat losses in the process, including both heat dissipated to the environment and heat taken away by the gases (both exhaust gases from combustion reaction and product gases from the shift reaction), and
- Incomplete reactions, including all three reactions involved.

Steam Reforming with Membrane Separation

Another elegant way of avoiding shift reaction and selective oxidation is to use a metal membrane to separate hydrogen from the reformate produced by steam reforming. These membranes allow only hydrogen to go through, so the product is high-purity hydrogen. Carbon monoxide and unconverted fuel and other byproducts of steam reformation are brought back into the combustion chamber (Figure 9-29), reducing the amount of fuel that has to be burned to produce the heat for the steam-reforming process. Such a process can be very efficient. However, filtration through the membrane requires relatively high pressure. For that reason, such a process is more suitable for liquid fuels, such as methanol, or for fuels that already come compressed, such as propane. Compression of natural gas would take a significant toll on the overall system efficiency.

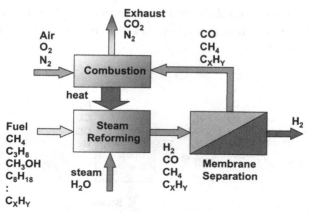

Figure 9-29 Schematic representation of the steam-reforming process with metal membrane separation.

9.3.3 Partial Oxidation and Autothermal Reforming

Unlike steam reforming, partial oxidation is an exothermic process. Essentially it is a combustion but with a less-than-stoichiometric amount of oxygen so that the products are carbon monoxide and hydrogen (instead of carbon dioxide and water vapor produced in full combustion). In a process similar to steam reforming, the reformate gas must go through a shift reaction to produce more hydrogen and through preferential oxidation to reduce the CO content to an acceptable level.

Partial oxidation produces less hydrogen per mole of fuel. In the case of methane, partial oxidation would generate only 2 moles of hydrogen per mole of methane, as opposed to 3 moles of hydrogen produced by steam reforming. In the case of octane, this ratio is even worse, that is, 9 moles vs. 17 moles.

In addition, unlike the product gas from steam reforming, which contains only hydrogen and carbon dioxide (with minute amounts of CO after preferential oxidation), the product gas from partial oxidation contains relatively large amounts of nitrogen. In other words, hydrogen content in the gas is much lower (about 40% vs. 80% from steam reforming), and that may have some impact on fuel cell performance. Figure 9-30 shows a schematic diagram of the partial oxidation process.

Because partial oxidation is an exothermic process (i.e., it produces heat), and steam reforming is an endothermic process (i.e., it needs heat), these two processes can actually be combined in the so-called *autothermal reforming process*. Instead of external combustion and heat transfer to the steam-reforming reactor, in an autothermal reformer heat is generated internally, by partial oxidation, and the heat is then carried by the reacting gases and partial oxidation products (hydrogen and carbon monoxide) to the steam-reforming zone (Figure 9-31).

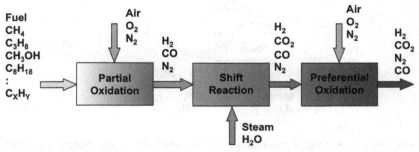

Figure 9-30 Schematic representation of the partial oxidation process.

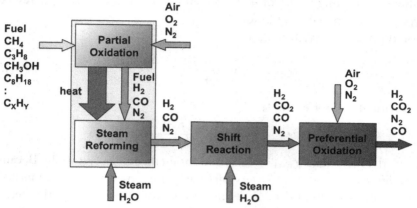

Figure 9-31 Schematic representation of an autothermal-reforming process.

The overall equation of an autothermal process with shift reaction is:

$$C_mH_nO_p + \chi O_2 + (2m - p - 2\chi)H_2O \;\rightarrow\; (2m + n/2 - p - 2\chi)H_2$$
(9-40)

where χ is the number of moles of oxygen per mole of fuel.

For methane, octane, and methanol, respectively, the overall equation is:

$$CH_4 + \chi O_2 + (2 - 2\chi)H_2O \;\rightarrow\; CO_2 + (4 - 2\chi)H_2 \qquad (9\text{-}41)$$

$$C_8H_{18} + \chi O_2 + (16 - 2\chi)H_2O \;\rightarrow\; 8CO_2 + (25 - 2\chi)H_2 \quad (9\text{-}42)$$

$$CH_3OH + \chi O_2 + (1 - 2\chi)H_2O \;\rightarrow\; CO_2 + (3 - 2\chi)H_2 \qquad (9\text{-}43)$$

The previous reaction may be endothermic, exothermic, or thermoneutral, depending on the amount of oxygen brought into the system, χ. Note that this is exactly the same overall equation as the overall equation for steam reforming with external combustion. However, the difference is that in the steam reforming, the two processes, combustion and reforming, are physically separated, whereas in autothermal reforming these processes follow each other in the same reactor.

If χ in the equation is selected so that no heat is needed and no heat is generated, this process is autothermal, and its theoretical efficiency is 100%. The efficiency, defined as a ratio between energy of hydrogen produced and energy of fuel consumed, are both expressed with their respective heating values:

$$\eta = \left(2m + \frac{n}{2} - p - 2\chi\right) \frac{\Delta H_{H_2}}{\Delta H_{fuel}} \qquad (9\text{-}44)$$

where the heating values of hydrogen and fuel, ΔH_{H_2} and ΔH_{fuel}, respectively, are expressed in kJ kmol^{-1}.

For 100% efficiency, $\eta = 1$, χ is:

$$\chi_{\eta=1} = m + \frac{n}{4} - \frac{p}{2} - \frac{\Delta H_{fuel}}{2\Delta H_{H_2}} \tag{9-45}$$

For natural gas, for example:

$$\chi_{\eta=1} = 1 + \frac{4}{4} - \frac{890.5}{2 \times 286} = 0.443$$

Similarly, for octane $\chi_{\eta=1} = 3.19$, and for methanol $\chi_{\eta=1} = 0.23$. Because χ is different for every fuel, very often an equivalence ratio is used instead. The equivalence ratio, ε, is defined as a ratio between the theoretical amount of oxygen needed for complete combustion and the actual amount of oxygen, χ:

$$\varepsilon = \frac{m + \dfrac{n}{4} - \dfrac{p}{2}}{\chi} \tag{9-46}$$

In that case the overall equation for autothermal reforming becomes:

$$C_m H_n O_p + \frac{m + \dfrac{n}{4} - \dfrac{p}{2}}{\varepsilon} O_2 + \frac{(\varepsilon - 1)(2m - p) - \dfrac{n}{2}}{\varepsilon} H_2 O$$

$$\rightarrow mCO_2 + \left(2m + \frac{n}{2} - p\right)\left(1 - \frac{1}{\varepsilon}\right) H_2 \tag{9-47}$$

The ideal efficiency of the autothermal reforming is then:

$$\eta = \left(2m + \frac{n}{2} - p\right)\left(1 - \frac{1}{\varepsilon}\right)\frac{\Delta H_{H_2}}{\Delta H_{fuel}} \tag{9-48}$$

Note that for $\varepsilon = 1$ the reaction becomes regular combustion and no hydrogen is being produced. As the equivalence ratio increases, the amount of hydrogen, and therefore the efficiency, increases. Figure 9-32 shows the theoretical efficiency of the reforming process as a function of the equivalence ratio for methane, octane, and methanol. However, there are both theoretical and practical limits to the equivalence ratio. The theoretical limit corresponds to the 100% efficiency, based on the higher heating value (for both hydrogen and fuel). An increase of the equivalence ratio means that less oxygen is used in the process. At the equivalent ratios approaching the

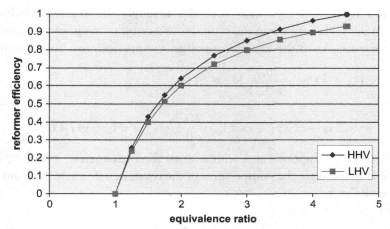

Figure 9-32 Relationship between reformer efficiency and equivalence ratio.

theoretical limit, there is a risk of carbon formation in the fuel processor through the following reactions:

$$CH_4 \rightarrow C + 2H_2 \tag{9-49}$$

$$2CO \rightarrow C + CO_2 \tag{9-50}$$

To reduce the risk of carbon formation, the equivalence ratio is always selected somewhat below the theoretical value, and steam is added to the reaction in excess of that theoretically required, typically at a rate corresponding to the steam-to-carbon ratio, σ, of 2.0–3.0 (moles of steam to moles of carbon in fuel).

The actual fuel reforming reaction is then:

$$C_mH_nO_p + \frac{m + \frac{n}{4} - \frac{p}{2}}{\varepsilon} O_2 + \frac{r_{O2}}{1 - r_{O2}} \frac{m + \frac{n}{4} - \frac{p}{2}}{\varepsilon} N_2 + \sigma H_2O \tag{9-51}$$

$$\rightarrow aC_mH_nO_p + bCH_4 + cCO_2 + dCO + eN_2 + fH_2 + gH_2O$$

Nitrogen appears in the reaction because it comes with air. Because nitrogen does not practically participate in the reaction, the coefficient e on the right side of the equation is equal to the nitrogen coefficient on the left side. Other coefficients on the right side, that is, the exact composition of the reformate gas, with given feed on the left side, will depend on the catalyst, reactor design, temperatures, pressure, and process control. Fuel appearing on the right side represents fuel slip through the reactor, obviously unwanted. CH_4 appears on the right side as a result of incomplete reactions.

For example, in the case of natural gas the previous reaction becomes:

$$CH_4 + \frac{2}{\varepsilon} O_2 + \frac{2}{\varepsilon} \frac{0.79}{0.21} N_2 + \sigma H_2O$$

(9-52)

$$\rightarrow bCH_4 + cCO_2 + dCO + \frac{2}{\varepsilon} \frac{0.79}{0.21} N_2 + fH_2 + gH_2O$$

With some assumptions and with known feed parameters, ε and σ, it is possible to calculate the composition of the resulting reformate gas from the species balance.

From carbon balance:

$$b + c + d = 1$$

(9-53)

From oxygen balance:

$$\frac{4}{\varepsilon} + \sigma = 2c + d + e$$

(9-54)

From hydrogen balance:

$$4 + 2\sigma = 4b + 2f + 2e$$

(9-55)

From nitrogen balance:

$$e = \frac{2}{\varepsilon} \frac{0.79}{0.21}$$

(9-56)

These are four equations, but there are six coefficients on the right side of Equation (9-52). The remaining two equations may be derived from the requirements for maximum CO content in the produced reformate gas (typically around 1% by volume in dry gas) and from the efficiency requirement.

The CO content (by volume.) in the dry reformate gas is:

$$r_{CO} = \frac{d}{b + c + d + e + f} = \frac{d}{1 + \frac{2}{\varepsilon} \frac{0.79}{0.21} + f}$$

(9-57)

Similarly, hydrogen content in the dry reformate gas is:

$$r_{H_2} = \frac{f}{b + c + d + e + f} = \frac{f}{1 + \frac{2}{\varepsilon} \frac{0.79}{0.21} + f}$$

(9-58)

The actual efficiency of the reforming process is:

$$\eta_{actual} = f \frac{\Delta H_{H_2}}{\Delta H_{fuel}} \qquad (9\text{-}59)$$

where the heating values of hydrogen and fuel, ΔH_{H_2} and ΔH_{fuel}, respectively, are expressed in kJ kmol^{-1}.

Hydrogen content in the reformate gas is directly related to the actual efficiency. From Equations (9-58) and (9-59) it follows:

$$\eta_{actual} = \frac{r_{H_2} \dfrac{\Delta H_{H_2}}{\Delta H_{fuel}} \left(1 + \dfrac{2}{\varepsilon}\dfrac{0.79}{0.21}\right)}{1 - r_{H_2}} \qquad (9\text{-}60)$$

Figure 9-33 shows the relationship between hydrogen content and the reformer efficiency for an autothermal natural gas reformer. It should be noted that for each equivalence ratio there is a limit that the efficiency can reach (from Equation 9.48 or Figure 9-32).

A typical composition of the reformate gas from an autothermal natural gas reformer is shown in Table 9-4. Similar analysis may be performed for other fuels.

The previous reactions include both the fuel reforming and shift reaction as they occur simultaneously, although they take place in separate reactors

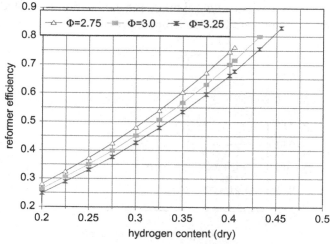

Figure 9-33 Relationship between hydrogen content in dry reformate gas (by volume) and autothermal natural gas reformer efficiency (CO content 1%).

using different types of catalysts. The high temperature in the fuel processor does not favor the shift reaction. Sometimes the shift reaction is divided into two steps: high-temperature shift reaction (400–500°C) and low-temperature reaction (200–250°C), also using different types of catalysts, typically iron and copper based. Recently, precious metal catalysts, such as Pt on ceria or gold on ceria, show more promise, not only because they are more tolerant to sulphur but, more important, because they allow higher space velocities, thus resulting in smaller reactors.

Shift reaction can reduce the CO content typically to below 1% (by volume in dry gas). This would still be too high for a PEM fuel cell, which has very low tolerance to CO (typically below 100 ppm, depending on the operating temperature). Further CO cleanup is achieved by an additional step: preferential or selective oxidation. This is essentially a controlled catalytic combustion of CO on a noble metal catalyst that at a given temperature has higher affinity to CO than to H_2. The reaction is simply:

$$CO + 1/2O_2 \rightarrow CO_2 \tag{9-61}$$

However, to ensure almost complete elimination of CO (below 100 ppm), oxygen must be provided in excess, typically 2.0–3.5 higher than the stoichiometric ratio. Excess oxygen in the presence of a noble metal catalyst reacts with hydrogen present in the reformate gas. The efficiency of the preferential oxidation process is then defined as a ratio between the amount of hydrogen at the outlet and the amount of hydrogen at the inlet. Figure 9-34 shows the efficiency of a typical preferential oxidation (PROX) reactor. In addition, air supply and PROX temperature must be carefully controlled

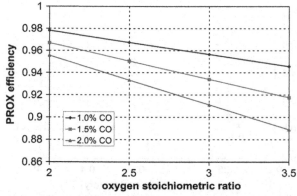

Figure 9-34 Efficiency of a PROX reactor as a function of oxygen stoichiometric ratio and CO content at the inlet.

to avoid hazardous situations, particularly if the gas flow rate is highly variable.

In general, control of the fuel processing is not an easy task. Fuel processing consists of a series of reactors, each operating at its own relatively narrow operating temperature window. The composition of the gas at the exit of each of those reactors depends highly on the composition of the inlet gas, plus oxygen and steam feed, and the reactor temperature. If any of the parameters is even slightly out of the operating window even for a very short period, the result is higher CO content at the outlet, which, as shown in a moment, may have a detrimental effect on fuel cell performance.

9.3.4 Effect of Reformate on Fuel Cell Performance

The composition of the reformate gas coming out of the reformer depends on the type of fuel, type of reformer, and the operating fuel processor efficiency. Table 9-6 gives approximate composition for some typical cases. In addition, not shown in the table, the reformate gas contains up to 100 ppm CO, unconverted fuel, and other hydrocarbons (methane, ethane

Table 9-5 Reformate Gas Composition for Various Equivalence Ratios and Efficiencies (for Natural Gas)

Equivalence ratio	3.0	3.0	3.20
Carbon to steam ratio	2.0	2.0	2.0
Theoretical efficiency (HHV)	0.85	0.85	0.88
Actual efficiency (HHV)	0.75	0.80	0.80
Actual efficiency (LHV)	0.70	0.75	0.75
CO_2	0.150	0.152	0.153
H_2	0.401	0.416	0.428
N_2	0.428	0.417	0.402
CO	0.010	0.010	0.010
CH_4	0.011	0.004	0.008

Table 9-6 Typical Reformate Compositions for Various Fuels and Various Reforming Processes (Including Preferential Oxidation)

	H_2	CO_2	N_2
Natural gas autothermal	0.42	0.16	0.42
Gasoline autothermal	0.40	0.20	0.40
Methanol autothermal	0.50	0.20	0.30
Natural gas steam reforming	0.75	0.24	0.01
Methanol steam reforming	0.70	0.29	0.01

ethylene), and it may contain traces of other byproducts, such as hydrogen sulfide, ammonia, aldehydes, and so forth.

The reformate gas has several effects on PEM fuel cell performance, such as:

- Loss of potential due to hydrogen's low content
- Catalyst poisoning with CO
- Catalyst and membrane poisoning with other constituents (such as ammonia and hydrogen sulfide), even at very small concentrations

Hydrogen content of 40% results in 2.5 times lower concentration than in pure hydrogen. The result is a lower cell voltage due to the Nernst equation. Typically, the voltage loss is less than 30 Mv (Figure 9-35).

Carbon monoxide has a much more severe impact, even at very low concentrations. At low temperatures (below 100°C), Pt has higher affinity toward CO than toward hydrogen, and as a result most of the catalyst sites get occupied with CO [15], even at concentrations as low as 100 ppm, resulting in practically no current-generating capability (Figure 9-36). It is well documented that in PEM fuel cells at 80°C, a PtRu catalyst has much higher tolerance to CO [16]. Figure 9-37 shows significant improvement in cell performance with a PtRu anode catalyst compared with a regular Pt catalyst. This is probably due to water activation by Ru and subsequent CO electrooxidation on a neighboring Pt atom [17] or due to weakening of the Pt/CO bond strength [15].

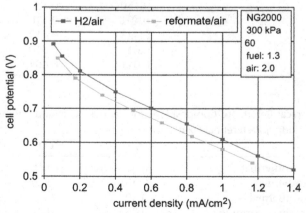

Figure 9-35 Effect of reduced hydrogen concentration in reformate on fuel cell performance [14].

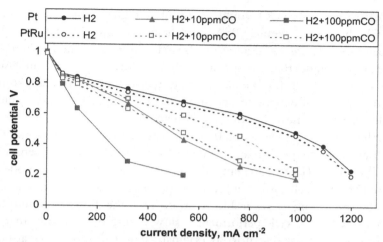

Figure 9-36 CO tolerance of Pt and PtRu catalyst (308 kPa, 80°C, 1.3/2.0 hydrogen/air stoichiometry) [15].

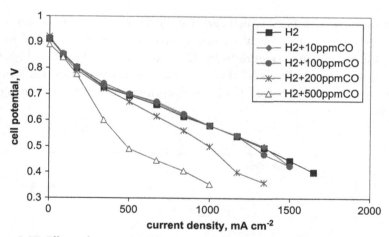

Figure 9-37 Effect of 2% air bleed on CO tolerance of PtRu catalyst (reformate/air, reformate composition 40% H_2, 40% N_2, 20% CO_2; 308 kPa, 60°C, 1.3/2.0 hydrogen/air stoichiometry) [18].

In addition, a little bit of air (typically about 2%) injected on the anode side of the fuel cell (called *air bleed*) helps oxidize the CO. With a PtRu catalyst and with 2% air bleed it is possible to operate a fuel cell with reformate containing up to 100 ppm with virtually no loss in performance (Figure 9-37) [18].

Better tolerance to CO would be possible at higher temperatures (>120–130°C); however, regular PSA membranes cannot operate at those temperatures. There are significant research efforts dedicated to development of membranes that can operate at higher temperatures. Although some show promise [19–21], none has been put to work in practical applications yet. An additional benefit of operation at a higher temperature would be smaller size of the heat rejection equipment.

Other possible byproducts in the reformate may have a detrimental effect on PEM fuel cell performance. Figure 9-38 shows what happens when a small quantity (13 to 130 ppm) of ammonia is injected in the anode fuel stream [22]. The cell current at given cell potential steadily declines. It seems that this poisoning is reversible, that is, after the flow of ammonia is disconnected, the cell performance slowly improves. Note that the performance did not return to its original value, even after several hours. Ammonia may be generated in the autothermal reformer under certain conditions (after all, the main constituents, hydrogen and nitrogen, are present).

Poisoning with H_2S, which also may be formed in the reformer if the fuel desulfurization step is not working properly, seems more fatal to the fuel cell performance (Figure 9-39) [23]. As low as 1–3 ppm of H_2S was sufficient to practically "kill" the cell, and performance does not seem to improve after the H_2S feed has been discontinued, that is, the damage is permanent.

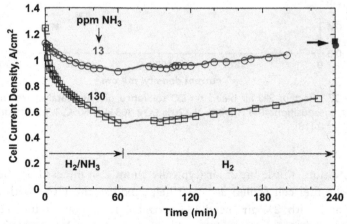

Figure 9-38 Effect of two concentrations of NH_3 on fuel cell performance at 80°C [22]. *(Reprinted by permission from the Electrochemical Society.)*

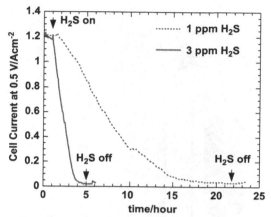

Figure 9-39 Effect of 1 ppm and 3 ppm H_2S burp on fuel cell performance operating at 0.5 V and 80°C. Humidified H_2/H_2S mix is injected directly to the anode [23]. *(Reprinted by permission from the Electrochemical Society.)*

9.3.5 System Integration

As discussed earlier, a fuel processor encompasses several processes, each taking place in a separate reactor or vessel. In addition, some of these processes require air supply, some require steam supply, and all require some kind of temperature control. The equipment for air supply, steam supply, and temperature control must be incorporated in the fuel processor subsystem. There are numerous ways the components and the flows between them can be arranged. The goal is to have a system with the highest possible efficiency (i.e., hydrogen yield), lowest possible CO content in the produced gas, minimum harmful byproducts (such as ammonia and hydrogen sulfide), and flexibility enough that it can respond to the fuel cell hydrogen needs. Figure 9–40 shows some possible reformer subsystem configurations. Although the two systems in Figure 9–40 contain the same components, they are arranged in slightly different ways. One, for example, uses the heat from the fuel cell tail gases burner to preheat the fuel and water, whereas the other uses this heat entirely to generate steam and uses the heat from the main reactor to preheat fuel. In the case of liquid fuel (such as gasoline or methanol), preheating also includes fuel evaporation. Both natural gas and gasoline, and particularly diesel fuel, contain sulphur, which must be removed, typically as the first step in fuel processing. Additional heat exchangers may be needed in the systems that intend to recuperate waste heat.

(a)

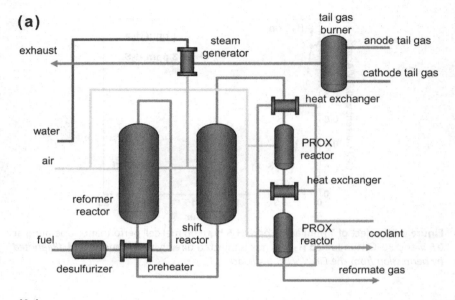

(b)

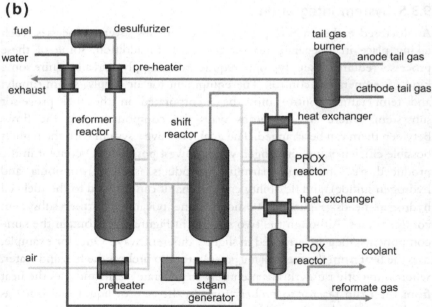

Figure 9-40 Schematic diagrams of two versions of an autothermal reformer.

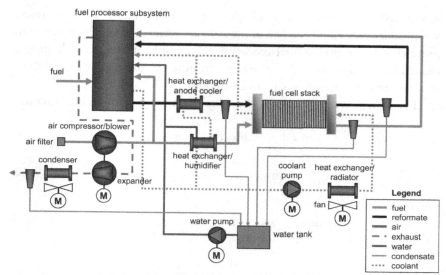

Figure 9-41 A complete fuel cell system integrated with fuel processor.

The fuel processor needs to be fully integrated with the fuel cell system. It not only provides fuel to the fuel cell and uses the heat from the fuel cell exhaust gases, but it also shares air, water, coolant, and control subsystems, as illustrated in Figure 9–41.

The reformate coming out of the fuel processor is hot and oversaturated with water. This eliminates the need for an anode humidifier, but it requires a cooler to bring the anode gas to the fuel cell operating temperature. Condensed water is separated before entering the fuel cell stack.

The fuel processor and the fuel cell typically operate at the same pressure, so they may share the same air supply. It is important to distribute the air flow to where it is needed, namely, the fuel cell stack, the fuel processor reactor, and the preferential oxidation reactors; an additional air bleed may be required at the stack anode inlet. To avoid expensive mass flow controllers (which also may require much higher pressures to operate properly), passive devices such as orifices are often used.

Water is needed by the fuel processor to generate steam for the fuel processor and shift reactors and by the fuel cell for humidification of the cathode inlet. Various cathode humidification schemes have been discussed. Typically, high-pressure water is required for both steam generation and injection in the humidifier. Fuel cell systems, regardless of application, may be designed to operate without need for makeup water. Water enters the system with ambient air and leaves the system with exhaust. Water is created as a product in the fuel cell and in the tail gas burner, and small quantities are produced in the preferential oxidation (as a result of unwanted hydrogen oxidation). At the same time, water is consumed in the fuel processor (both in steam reforming and in gas shift reactions).

The system water balance is therefore given by the following equation:

$$\dot{m}_{H_2O}\Big|_{in}^{air} + \dot{m}_{H_2O}\Big|_{gen}^{FC} + \dot{m}_{H_2O}\Big|_{gen}^{TGC} + \dot{m}_{H_2O}\Big|_{gen}^{PROX}$$

$$= \dot{m}_{H_2O}\Big|_{cons}^{FP} + \dot{m}_{H_2O}\Big|_{out}^{exh} \tag{9-62}$$

where:

$\dot{m}_{H_2O}\Big|_{in}^{air}$ = amount of water (humidity) that enters the system with ambient air

$$\dot{m}_{H_2O}\Big|_{in}^{air} = \dot{m}_{Airin} = \dot{m}_{Airin}\frac{M_{H_2O}}{M_{air}}\frac{\varphi_{amb}P_{vs}(T_{amb})}{P_{amb} - \varphi_{amb}P_{vs}(T_{amb})} \tag{9-63}$$

where $\dot{m}_{Airin}$ is the total amount of dry air needed by the fuel cell system, which includes air for the fuel cell, air for the fuel processor, air for preferential oxidation, and air for air bleed:

$$\dot{m}_{Airin} = \dot{m}_{Air}\Big|_{in}^{FC} + \dot{m}_{Air}\Big|_{in}^{FP} + \dot{m}_{Air}\Big|_{in}^{PROX} + \dot{m}_{Air}\Big|_{Airbleed} \tag{9-64}$$

Air for the fuel cell is given by Equation (5-42):

$$\dot{m}_{Air}\Big|_{in}^{FC} = \frac{S_{O2}}{r_{O2}}\frac{I\cdot n_{cell}}{4F}M_{Air} \tag{9-65}$$

Air for the fuel processor may be calculated from Equations (5-39, 9-51, and 9-59):

$$\dot{m}_{Air}\Big|_{in}^{FP} = S_{H_2}\frac{I\cdot n_{cell}}{2F}\frac{\Delta H_{H_2}}{\eta_{FP}\Delta H_{fuel}}\frac{m + n/4 - p/2}{\varepsilon}\frac{M_{Air}}{r_{O2}} \tag{9-66}$$

Air for preferential oxidation may be calculated from Equations (5-39, 9-51, 9-57, 9-59, and 9-61):

$$\dot{m}_{Air}\Big|_{in}^{PROX} = S_{H_2}\frac{I\cdot n_{cell}}{2F}\frac{\Delta H_{H_2}}{\eta_{FP}\Delta H_{fuel}}r_{CO}$$

$$\left(1 + \frac{m + n/4 - p/2}{\varepsilon}\frac{1 - r_{O2}}{r_{O2}} + \eta_{FP}\frac{\Delta H_{fuel}}{\Delta H_{H_2}}\right)\frac{S_{PROX}}{2}\frac{M_{Air}}{r_{O2}} \tag{9-67}$$

Air bleed is usually about 2% of the anode flow:

$$\dot{m}_{Air}\big|_{Airbleed} = 0.02 \cdot \frac{S_{H_2}}{r_{H_2}} \frac{I \cdot n_{cell}}{2F} M_{Air} \tag{9-68}$$

$\dot{m}_{H_2O}\big|_{gen}^{FC}$ = amount of water generated in fuel cell (from Equation 5-7):

$$\dot{m}_{H_2O}\bigg|_{gen}^{FC} = \frac{In_{cell}}{2F} M_{H_2O} \tag{9-69}$$

$\dot{m}_{H_2O}\big|_{gen}^{TGC}$ = amount of water generated in tail gas combustor:

$$\dot{m}_{H_2}\bigg|_{gen}^{TGC} = (S_{H_2}-1) \frac{I \cdot n_{cell}}{2F} M_{H_2O} \tag{9-70}$$

$\dot{m}_{H_2O}\big|_{gen}^{PROX}$ = amount of water generated in preferential oxidation of CO (from Equations 9-51, 9-57, 9-59, and 9-61):

$$\dot{m}_{H_2O}\bigg|_{gen}^{PROX} = S_{H_2}\frac{I \cdot n_{cell}}{2F} \frac{\Delta H_{H_2}}{\eta_{FP}\Delta H_{fuel}} r_{CO}$$

$$\left(1 + \frac{m+n/4-p/2}{\varepsilon}\frac{1-r_{O2}}{r_{O2}} + \eta_{FP}\frac{\Delta H_{fuel}}{\Delta H_{H_2}}\right)(S_{PROX}-1)M_{H_2O} \tag{9-71}$$

$\dot{m}_{H_2O}\big|_{cons}^{FP}$ = amount of water consumed in the fuel processor (from Equations 9-47, 9-57, and 9-59):

$$\dot{m}_{H_2O}\bigg|_{cons}^{FP} = S_{H_2}\frac{I \cdot n_{cell}}{2F} \frac{\Delta H_{H_2}}{\eta_{FP}\Delta H_{fuel}} \frac{(\varepsilon-1)(2m-p)-n/2}{\varepsilon} M_{H_2O} \tag{9-72}$$

$m.{H_2O}\big|_{out}^{exh}$ = amount of water that exits the system with exhaust gases

$$\dot{m}_{H_2O}\bigg|_{out}^{exh} = \dot{m}_{exh}\frac{M_{H_2O}}{M_{exh}}\frac{P_{vs}(T_{exh})}{P_{amb}-P_{vs}(T_{exh})} \tag{9-73}$$

where $\dot{m}_{exh}$ is the mass flow of exhaust gases, which may be easily calculated from the system mass balance:

$$\dot{m}_{exh} = \dot{m}_{fuelin} + \dot{m}_{Airin} - \dot{m}_{H_2O}\bigg|_{gen}^{FC} - \dot{m}_{H_2O}\bigg|_{gen}^{TGC} - \dot{m}_{H_2O}\bigg|_{gen}^{PROX}$$

$$+ \dot{m}_{H_2O}\bigg|_{cons}^{FP} \tag{9-74}$$

$$\dot{m}_{fuelin} = S_{H_2}\frac{I \cdot n_{cell}}{2F}\frac{\Delta H_{H_2}}{\eta_{FP}\Delta H_{fuel}}M_{fuel} \tag{9-75}$$

Note that in Equations (9-65) through (9-75), the heating values of hydrogen and fuel, ΔH_{H_2} and ΔH_{fuel}, respectively, are expressed in kJ kmol^{-1}. Either lower or higher heating value may be used, but consistently, depending on how the fuel processor efficiency, η_{FP}, is expressed. Also in Equations (9-65) through (9-75), $I \cdot n_{cell}$ may be replaced by W/V_{cell}, where W is the fuel cell power output.

Equation (9-62) is satisfied and neutral water balance is accomplished by condensing the exhaust gases to a required temperature, T_{exh}. In systems where neutral water balance is not critical, the condenser at the exhaust may be omitted.

Water presence in the system makes the fuel cell systems susceptible to freezing if used in a cold environment. Various engineering solutions have been proposed and a few tried in the field to prevent freezing, ranging from draining the water from the system at shutdown to keeping the system warm, either by electric heaters or having the system run (constantly or periodically) at very low power. The former requires changes in both system and stack design, whereas the latter may have a significant impact on overall system efficiency.

Deionized water may be used as coolant, but in the systems susceptible to freezing, the coolant loop is separated from the water loop, which allows for antifreeze coolants to be used, such as aqueous solutions of ethylene-glycol or propylene-glycol. Another option is to have an air-cooled system, which is typically employed for low-power applications (below 3–4 kW).

In addition to taking the heat away from the fuel cell stack, the cooling loop also cools down the reformate gas coming from the fuel processor and provides heat, if needed, in the air humidification process. Moreover, temperature control of the preferential oxidation process is needed (as shown in Figures 9-40 and 9-41). The heat is then rejected from the system to the environment by the heat exchanger, typically of a radiator type. Some heat may be rejected from the system in the condenser just before the exhaust gases leave the system. The condenser is needed to condense and save the water in the exhaust to maintain a neutral water balance in the system.

The heat load distribution on the radiator and condenser depends on the operating pressure and temperature. Figure 9-42 shows the heat loads on the radiator and condenser as a function of operating pressure and

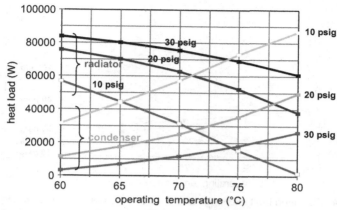

Figure 9-42 Radiator and condenser heat load as a function of operating pressure and temperature of a 50-kW$_e$ net system operating at 32.5% system efficiency [24].

temperature [24]. Higher operating temperatures and lower operating pressures result in the shifting of heat load from the radiator to the condenser. Though at high pressure (300 kPa) and low temperature (60°C) almost all the heat is rejected in the radiator, at low pressure (170 kPa) and high temperature almost all the heat is rejected in the condenser. The major difference between the radiator and the condenser is that the radiator is a liquid/gas heat exchanger and the condenser is a gas/gas heat exchanger. (Some heat transfer is used for phase change, that is, condensation of water, but the amount of water is orders of magnitude lower than the amount of gas.) The heat transfer coefficients are significantly lower for gas/gas heat exchangers, which means that for the same amount of heat load these heat exchangers require a much larger heat exchange area.

Figure 9-43 compares the heat exchange areas for the radiator and the condenser at different operating temperatures and pressures [24]. Although the automotive heat exchangers are typically densely packed (>1000 m²/m³), the size of the heat exchangers may very well be a limiting factor for the automotive fuel cell systems. Higher operating temperature does not necessarily mean smaller heat exchanger size if water balance is to be maintained. For this particular analysis [24], the smallest heat exchangers resulted in a system that operates at high pressure (308 kPa) and low temperature (60°C). At an operating temperature below 60°C at sufficiently high operating pressure (above 300 kPa), there is no need for a condenser, and all the heat can be rejected in the radiator.

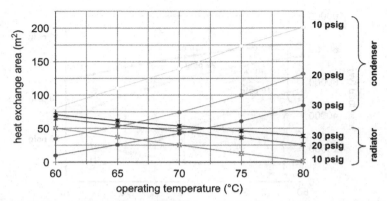

Figure 9-43 Required heat exchange area for a 50-kW$_e$ net system operating at 32.5% system efficiency (liquid/gas heat transfer coefficient = 60 W m^{-2} K, air/air heat transfer coefficient 15 W m^{-2} K, fin area/tube area = 10) [24].

9.4 ELECTRICAL SUBSYSTEM

An electrical subsystem is needed to deliver power produced by a fuel cell to the user (load); this subsystem typically does not just deliver, it also modifies the fuel cell electrical output so that it matches the load requirements in terms of voltage, type of current, power quality, and transients. Obviously the configuration and characteristics of such a subsystem strongly depend on the load requirements, which vary with application.

Fuel cells generate direct current (DC) at a voltage that depends on the number of cells connected in series and on the current being generated. (Remember, fuel cell voltage and current are related by the fuel cell's polarization curve.) The fuel cell voltage, therefore, changes with current. The fuel cell open circuit voltage is around 1 V per cell (a little bit above 1 V for hydrogen/oxygen and a little bit below 1 V for hydrogen/air systems). The cell potential at nominal power is a design variable and is typically selected at a value between 0.6 V and 0.7 V. Therefore, any fuel cell stack will have a voltage swing between 0.6:1.0 and 0.7:1.0. Very few loads could tolerate such voltage swings.

Voltage regulation is one of the most common functions of the electrical subsystem. Depending on where the fuel cell swing potential is relative to the required load potential, the electrical subsystem must either reduce the fuel cell voltage (a so-called *buck converter*) or increase it (a so-called *boost converter*). If the load voltage is inside the fuel cell voltage range, a buck-boost converter is needed (Figure 9-44).

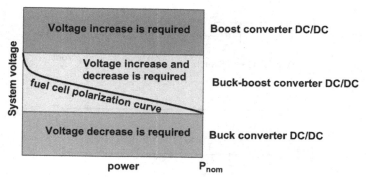

Figure 9-44 Fuel cell polarization curve determines the type of converter needed to match the load voltage.

Voltage regulation is accomplished using switching or chopping circuits [25] that use electronic switches, such as:

- Thyristors and gate–turn–off (GTO) thyristors
- A metal oxide semiconductor field effect transistor (MOSFET), used typically in low-voltage systems up to approximately 1 kW
- An insulated gate bipolar transistor (IGBT) for high-current applications (>50 A)

When arranged in electronic circuits with diodes, capacitors, and inductors (as shown in Figure 9-45), these electronic switches can provide the desired voltage output of the circuit by switching on and off certain parts of the circuit. The output voltage is a function of input voltage and switching time.

For a step-down or buck converter:

$$V_{out} = DV_{in} \qquad (9\text{-}76)$$

where D is defined as a switching function, $D = \dfrac{t_{on}}{t_{on} + t_{off}}$

For a step-up or boost converter:

$$V_{out} = \frac{V_{in}}{1 - D} \qquad (9\text{-}77)$$

The actual voltage is always somewhat lower than that given by Equations (9.76) and (9.77) because of energy losses in the circuit, which include switching losses, power loss in the switch while it is turned on, power loss due to the resistance of the inductor, and losses in the diode. In practice, these losses are very low [25]. The efficiencies of step-down converters are typically more than 90%, and in high-voltage systems ($V_{in} > 100$ V) they can be as high as 98%. The efficiencies of the step-up converters are somewhat

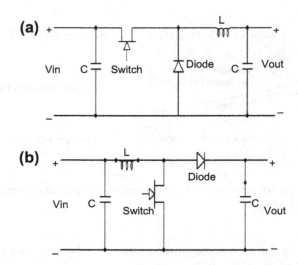

Figure 9-45 Switching circuits used in DC converters: (a) step-down or buck converter; (b) step-up or boost converter.

lower because of a different duty cycle (in high-voltage systems, they can be as high as 95%). It should be mentioned that the efficiency is a function of current through different elements of the circuit as well as of the switching function, and therefore it is not constant throughout the power range.

Fuel cells generate direct current (DC), but some loads or applications require alternating current (AC). For those applications, the electrical subsystem must also include a step of converting DC to AC of a fixed frequency (60 Hz or 50 Hz). This is also accomplished by the electronic switching devices arranged in electrical circuits as shown in Figure 9-46. The fuel cell system may be connected to a single AC voltage (typically in domestic applications) or to a three-phase supply (in larger industrial systems).

The resulting current from AC inverters is in a square-wave shape. For some applications this may be sufficient; however, for most applications, particularly the grid-connected applications, the square-wave output is not acceptable. In that case modulation is required to generate the output closer to the pure sine wave. Typically, pulse-width modulation (PWM), or more recently the tolerance-bend pulse method, is used. The efficiency of commercially available DC/AC inverters varies between 70% and 90% and is a strong function of the required power quality (i.e., how close to the true sine wave the output has to be).

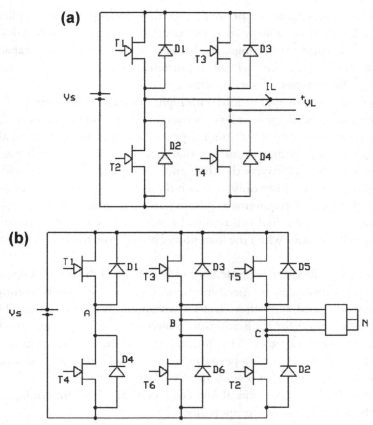

Figure 9-46 Configurations of typical DC/AC inverters: (a) H-bridge for single phase; (b) three-phase inverter.

A fuel cell system has several electric power-driven components, such as pumps, fans, blowers, solenoid valves, instruments, and so on. These components may operate with either DC or AC. The electrical subsystem must also provide power for these components at certain voltage and current.

Very often a fuel cell system is equipped with a battery for startup or for peaking. A startup battery must match the voltage of the critical components to be started before the fuel cell is operational or an additional voltage converter may be needed. The batteries or other peaking devices, such as ultracapacitors, can respond to sudden changes in power much faster than can a fuel cell system. A fuel cell itself can respond to the changes in load as quickly as a battery providing that it has sufficient supply of reactant gases.

The gases, particularly air or fuel other than hydrogen, are supplied by mechanical devices, which, by their nature, cannot respond as quickly as electrical demand may change. In that case the batteries or ultracapacitors provide power until the fuel cell can match it. The fuel cell then can also recharge the batteries or ultracapacitors.

In some cases, in applications with highly variable power, the fuel cell is intentionally sized at a power level between average and peak power. A fuel cell provides power up to its nominal power. An increase in demand above this power level further drops the fuel cell potential, in which case the battery steps in and covers the difference in power. If that happens, the battery must be sized not only to match the power requirements but also to match the energy requirements during those periods when the power demand exceeds the fuel cell nominal power. The fuel cell automatically recharges the battery when the load power goes below the fuel cell nominal power.

A control subsystem is needed not only to control the fuel cell operating parameters (flow rates, temperature, humidity, etc.) but also to communicate with the load and other electrical components of the system. This is particularly critical for applications in which a fuel cell operates integrated within an electrical grid. The functioning of the electric and control subsystems depends on application (e.g., standalone, grid integrated, combined with another power source, backup power).

Figure 9-47 shows a typical fuel cell electrical subsystem configuration with the following main components:

- A fuel cell stack that generates DC current at certain voltage
- A converter to step down or to boost up the voltage level that is produced by the fuel cell, depending on the load requirements
- A battery or peaking device, such as an ultracapacitor, that supplies power during periods when the demand exceeds fuel cell capabilities
- A diode to prevent current flow back to the fuel cell
- A capacitor just before the DC/DC converter to filter out the ripple
- A controller to ensure efficient and reliable working of the system by monitoring current and voltages in the subsystem and determining and providing respective control signals
- A power supply (DC) for auxiliary fuel cell equipment (fan, blower, pumps, solenoid valves, instruments, etc.)

If the load requires AC current (120/240 V) at a certain frequency (50 or 60 Hz), an additional component of the electrical subsystem—an inverter that generates AC current from a DC input—is needed (Figure 9-48).

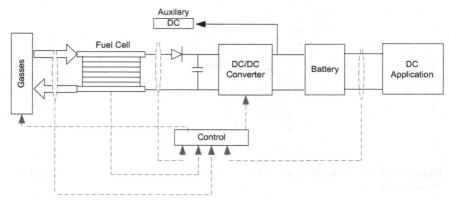

Figure 9-47 Schematic diagram of a fuel cell electric subsystem for DC power application.

Depending on the application, the system may operate as standalone (as shown in Figure 9-48) or parallel with the grid. In the latter case, an extra unit—a transfer switch—is needed to be connected to and to guarantee synchronizing with the grid (Figure 9-49).

There are certain regulations for power-generating devices connected to the grid. For example: (1) they must have the same voltage level and frequency as the grid, (2) they must be in phase with the grid, (3) they must have total harmonic distortion as good or better than the grid, and (4) for safety reasons, they must automatically disconnect themselves from the grid if the grid goes down. In addition, for a system that "exports" at least a portion of its power to the grid, net metering may also be included.

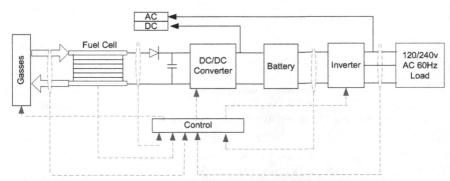

Figure 9-48 Schematic diagram of a fuel cell electric subsystem for a standalone AC power application.

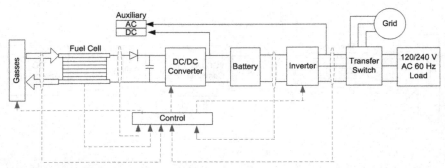

Figure 9-49 Schematic diagram of a fuel cell electric subsystem for a grid parallel AC power application.

9.5 SYSTEM EFFICIENCY

The fuel cell theoretical efficiency is 83% (Equation 2-43), based on hydrogen's higher heating value. The actual fuel cell efficiency in operation is much lower because of various losses (heat, electrode kinetics, electric and ionic resistance, mass transport). Additional components, such as fuel processor, power conditioning, and balance of plant, cause additional losses on the system level. Figure 9-50 shows a block diagram of a very generic fuel cell system with the flows of energy used to define the efficiencies of the individual steps.

The system efficiency is defined as a ratio between the output electrical energy, E_{net}, and the energy in fuel fed to the system, F_{in}:

$$\eta_{sys} = \frac{E_{net}}{F_{in}} \tag{9-78}$$

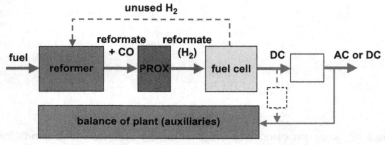

Figure 9-50 Simplified block diagram of a generic fuel cell system.

This efficiency is actually a product of several efficiencies of the individual components:

$$\eta_{sys} = \frac{H}{F_{in}} \frac{H_{cons}}{H} \frac{E_{FC}}{H_{cons}} \frac{E_{net}}{E_{FC}} \tag{9-79}$$

where:

H = energy of hydrogen produced by the fuel processor (product of flow rate and heating value resulting in watts)

H_{cons} = energy of hydrogen consumed in fuel cell electrochemical reaction (also a product of flow rate and heating value resulting in watts)

E_{FC} = power produced by fuel cell (W)

Note that by definition, hydrogen fed to the fuel cell and hydrogen consumed in the fuel cell are related:

$$H = S_{H_2} H_{cons} \tag{9-80}$$

Reformer efficiency is defined as a ratio between the energy in produced hydrogen and the energy in fuel fed to the system.

$$\eta_{ref} = \frac{H}{F_{in}} = \frac{S_{H_2} H_{cons}}{F_{in}} \tag{9-81}$$

The reformer efficiency typically does not include the efficiency of the preferential oxidation (PROX), which, as shown previously, is a function of the amount of air used in the process. Optimum oxygen stoichiometry that results in the lowest CO output is between 2 and 3. As shown in Figure 9-34, the PROX efficiency in that case is about 95%.

Fuel cell efficiency has two parts: fuel efficiency and voltage efficiency. Fuel efficiency is defined with the stoichiometric ratio. Because hydrogen is diluted at the fuel cell entrance, it has to be supplied in excess of the stoichiometric amount so that hydrogen concentration at the exit is higher than zero. Typically, the reformate fuel cells operate with hydrogen stoichiometric ratios between 1.1 and 1.2. The fuel efficiency, defined as the ratio between hydrogen consumed in the fuel cell and hydrogen actually supplied to the fuel cell, is:

$$\eta_{fuel} = \frac{H_{cons}}{H} = \frac{H_{cons}}{S_{H_2} H_{cons}} = \frac{1}{S_{H_2}} \tag{9-82}$$

Unused hydrogen, however, is rarely just discarded into the atmosphere. Typically, unused hydrogen from the fuel cell stack is used to generate power in a turbine/expander or in the fuel processing process (to preheat the fuel or to generate steam, or, in the steam-reforming process, it may be used as fuel, thus reducing the amount of fuel needed, that is, improving the overall efficiency). Fuel consumption is:

$$F_{in} = \frac{S_{H_2} H_{cons}}{\eta_{ref}} \tag{9-83}$$

Fuel consumption reduced by the recirculated hydrogen is:

$$F_{in} = \frac{S_{H_2} H_{cons}}{\eta_{ref}} - (S_{H_2} - 1)H_{cons} \tag{9-84}$$

The effective reformer efficiency, defined as shown previously as a ratio between the energy in produced hydrogen and energy in fuel fed to the system, is:

$$\eta_{ref}^{eff} = \frac{H}{F_{in}} = \frac{S_{H_2} H_{cons}}{\dfrac{S_{H_2} H_{cons}}{\eta_{ref}} - (S_{H_2} - 1)H_{cons}} = \frac{1}{\dfrac{1}{\eta_{ref}} + \dfrac{1}{S_{H_2}} - 1} \tag{9-85}$$

The voltage efficiency of a fuel cell is:

$$\eta_{FC} = \frac{E_{FC}}{H_{cons}} = \frac{V_{cell}}{1.482} \tag{9-86}$$

Power produced by a fuel cell goes through power conditioning, where it is brought to a desired constant voltage and a desired current type, that is, direct or alternate. Some power is needed to run the system's ancillary components. This equipment may use either AC or low-voltage DC (typically 12 V, 24 V, or 42 V), in which case an additional DC/DC converter is needed, as shown in Figure 9-50. The resulting power conversion and parasitic loss efficiency is:

$$\eta_{PC} = \frac{E_{net}}{E_{FC}} = (\eta_{DC} - \xi) \tag{9-87}$$

for the case where the parasitic load runs on same voltage and current as the main load, with the efficiency of DC/DC or DC/AC power conversion η_{DC}. Coefficient ξ is a ratio between power needed to run the parasitic load and fuel cell gross power:

$$\xi = \frac{E_{aux}}{E_{FC}} \tag{9-88}$$

Table 9-7 Range of Efficiencies and Operating Parameters of Fuel Cell and Supporting Components and Resulting System Efficiencies

	η_{ref}	η_{PROX}	S_{H_2}	V_{cell}	η_{DC}	ξ	η_{sys} (HHV)	η_{sys} (LHV)
Hydrogen/oxygen systems								
Low			1.01	0.8	0.93	0.05	0.48	0.56
High			1.01	0.85	0.96	0.03	0.53	0.62
Hydrogen/air systems								
Low			1.05	0.7	0.93	0.10	0.37	0.44
High			1.01	0.8	0.96	0.05	0.49	0.57
Systems with fuel processor								
Low	0.80	0.95	1.2	0.7	0.93	0.10	0.25	0.27
High	0.90	0.97	1.1	0.8	0.96	0.05	0.40	0.44

$$\eta_{PC} = \frac{E_{net}}{E_{FC}} = \eta_{DC}\left(1 - \frac{\xi}{\eta_{DCaux}}\right) \tag{9-89}$$

for the case where the parasitic load runs on low-voltage DC and requires an additional power conversion with the efficiency η_{DCaux}. If the two power conversions have the same efficiency, Equation (9-89) becomes Equation (9-87).

The overall system efficiency is then:

$$\eta_{sys} = \eta_{ref}\eta_{PROX}\eta_{fuel}\eta_{FC}\eta_{PC} \tag{9-90}$$

Obviously, hydrogen/oxygen and hydrogen/air systems do not include the reforming step. In that case the system efficiency is:

$$\eta_{sys} = \eta_{fuel}\eta_{FC}\eta_{PC} \tag{9-91}$$

So, what would be a reasonable fuel cell system efficiency? Table 9-7 lists some ranges of components' efficiency and the resulting system efficiency. "Low" efficiencies should be achievable with today's technology, whereas "high" efficiencies may require some advances in reforming technology, power conditioning, and system integration.

An obvious solution for improvement of the system efficiency is to operate a fuel cell at a higher voltage. Because of the power-voltage fuel cell characteristic, higher voltage means lower power density, which in turns means a larger fuel cell for the same power output. Improvements in fuel cell catalysts, catalyst layer design, membrane materials, and fuel cell design should result in higher and less steep polarization curves, which should allow selection of higher operating voltages while still operating with considerable power densities. There seems to be room for improvement of the fuel cell efficiency, that is, operating potential, because current operating potentials are far below the theoretical potential.

PROBLEMS

1.

    a. Calculate the temperature at the end of compression in a single-stage compressor used to provide air for a fuel cell system. Ambient air is at 15°C and 101.3 kPa, and (for the sake of this exercise) assume it is completely dry. Delivery pressure is 300 kPa. The fuel cell generates 50 kW at 0.7 V/cell and operates with oxygen stoichiometric ratio of 2. Assume adiabatic compression with efficiency of 0.7.

    b. Calculate the power (in kW) needed to run the compressor. Mechanical efficiency of the compressor is 93% and the efficiency of the electric motor and controller is also 93%.

    c. How much power can be recovered if the exhaust air is run through a turbine? The exhaust gas is fully saturated with water vapor at 60°C after liquid water has been removed. The pressure drop through the fuel cell is 25 kPa. The turbine efficiency is 70%.

2. For the fuel cell system from Problem 1, calculate:

    a. The amount of water (in g s^{-1}) needed to fully saturate the air at the fuel cell entrance at 60°C.

    b. Heat (in watts) needed to be brought in by the air humidification process (other than the heat of the incoming fluids—i.e., air and water—assume water for humidification is available at 55°C).

3. A 10-kW hydrogen/air fuel cell operates at 0.7 V/cell at 70°C and ambient pressure, with oxygen stoichiometry of 2.25. Liquid water is separated from the cathode exhaust. Ambient conditions are 23°C, 101.3 kPa, and 75% relative humidity. Calculate how much water would have to be stored for seven days' operation. Propose a solution to make this system water neutral (and support your proposal with calculations showing the resulting water balance).

4. Calculate the maximum theoretical LHV efficiency for reforming isooctane C_8H_{18}. What equivalence ratio does this efficiency correspond to?

5. Calculate composition (in volume % of dry gas) of a reformate gas obtained by reforming methane CH_4 in an autothermal fuel processor with gas shift reactors, operating with equivalence ratio of 3.0 and LHV efficiency of 75%. CO content is 1% (by volume dry).

6. Hydrogen content in a reformate gas from a gasoline (assume octane) autothermal reformer is 41%. CO content is 1.2. Assume that in addition to products (H_2, CO, CO_2), the only other byproduct is CH_4.

   a. Calculate the reformer efficiency if the equivalence ratio is 3.13.

   b. Calculate the reformate generation rate (in standard liters per minute) for a 50-kW fuel cell operating at 0.66 V/cell with H_2 stoichiometry of 1.15.

   c. Calculate the fuel consumption rate (in $g\ s^{-1}$).

7. The goal for the efficiency of an 85-kW fuel cell system using gasoline as fuel is 40% or higher (based on the lower heating value). The available reformer is 90% efficient, and the preferential oxidation has efficiency of 97%. The system uses power conditioning that is 94% efficient for both main and parasitic loads. The ancillary components operate at 24 volts DC and consume 7500 W. What should be the nominal fuel cell operating potential (in volts per cell)? How much power would the fuel cell generate? What would be fuel consumption (in g/s)? Discuss what would happen with the system efficiency at partial load, that is, 15-kW net power output.

QUIZ

1. In a fuel cell system, compressor power consumption:
   a. Is higher at higher fuel cell voltage
   b. Is lower at higher fuel cell voltage
   c. Does not depend on fuel cell voltage

2. A positive displacement compressor:
   a. Needs certain backpressure to operate
   b. Delivers air at a pressure higher than the fuel cell backpressure
   c. Delivers air at a pressure equal to the fuel cell backpressure

3. Humidification of air in a fuel cell system:
   a. Always needs additional heat
   b. Does not need additional heat in pressurized systems
   c. May need additional heat depending on operating pressure and temperature

4. An expander may generate enough power to run the compressor:
    a. Only if the fuel cell system operates at high pressure and high temperature
    b. If the pressure drop is sufficiently low
    c. Only if additional hydrogen is burned and this hot gas is used to run the expander
5. In a fuel cell system with hydrogen recirculation:
    a. Hydrogen flow rate at the stack inlet has stoichiometric ratio of 1
    b. Hydrogen flow rate at the stack inlet has stoichiometric ratio higher than 1
    c. Hydrogen consumption is higher than in dead-end mode operation
6. The difference between reformate produced by steam reforming and reformate produced by autothermal reforming is:
    a. Reformate from steam reforming does not contain nitrogen
    b. Reformate from authothermal reforming contains no water vapor
    c. There is no difference
7. A shift reactor:
    a. Improves the efficiency of the reforming process
    b. Improves the yield of hydrogen
    c. Improves both yield and the efficiency
8. An equivalence ratio is:
    a. A ratio of air to fuel
    b. A ratio of theoretical air flow rate in the reformer (needed for complete reformation) and actual air flow rate
    c. A ratio of theoretical air-to-fuel ratio at the reformer inlet (needed for complete combustion) and actual air-to-fuel ratio
9. Air bleed is the process of adding air to eliminate CO:
    a. At the stack inlet
    b. In the last stage of the reforming process
    c. During the reformer process
10. Unused hydrogen from the fuel cell outlet may be used to reduce the fuel requirement at the steam reformer inlet, thus:
    a. Resulting in higher fuel processing efficiency
    b. Resulting in higher hydrogen concentration at the outlet
    c. Having no effect on the reformer efficiency

REFERENCES

[1] Justi EW. A Solar-Hydrogen Energy System. New York: Plenum Press; 1987.

[2] Winter C-J, Nitsch J, editors. Hydrogen as an Energy Carrier. Berlin Heidelberg: Springer-Verlag; 1988.

[3] McCarty RD, Hord J, Roder HM. Selected Properties of Hydrogen (Engineering Design Data), U.S. Department of Commerce, National Bureau of Standards, Monograph 168. Washington, DC: U.S. Government Printing Office; 1981.

[4] Quantum Technologies website: www.qtww.com/core_competencies/gf_storage. shtml (accessed February 2005).

[5] Blomen LJMJ, Mugerwa MN. Fuel Cell Systems. New York: Plenum Press; 1993.

[6] Dynetek Industries website: www.dynetek.com (accessed February 2005).

[7] Michel F, Fieseler H, Meyer G, Theissen F. Onboard Equipment for Liquid Hydrogen Vehicles. In: Veziroglu TN, Winter C-J, Baselt JP, Kreysa G, editors. Proc. of the 11th World Hydrogen Energy Conference. Hydrogen Energy Progress XI, 2. Germany: Stuttgart; 1996. p. 1063–78.

[8] Perry Jr JH, Person A, Misiaszek SM, Alessi Jr DP. Closed Loop Reactant/Product Management System for Electrochemical Galvanic Energy Devices; 1991. U.S. Patent 5,047,298.

[9] Li P-W, Zhang T, Wang Q-M, Schaefer L, Chyu MK. The Performance of PEM Fuel Cells Fed with Oxygen through the Free-Convection Mode. Journal of Power Sources 2003;114(1):63–9.

[10] PaxiTech website: www.paxitech.com (accessed February 2005).

[11] Wilkinson DP, St-Pierre J. In-Plane Gradients in Fuel Cell Structure and Conditions for Higher Performance. Journal of Power Sources 2003;113(1):101–8.

[12] Barbir F, Nadal M, Fuchs M. Fuel Cell Powered Utility Vehicles. In: Buchi F, editor. Proc. of the Portable Fuel Cell Conference. Switzerland: Lucerne; June 1999. p. 113–26.

[13] Weast RC. CRC Handbook of Chemistry and Physics. Boca Raton, FL: CRC Press; 1988.

[14] Neutzler J, Barbir F. Development of Advanced, Low-Cost PEM Fuel Cell Stack and System Design for Operation on Reformate Used in Vehicle Power Systems, Fuel Cells for Transportation, FY1999 Annual Progress Report. Washington, DC: U.S. Department of Energy, Office of Advanced Automotive Technologies; October 1999. 71–76.

[15] Ralph TR, Hogarth MP. Catalysis for Low Temperature Fuel Cells. Platinum Metal Review 2002;46(3):117–35.

[16] Gottesfeld S, Zawodzinski TA. Polymer Electrolyte Fuel Cells. In: Alkire RC, Gerischer H, Kolb DM, Tobias CW, editors. Advances in Electrochemical Science and Engineering, 5. New York: Wiley-VCH; 1997. p. 219.

[17] Gasteiger HA, Markovic N, Ross PN, Cairns EJ. CO Electrooxidation on Well-Characterized Pt-Ru Alloys. Journal of Physical Chemistry 1994;98(2):617–25.

[18] Barbir F. Short Course on Automotive Fuel Cell System Engineering. In: Proc. of the Intertech Conference Commercializing Fuel Cell Vehicles, 2000; April 2000. Berlin, Germany.

[19] Sumner MJ, Harrison WL, Weyers RM, Kim YS, McGrath JE, Riffle JS, et al. Novel Proton Conducting Sulfonated Poly(Arylene Ether) Copolymers Containing Aromatic Nitriles. Journal of Membrane Science 2004;239(2):199–211.

[20] Pawlik J, Henschel C. Membrane-Electrode-Assemblies for Reformed Hydrogen Fuel Cells. In: Proc. of the 2004 Fuel Cell Seminar. TX: San Antonio; 2004. p. 93–6.

[21] Herring A, Dec S, Malers J, Meng F, Horan J, Turner J, et al. The Use of Heteropoly Acids in Composite Membranes for Elevated Temperature PEM Fuel Cell Operation:

Lessons Learned from Three Different Approaches. In: Proc. of the 2004 Fuel Cell Seminar. TX: San Antonio; , 2004. p. 97–100.

[22] Uribe FA, Gottesfeld S, Zawodzinski Jr TA. Effect of Ammonia as Potential Fuel Impurity on Proton Exchange Membrane Fuel Cell Performance. Journal of the Electrochemical Society 2002;149. p. A293.

[23] Uribe FA, Zawodzinski TA. Effects of Fuel Impurities on PEM Fuel Cell Performance. In: The Electrochemical Society Meeting Abstracts, 2001–2. San Francisco; September 2001. Abstract 339.

[24] Barbir F, Balasubramanian B, Neutzler J. Trade-off Design Study of Operating Pressure and Temperature in PEM Fuel Cell Systems. Advanced Energy Systems 1999;39:305–15. ASME.

[25] Larminie J, Dicks A. Fuel Cell Systems Explained. Second ed. Chichester, England: John Wiley; 2003.

CHAPTER TEN

Fuel Cell Applications

Fuel cells can generate power from a fraction of a watt to hundreds of kilowatts. For that reason they may be used in almost every application that requires local electricity generation. Applications such as automobiles, buses, utility vehicles, scooters, bicycles, and submarines have been demonstrated. Fuel cells are suitable for distributed power generation at the level of individual home, building, or community, offering tremendous flexibility in power supply. In some cases both power and heat produced by a fuel cell may be utilized, resulting in very high overall efficiency. As a backup power generator, fuel cells offer several advantages over either internal combustion engine generators (noise, fuel, reliability, maintenance) or batteries (weight, lifetime, maintenance). Small fuel cells are attractive for portable power applications, either as replacements for batteries (in various electronic devices and gadgets) or as portable power generators.

Fuel cell system design is not necessarily the same for each of these applications. On the contrary, each application, in addition to power output, has its own specific requirements, such as efficiency, water balance, heat utilization, quick startup, long dormancy, size, weight, and fuel supply (as shown in Table 10-1). Some of those application-specific requirements and corresponding design variations are discussed in this chapter.

10.1 TRANSPORTATION APPLICATIONS

10.1.1 Automobiles

Almost all major car manufacturers have demonstrated prototype fuel cell vehicles, and some have produced small series of fuel cell vehicles and announced plans for production and commercialization in the near future (see Table 10-1 and Figure 10-1 [1,2]). The race to develop a viable fuel cell vehicle and bring it to market began during the 1990s and continues in the first decades of the 21st century.

The major drivers for development of automotive fuel cell technology are these vehicles' efficiency, low or zero emissions, and fuel that could be produced from indigenous sources rather than imported. The main obstacles

Table 10-1 Summary of Market Requirements for Fuel Cell Systems

	Automotive	Stationary (Primary Power)	Stationary (Backup Power)
Power output	50−100 kW	1−10 kW & 200 kW	1−10 kW
Fuel	Hydrogen	Reformate	Hydrogen
Life (operational)	5000 hours	>40,000 hrs	<2000 hrs
High efficiency	Critical	Critical	Not critical
Instant start	Very important	Not important	Very important
Output mode	Highly variable	Variable	~Constant
Operation	Intermittent	Constant	Intermittent
Preferred voltage	>300 V	110/220 VAC	24 or 48 VDC or 110/220 VAC
Heat recovery	Not needed	Very important	Not needed
Water balance	Very important	Very important	Not critical
Size and weight	Critical	Not critical	Not critical
Extreme conditions	Critical	Not critical	Important
Cost	<$50/kW	<$1000/kW	<$5000/kW

for fuel cell commercialization in automobiles are the cost of fuel cells and the cost and availability of hydrogen.

Configurations of Automotive Fuel Cell Systems

The fuel cell may be connected to the propulsion motor in several ways [3]:

1. The fuel cell is sized to provide all the power needed to run the vehicle. A battery may be present, but only for startup (such as a 12-V battery). This configuration is typically possible only with direct hydrogen fuel cell systems. A system with a fuel processor would not have as good a dynamic response. Also, a small battery would not be sufficient to start up a system with a fuel processor.

2. The fuel cell is sized to provide only the base load, but the peak power for acceleration of the vehicle is provided by the batteries or similar peaking devices (such as ultracapacitors). This may be considered a parallel hybrid configuration because the fuel cell and the battery operate in parallel—the fuel cell provides cruising power, and the battery provides peak power (such as for acceleration). The presence of a battery in the system results in much faster response to load changes. The vehicle can be started without preheating of the fuel cell system, particularly the fuel processor, and operated purely as a battery-electric vehicle until the fuel cell system becomes operational. A battery allows for recapturing the braking energy, resulting in a more efficient system. The disadvantages of having the battery are extra cost, weight, and volume.

Figure 10-1 Fuel cell vehicles demonstrated by major car manufacturers (clockwise from top left: Honda FCX Clarity, Mercedes B-Class F-Cell, General Motors Equinox, Hyundai Tucson ix35 FCEV, Nissan X-Trail, VW Tiguan HyMotion, Kia Norego FCEV, Toyota FCHV Adv)

3. The fuel cell is sized only to recharge the batteries. The batteries provide all the power needed to run the vehicle. This may be considered a serial hybrid configuration (the fuel cell charges the battery and the battery drives the electric motor). The same advantages and disadvantages of having a battery apply as for the parallel hybrid configuration. The fuel cell nominal power output depends on how fast the batteries would have to be recharged. A smaller battery would have to be recharged faster and would result in a larger fuel cell.

4. The fuel cell serves only as an auxiliary power unit, that is, another engine is used for propulsion, but the fuel cell is used to run part of or the

entire vehicle electrical system [4]. This may be particularly attractive for trucks because it would allow operation of an air-conditioning or refrigeration unit while the vehicle is not moving and without the need to run the main engine.

Power Demand and Efficiency (Fuel Economy)

The power requirement of an automotive engine depends on many parameters, such as vehicle mass, frontal cross-sectional area, drag coefficient, rolling resistance coefficient, and the efficiency of the drive train, and it changes with the vehicle speed, acceleration, and road slope.

The efficiency of an automobile engine is more often expressed as specific fuel consumption (g kWh^{-1}):

$$f_c = \frac{3.6 \times 10^6}{\eta_{sys} H_{LHV}} \tag{10-1}$$

where:

H_{LHV} = lower heating value of fuel (kJ/kg)

η_{sys} = the vehicle efficiency, which is a product of the fuel cell system efficiency (including the fuel processor, if any, fuel cell, and power converter), traction efficiency (typically about 93%), and electric drive efficiency (typically 90% or higher).

For a gasoline internal combustion engine at its most favorable point of operation, the specific fuel consumption is about 240 g kWh^{-1} [5], which corresponds to the system efficiency of 34%. Diesel engines have a higher efficiency of about 40%. A fuel cell engine, depending on system configuration, may have the efficiency at its most favorable operating point well above 50%, corresponding to a specific fuel consumption of below 60 g kWh^{-1}. Note that 1 g of hydrogen contains the same amount of energy as 2.73 g of gasoline, based on its lower heating value. It should also be noted that the lower heating value efficiencies are used for comparison with internal combustion engines.

The efficiency of fuel cells vs. an internal combustion engine should not be compared at their most favorable operating points. These two technologies are intrinsically different and have very different efficiency-power characteristics. Although an internal combustion engine has the maximum efficiency at or near its maximum power [6], a fuel cell system has its maximum efficiency at partial load [7] (Figure 10-2). For this reason, the efficiency of a hydrogen-fueled fuel cell propulsion system in a typical driving schedule (such as the one shown in Figure 10-3), where an

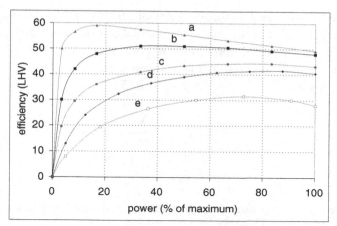

Figure 10-2 Comparison of the efficiency of fuel cells and internal combustion engines: a) fuel cell system operating at low pressure and low temperature; b) fuel cell system operating at high pressure and high temperature; c) fuel cell system with an onboard fuel processor; d) compression ignition internal combustion engine (diesel); e) spark ignition internal combustion engine (gasoline). *(Compiled from [6] and [7].)*

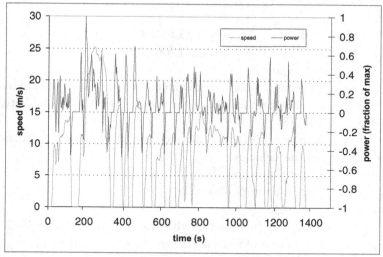

Figure 10-3 U.S. EPA Urban Dynamometer Driving Schedule (UDDS) [8]: average speed 8.7 ms^{-1}, average power 12% of maximum power.

automobile engine operates most of the time at partial load, can be about twice that of an internal combustion engine [8–10]. The hydrogen fuel cell system efficiency in a driving schedule can be upward of 40% and even above 50%. The efficiency of a fuel cell propulsion system with an onboard fuel processor is lower than the efficiency of a hydrogen fuel cell system, but it's still higher than the efficiency of an internal combustion engine. The fuel cell efficiency advantage diminishes if both a fuel cell and an internal combustion engine are used in a hybrid configuration.

The vehicle fuel economy (in $g\,m^{-1}$) is [5]:

$$B_e = \frac{\int f_c\, v \sum Fdt}{3.6 \times 10^6 \int vdt} \qquad (10\text{-}2)$$

where:

f_c = specific fuel consumption in $g\,kWh^{-1}$ as defined by Equation (10-1)

v = vehicle speed (ms^{-1})

ΣF = sum of the forces to be overcome (N), such as:

- Air resistance $F_a = 0.5\, C_d A_v \rho v^2$
- Rolling resistance: $F_r = C_r m_v g$
- Inertial forces: $F_i = m_v a$
- Gravitational force: $F_g = m_v g sin\theta$

where:

C_d = drag coefficient (typically 0.25 to 0.4)

A_v = vehicle frontal area (typically 1.5 to 2.5 m^2)

ρ = air density (kgm^{-3})

C_r = rolling resistance coefficient, which is a function of tire radius and tire distortion and thus a function of vehicle speed (typically between 0.008 and 0.02) [5]

m_v = vehicle mass (varies from 1,000 to more than 2,000 kg, depending on the class of the vehicle)

a = vehicle acceleration (ms^{-2})

g = gravity acceleration $(9.81\ ms^{-2})$

θ = road slope angle

For a given driving schedule, such as the one shown in Figure 10-3, the vehicle speed is a function of time, and so are the system efficiency, η_{sys}, and the sum of forces, ΣF.

Emissions

A hydrogen fuel cell does not generate any pollution. The only byproduct is pure water, which leaves the system as both liquid and vapor, depending on the operating conditions (temperature and pressure) and system configuration. The amount of water produced by a fuel cell propulsion system in a typical driving schedule is comparable to the amount of water produced by an internal combustion engine. If another fuel is used (such as methanol or gasoline) and reformed onboard, the propulsion system has some emissions generated in the reforming process, but those emissions are in general still much lower than the emissions from an internal combustion engine, and such vehicles would typically qualify as ultralow emission vehicles (ULEVs).

In an emission analysis, it is important to take into account the entire fuel cycle (well to wheels); otherwise, skewed results may be generated. In general, if hydrogen is produced from fossil fuels, the emissions resulting from that process (particularly the CO_2 emissions) should be taken into account, regardless of whether the hydrogen generation takes place in a refinery, at the refueling station, or on the vehicle. Figure 10-4 shows the results of one such well-to-wheels life-cycle study [12]. Note that the fuel cell vehicles generate significantly less greenhouse emissions than the comparable gasoline-, diesel-, or methanol-powered internal combustion engine vehicles. The lowest emissions in this study were attributed to the hydrogen-powered internal combustion engines. This is the result of higher emissions assigned to the process of fuel cell vehicle production. Hydrogen production from renewable energy sources and water does not generate any direct emissions (Figure 10.4 does show some emissions, which are the result of manufacturing of the equipment for utilization of renewable energy), and

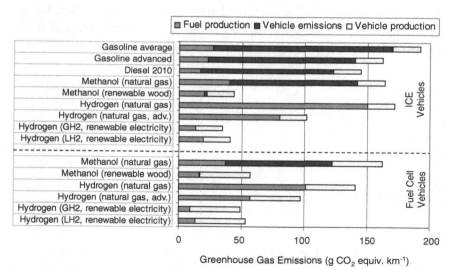

Figure 10-4 The results of a life-cycle analysis: greenhouse gas emissions for different power train and fuel options. *(Adapted from [12].)*

that is the path that takes full advantage of the fuel cell technology. More about hydrogen as a fuel of the future is included in Chapter 12.

Cost

Automobile engines, although relatively complex, particularly compared with fuel cells, are relatively inexpensive ($35–50 per kW). This is mainly because of mass production techniques employed in their manufacturing. There are millions of cars (and of course, engines) produced annually. Fuel cells, being an immature technology and manufactured on a prototype level, are far more expensive than the internal combustion engines. However, studies conducted by or on behalf of the major car manufacturers have shown that the fuel cells could be produced cost-competitively, assuming mass-production manufacturing techniques are applied [13]. Figure 10-5 shows cost estimates for the entire fuel cell system as a function of manufacturing rate, demonstrating that for a manufacturing rate of 500,000 units per year the achievable manufacturing cost with state-of-the-art technology (2012) would be $49 per kW [13].

The major high-cost contributing components are the catalyst (precious metal, Pt, or Pt-alloys) and the ionomer membrane.

Typical platinum loading in PEM fuel cells is about 0.3 mg per cm^2 of each electrode active area. Assuming power density of 0.7 W cm^{-2} (for example 0.7 V and 1 A cm^{-2}), this corresponds to approximately 0.85 mg W^{-1}, or 0.85 g kW^{-1}. The price of platinum varies on the market, but at, for example, $1,500 per ounce or $50 per gram, this corresponds to about $40 per kW, a significant portion of the total fuel cell system cost allowance.

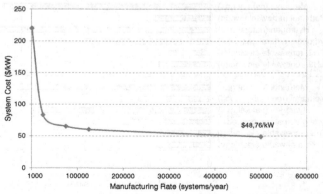

Figure 10-5 Fuel cell system cost estimates as function of manufacturing rate. (*Adapted from [13].*)

Lower Pt loading (<0.1 mg per cm^2) and higher performance (>1 W cm^{-2}) would result in 0.1 g Pt, or \$5 per kW, which would be acceptable.

The other expensive component is the ionomer membrane (Nafion or similar fluoropolymers). At \$500 per m^2 this is prohibitively expensive, resulting in more than \$50–70 per kilowatt. Membrane manufacturers estimate that for every two orders of magnitude increase in manufacturing volume the price may be cut in half.

The cost targets (\$35–50 per kW) therefore require improvements in fuel cell performance (more watts per unit active area); reduction in catalyst loading or alternative, less expensive catalysts without sacrificing the performance; and novel, less expensive membranes.

Fuel Issues (Availability, Cost, Storage)

An automotive fuel cell system configuration greatly depends on the choice of fuel. Possible fuels for the fuel cell vehicles are hydrogen, gasoline, or methanol, each with its own advantages and disadvantages. The selection of the fuel depends on several factors:
- Fuel supply infrastructure and the cost of establishing a new one
- The cost of fuel per energy content, but more important, per mile
- Environmental implications (i.e., well-to-wheels emissions)
- Complexity and the cost of onboard storage and processing
- Safety, both real and perceived
- National security and related national energy policy

Lack of hydrogen infrastructure is considered the biggest obstacle for introduction of fuel cell vehicles, although this may be a classic chicken-and-egg problem (i.e., there are no fuel cell vehicles because there are no hydrogen refueling stations, but there are no hydrogen refueling stations because there is no demand for hydrogen as transportation fuel). Establishing a necessary hydrogen infrastructure (hydrogen production and distribution) would require significant capital. Nevertheless, there are already hundreds of hydrogen refueling stations, mainly in the United States, Japan, and Germany. California already has a network of hydrogen refueling stations that allows deployment of hydrogen vehicles. Germany announced plans for establishing a network of hydrogen refueling stations as a necessary condition for widespread deployment and commercialization of hydrogen vehicles [14].

Hydrogen produced from natural gas, either in a central facility or at the refueling station, can be cheaper than gasoline. The wholesale price of natural gas is 2–3 times lower than the retail price of gasoline, so at the efficiency of 70% to 80% hydrogen could be produced cost-competitively. However, the retail price of gasoline includes a hefty tax burden, but a discussion about the tax policy is certainly beyond the scope of this book. Hydrogen produced from water and electricity (via electrolysis) is in general more expensive than gasoline, unless low-cost off-peak electricity is used from the power plants that have low or zero fuel cost (nuclear and renewable power plants).

Hydrogen is the only fuel that results in a zero-emissions vehicle, particularly if hydrogen is produced from renewable energy sources. Also, hydrogen from renewable energy sources as a transportation fuel could reduce dependency on imported oil. A fuel cell system that runs on pure hydrogen is relatively simple, has the best performance, runs more efficiently, and has the longest stack life. Hydrogen is nontoxic and, despite its reputation, has some very safe features (as discussed in Section 12.3.5, "Safety Aspects of Hydrogen as Fuel").

One of the biggest problems related to hydrogen use in passenger vehicles is its onboard storage. Hydrogen can be stored as compressed gas, as a cryogenic liquid, or in metal hydrides. Tanks for compressed gaseous hydrogen are bulky, even if hydrogen is compressed to 700 bar, which nowadays seems to be accepted as a standard for vehicle hydrogen tanks. At this pressure it takes about 26 liters of space to store 1 kg of hydrogen (energy content of 1 kg of hydrogen is approximately equivalent to energy content of 1 gallon or 3.785 l of gasoline). The amount of fuel to be stored onboard depends on the vehicle fuel efficiency and required range. The fuel efficiency of vehicles is measured in miles per gallon (in the United States) or in liters per 100 km. The average fuel efficiency of new cars is between 20 and 30 mpg (7.9 and 11.8 l/100 km) and even better in smaller European and Japanese cars. The range is typically above 300 miles (480 km). A typical vehicle thus has about 10 to 16 gallons of gasoline onboard, taking space of 40 to 60 liters. Assuming that hydrogen fuel cell vehicles have twice the efficiency of gasoline vehicles, they will have to store at least 5 kg of hydrogen, which would take about 130 liters of space, several times larger than current gasoline tanks. Further improvements in vehicle design and use of lightweight materials could improve the fuel efficiency by a factor of 2 (bringing hydrogen onboard storage requirement down to 2.5–4.0 kg), but that would apply to both hydrogen and gasoline vehicles.

Liquid hydrogen tanks take about the same volume as 700 bar compressed hydrogen tanks (about 30 liters per 1 kg of hydrogen), but liquid hydrogen is a cryogenic fuel (at 20K) and its use would be associated with significant handling challenges. Hydrogen liquefaction is an energy-intensive process, requiring energy equal to about 30% of liquefied hydrogen's higher heating value and therefore more expensive than compressed hydrogen. (It should be mentioned that hydrogen compression to 700 bars consumes energy in the amount equal to about 12% of compressed hydrogen's higher heating value.) In addition, storing hydrogen at such a low temperature would result in significant boil-off losses. Well-insulated, large-volume liquid hydrogen containers (such as those used by NASA) typically have loss rates of less than 0.1% per day [15]. Modern automotive liquid hydrogen tanks can realize energy densities of 22 MJ/kg and evaporation rates of approximately 1% per day [16].

The difficulty of storing hydrogen onboard a vehicle, as well as lack of hydrogen infrastructure, has forced car manufacturers to consider other, more conveniently supplied fuels for fuel cells. In that case a fuel cell system would have to be integrated with a fuel processor. From the infrastructure point of view, gasoline would be a logical choice, because it would allow quick penetration of fuel cell vehicles into the market. However, gasoline is not an easy fuel to reform. Because gasoline is a heavily processed fuel optimized specifically for combustion in the internal combustion engine, gasoline suppliers considered alternative and relatively easy-to-reform fuels such as hydrotreated naphta, hydrocrackate, alkylate/isomerate, or liquid fuels generated from natural gas such as methanol [17]. Nevertheless, onboard reforming is not easy, and in spite of several more or less successfully demonstrated prototype vehicles, it does not appear to be a likely path for bringing fuel cell vehicles to market. There are numerous engineering issues with successfully integrating a fuel processor with a fuel cell:

1. Vehicles with an onboard fuel processor are not zero-emissions vehicles.
2. Onboard reforming reduces the efficiency (fuel to wheels) of the propulsion system. The reformers are typically 80% to 90% efficient (fuel to hydrogen). Use of diluted hydrogen reduces fuel cell voltage, which has a direct impact on either the efficiency or the size of the fuel cell. In addition, fuel utilization of diluted hydrogen is much lower, which further reduces the system efficiency.
3. Onboard reforming increases complexity, size, weight, and cost of the entire propulsion system.
4. Fuel processors need time to start producing hydrogen, that is, to warm up to their operating temperature. In the first prototypes the startup time was between 15 and 30 minutes, clearly not acceptable for practical applications. Significant engineering efforts are required to reduce the

startup time [18]. This issue may be avoided in hybrid configurations because the batteries may provide power during the fuel processor warmup.

5. Controlling relatively small reformer devices in highly dynamic operation, while at the same time maintaining high purity of hydrogen at all times is not an easy task. The reformer, as shown in Figure 9-40, is a series of chemical reactors, each operating within a relatively narrow window of operating temperatures. Because of relatively small size, the whole system is very sensitive to changes in temperature resulting from disturbances such as sudden change in fuel input. As a result, it is difficult to keep CO concentration at low, fuel-cell-safe levels during these transitions.

6. The long-term effects of fuel impurities on reformer lifetime and reformer byproducts on fuel cell lifetime are not well known at present. Researchers at Los Alamos National Laboratory have reported on negative impacts of small concentrations of NH_3 and H_2S possibly present in the reformate from partial oxidation (as shown in Figures 9-38 and 9-39 [19,20]).

Because of these problems with onboard reforming of liquid hydrocarbon fuels, the car companies have given up on the idea of onboard reforming and are concentrating on bringing hydrogen-fueled fuel cell vehicles to market.

Lifetime

The average lifetime of a vehicle is about 10 to 12 years, but the actual operating time is only on the order of 3,000 to 5,000 hours. The fuel cells for automotive applications are expected to have a similar lifetime. Limited laboratory testing to date has confirmed that the PEM fuel cells can meet these expectations. Continuous operation for 5,000 hours is not a problem, but operation with highly variable power, numerous startups and shutdowns, operation in various ambient conditions, impurities in fuel and in air, and the like may have dramatic effects on reducing the fuel cell operating lifetime. The problem with corrosion of catalyst carbon support during startup and shutdown, when the fuel cell voltage is high, is described in Chapter 11.

Heat Rejection, Water Balance, and Freezing

Automotive fuel cells must survive and operate in extreme weather conditions ($-40°C$ to $+40°C$). This requirement has a tremendous effect on system design. Survival and startup in extremely cold climates require specific engineering solutions, such as use of antifreeze coolant and water management. Water cannot be completely eliminated from the system, because water is essential for the membrane ionic conductivity.

The fuel cell heat-rejecting equipment (radiator and condenser) must be sized for heat rejection in extremely hot weather (typically from 32°C to 40°C[6]). Although a fuel cell system is more efficient than an internal combustion engine, it has similar or larger cooling loads. More important, because of the fuel cell's low operating temperature (60°C to 80°C), the heat-rejection equipment is typically much larger than that for a comparable internal combustion engine [6,21].

The water balance requirement results in additional cooling loads [6,21]. Although water is produced in a fuel cell, water is needed for humidification of reactant gases and for fuel processing (in case of an onboard fuel processor), and it has to be reclaimed from the exhaust gases.

Size and Weight

An automotive fuel cell system replacing an internal combustion engine must be of a similar size and weight. The size of the stack greatly depends on the selected nominal cell voltage, that is, the stack voltage efficiency. The stack specific volume (m^3 per kW) is:

$$V_s = 0.1 \frac{n_{cells}\, d_{cell} + 2d_{ep}}{\alpha_{act} n_{cells} V_{cell} i} \tag{10-3}$$

where:

n_{cells} = number of cells in a stack

d_{cell} = individual cell thickness, including the cooling arrangements (m)

d_{ep} = thickness of the end plates and the bus plates if their function is not included in the bus plates (m)

α_{act} = ratio of cell active area and bipolar plate area, including the perimeters reserved for manifolds and seals; typical values for larger stacks are in the range of 0.80 to 0.86 [5]

V_{cell} = cell potential at nominal power (V)

i = current density at nominal power (A cm^{-2})

Note that the cell potential, V_{cell}, and current density, i, are connected by the cell polarization curve. The product of cell potential and current density is the fuel cell-specific power (per unit active area):

$$P_f = V_{cell}i(W\ cm^{-2}) \text{ or } P_f = 10V_{cell}i(kW\ m^{-2}) \tag{10-4}$$

Obviously, a lower stack–specific volume results in thinner cells, better utilization of bipolar plate area, and higher performance.

The stack-specific mass (in kg kW^{-1}) is then simply:

$$m_s = V_s \rho_s \qquad (10\text{-}5)$$

where ρ_S is the stack density (kg m^{-3}), a function of stack design and materials used. Typical fuel cell stacks have density below 2000 kg m^{-3}.

From the parameters in Equations (10-4) and (10-5), a new parameter may be derived, the so-called design factor, D_f (kg m^2), which is simply the stack mass divided by the active area. This factor is useful in comparisons of various stack designs. A good fuel cell stack design, which means a stack with thin cells, high utilization of bipolar plate area, and selection of lightweight materials, should result in a design factor below 10 kg m^{-2}. From a known design factor and specific performance it is easy to calculate the stack-specific mass:

$$m_s = \frac{D_f}{P_f} \qquad (10\text{-}6)$$

The fuel cell stacks with weight and volume less than 1 kg kW^{-1} and 1 l kW^{-1}, respectively, have already been demonstrated. The complete fuel cell systems are of course heavier and bulkier, approximately 2.5 kg kW^{-1} and 2.5 l kW^{-1}, but the goals for the complete fuel cell system, excluding hydrogen storage and electric motor, are 1.54 kg kW^{-1} and 1.18 l kW^{-1} (from Table 10-2). Table 10-3 lists the U.S. DOE technical targets [22] for an integrated direct hydrogen automotive fuel cell system (stack, air management, thermal subsystem).

10.1.2 Buses

Buses for city and regional transport are considered the most likely type of vehicles for an early market introduction of fuel cell technology. Most of the issues discussed in the previous section on automobiles also apply to fuel cell applications in buses. The major differences are in hydrogen storage and refueling sites.

Buses require more power than passenger automobiles, typically about 150 kW or more. They operate in a more demanding operating regimen with frequent starts and stops. Nevertheless, the average fuel economy of a bus fuel cell system is roughly 15% better than that of a diesel engine [23]. Buses are almost always operated in a fleet and refueled in a central facility. This makes refueling with hydrogen much easier. In addition, storing larger quantities of hydrogen (typically above 20 kg) onboard is less

Table 10-2 Prototype Fuel Cell Vehicles by Major Car Manufacturer (compiled from HyWeb [1] and Fuel Cells 2000 [2])

Manufacturer	Model	Year	Fuel cell	Batteries	H2 storage	H2 pressure	Range (km)
Honda	FCX Clarity	2005	100 kW	Li-Ion	5 kg; 2 tanks	350 bar	560
GM	Equinox	2006	115 kW	NiMH; 35 kW	4.5 kg	700 bar	300
Volkswagen	Tiguan HyMotion	2007	80 kW	Li-Ion, 22 kW	3.2 kg	700 bar	
Toyota	FCHV Adv	2008	90 kW	NiMH	156 l	700 bar	690
KIA	Borego	2008	100 kW	Supercapacitor, 450 V	3x76 l tanks	700 bar	600
Mercedes	B-Class F-Cell	2009	100 kW	Li-Ion; 35 kW; 1,4 kWh	3 tanks	700 bar	400
Nissan	X-Trail	2009	90 kW	Li-Ion		700 bar	
Hyundai	Tucson ix35	2010	100 kW	Li-Poly, 21 kW	5.6 kg; 2 tanks	700 bar	650

Table 10-3 U.S. DOE Technical Targets[a] for Automotive Applications: 80-kWe (net) Integrated Transportation Fuel Cell Power Systems Operating on Hydrogen

Characteristic	Units	2011 Status	2017 Targets	2020 Targets
Energy efficiency[b] @ 25% of rated power	%	59	60	60
Power density	W/L	400[c]	650	850
Specific power	W/kg	400[c]	650	650
Cost[d]	$/kW$_e$	49[e]	30	30
Cold start-up time to 50% of rated power	seconds	20[f]	30	30
@−20°C ambient temp	seconds	<10	5	5
@+20°C ambient temp				
Start up and shut down energy[g]	MJ	7.5	5	5
from −20°C ambient temp	MJ	–	1	1
from +20°C ambient temp				
Durability in autotive drive cycle	hours	2,500[h]	5,000[i]	5,000[i]
Assisted start from low temperatures[j]	°C	–	−40	−40
unassisted start from low temperatures[j]	°C	−20[f]	−30	−30

[a]Targets exclude hydrogen storage, power electronics, and electric drive.
[b]Ratio of DC output energy to the lower heating value of the input fuel (hydrogen).
[c]Estimates
[d]Cost projected to high-volume production (500,000 systems per year).
[e]The projected cost status is from a 2011 DTI study (ref. and refer to Figure)
[f]Based on average of status values reported at 2010 SAE World Congress. These systems do not necessarily meet other system-level targets.
[g]H$_2$ fuel energy (Lower Heating Value) to include the fuel energy required to account for the electrical energy consumed from cold start.
[h]Projected time to 10% voltage degradation from the Technology Validation activity.
[i]Based on U.S. DRIVE Fuel Cell Tech Team Cell Component Accelerated Stress Test and Polarization Curve Protocols (http://www.uscar.org/guest/view_team.php?teams_id=17), Table 6, <10% drop in rated power after test.
[j]8-hour soak at stated temperature must not impact subsequent achievement of targets.

of a problem. Fuel cell buses typically store hydrogen in composite compressed gas cylinders at 350 bar, located on the roof. Since availability of space on the bus is not an issue, there is no need to use 700 bar storage tanks, so almost all the buses built today use 350 bar compressed hydrogen tanks. Because hydrogen is much lighter than air, the roof location is considered very safe.

Table 10-4 presents characteristics of the most recent fuel cell buses developed and demonstrated all over the world [1], some of which are shown in Figure 10-6.

The hydrogen fuel cell buses have a major advantage over their competition (diesel buses) because they produce zero emissions. This is particularly important in already heavily polluted, densely populated cities. Under the Clean Urban Transport for Europe (CUTE) program, hydrogen fuel cell buses have been employed in major European cities such as Amsterdam, Barcelona, Hamburg, London, Luxembourg, Madrid, Porto, Reykjavik, Stockholm, and Stuttgart [24]. Sunline Transit Authority in Palm Springs, California, has been operating fuel cell buses in regular service for several years. The United Nations Development Program (UNDP) and Global Environment Facility (GEF) are funding and coordinating an international program that aims to introduce fuel cell buses in major cities around the world, such as Sao Paolo, Mexico City, New Delhi, Cairo, Shanghai, and Beijing; some of these have already been developed and demonstrated. When hydrogen is produced from clean, renewable energy sources, the fuel cell buses could make a significant contribution toward cleaner air in these cities.

The main obstacles for commercialization of fuel cell buses are fuel cell cost and durability. Because of the smaller manufacturing series, the cost of the bus engines per kilowatt is somewhat higher than the cost of the automobile engines. The expected lifetime is also higher because a typical city bus may be in operation more than 6,000 hours per year. This, combined with highly intermittent operation with many starts and stops, poses a challenge to fuel cell durability with current technology.

10.1.3 Utility Vehicles

Utility vehicles such as material-handling industrial vehicles, airport ground-support tow vehicles, airport people movers, golf carts, lawn maintenance vehicles, and similar vehicles may be another early adapter of fuel cell technology. These applications are not as demanding as passenger vehicles or buses. The competing technology is typically batteries, most often lead acid

Table 10-4 Fuel Cell Buses Demonstrated to Date (Compiled from Hy Web [1] and Fuel Cell 2000 [26])

Manufacturer	Model	Year	Passengers	Fuel cell	Provider	Batteries	Electric motor	H2 storage	H2 storage pressure	Status
Toyota/Hino	Blue Ribbon	2010	26	180 kW	Toyota	Ni-MH 84 kw	2x80 kW		350 bar	In service connecting downtown Tokyo with Haneda and Narita Airports
ISE Corp./ Wrightbus	Pulsar H	2010		75 kW	Ballard	Supercaps 2x85		6 tanks	350 bar	8 buses in operation in London
APTS	Phileas	2010		150 kW (130 net)	Ballard	Ni-MH 100 kW; supercaps 100 kW	240 kW	36.4 kg in 8 tanks (1640 l)	350 bar	2 buses at RVK (Cologne) and 2 at GVB (Amsterdam)
HAN Automotive	VDL Ambassador ALE 120–205/ 225	2010			Nedstack			3 tanks	300 bar	demonstration
SAIC	SWB6129FC	2010	49	80 kW net	Ballard	Li-iron-phosphate 100 kW			350 bar	demonstration at EXPO 2010 Shangai
Fbus	Battery City Bus	2009	22	38 kW	Ballard	50 Ni-Cd (100 Ah)	130 kW	15.5 kg in 2 tanks (700 l)	350 bar	demonstration
VanHool	A300L new	2009		120 kW	UTC Fuel Cells	yes	2x85 kW		350 bar	12 buses for AC Transit (CA) 4 buses for CT Transit (CT)
Hyundai	Super Aero City	2009	28	200 kW (160 net)	Hyundai/ KIA	Supercaps 100 kW	3x100 kW	27.3 kg in 6 tanks (1260 l)	350 bar	test runs in Seoul and other Korean cities
Daimler AG	Mercedes Citaro E4	2009	77	150 kw (120 net)	AFCC	Li-Ion 250 kW		32 kg in 8 tanks (1435 l)	350 bar	small series

Figure 10-6 Some prototype fuel cell buses (clockwise from top left: ISE Corp. Wrightbus, Han Automotive HyMove, APTS Phileas, Mercedes Citaro) *(compiled from [1])*

batteries that require frequent and lengthy charging and pose significant maintenance problems.

Early demonstrations of fuel cell-powered utility vehicles have shown that such vehicles offer lower operating cost, reduced maintenance, lower downtime, and extended range.

One emerging market is in material-handling equipment (also known as lift trucks, which include forklifts, pallet jacks, and stock pickers). Fleets of fuel cell material-handling equipment are already in use at dozens of warehouses, distribution centers, and manufacturing plants. With increased performance and decreased maintenance needs, fuel cells are competitive with batteries on a life-cycle basis. Like batteries, hydrogen PEM fuel cells—whose only byproducts are water and heat—do not emit any harmful air pollutants while in use, so they are suitable for indoor use. Unlike batteries, however, fuel cells can be rapidly refueled, boosting productivity by eliminating the time and cost associated with battery change-outs. Hydrogen fuel cell lift trucks can be refueled in 2–3 minutes on average and can operate for over 8 hours on a single fill. In addition, fuel cells maintain full power capability between refueling, whereas batteries typically have some power loss between charges. Growing numbers of lift truck installations are demonstrating the economic, environmental, and performance benefits of fuel cells [25]. Figure 10-7 shows a prototype of a fuel cell-powered forklift.

Figure 10-7 Fuel cell powered forklift *(Courtesy of UNIDO-ICHET)*.

10.1.4 Scooters and Bicycles

Scooters and bicycles may be a significant market for fuel cell technologies, particularly in developing countries. Despite the stringent requirements relating to weight, size, and low cost, fuel cells have been successfully demonstrated in various scooters and bicycles. The power requirement is considerably less than that for automobiles—up to 3 kW for scooters and a few hundred watts for bicycles. Although the range may be smaller than for automobiles too, the volume of hydrogen storage is one of the critical issues. The fuel cells for scooters and bicycles are almost always air cooled. The refueling issue of these vehicles in mass markets is as complex as the issue of automobile refueling. However, because of significantly smaller quantities of hydrogen to be stored onboard, additional options, such as distribution of metal hydride tanks or home refueling devices (such as electrolyzers), are possible.

10.2 STATIONARY POWER

Although development and demonstrations of fuel cells in automobiles usually draw more attention, applications for stationary power generation offer even greater market opportunity. The drivers for both market sectors are similar: higher efficiency and lower emissions. The system design for both applications is also similar in principle. The main differences are in the choice of fuel, power conditioning, and heat rejection [26]. There are also some differences in requirements for automotive and stationary fuel cell systems. For example, size and weight requirements are very important in automotive applications but not so significant in stationary applications. The acceptable noise level is lower for stationary applications, especially if the unit is to be installed indoors. The fuel cell itself, of course, does not generate any noise; noise may be coming from air- and fluid-handling devices. Automobile systems are expected to have a very short startup time (a fraction of a minute), whereas the startup of a stationary system is not time limited unless the system is operated as a backup or emergency power generator. Both automotive and stationary systems are expected to survive and operate in extreme ambient conditions, although some stationary units may be designed for indoor installation only. Finally, the automotive systems for passenger vehicles are expected to have a lifetime of 3,000 to 5,000 operational hours, and systems for buses and trucks somewhat longer, but the stationary fuel cell power systems are expected to operate 40,000 to 80,000 hours (5 to 10 years). Table 10-5 shows technical targets for small (1–10 kW) residential and distributed generation fuel cell systems operating on natural gas.

Stationary fuel cell power systems will enable the concept of distributed generation, allowing the utility companies to increase their installed capacity following the increase in demand more closely rather than anticipating the demand in huge increments by adding gigantic power plants. Currently, obtaining permits and building a conventional power plant have become very difficult tasks. Fuel cells, on the other hand, do not need special permitting and may be installed virtually everywhere—inside residential areas, even inside residential dwellings. To end users the fuel cells offer reliability, energy independence, "green" power, and, ultimately, lower energy costs.

10.2.1 Classification of Stationary Fuel Cell Systems

A variety of stationary fuel cell systems are being developed. Design options of stationary fuel cell power systems with respect to application, grid connection, nominal power output, load following, choice of fuel, installation, and cogeneration capabilities are discussed next.

Application and Grid Connection

Stationary fuel cells may be used in a variety of applications:
- As the only power source, thus competing with or replacing the grid, or providing electricity in areas not covered by the grid

Table 10-5 U.S. DOE Technical Targets: 1–10 kWe Residential Combined Heat and Power and Distributed Generation Fuel Cell Systems Operating on Natural Gas[a]

Characteristic	2011 Status	2015 Targets	2020 Targets
Electrical efficiency at rated power[b]	34–40%	42.5%	>45%[c]
CHP energy efficiency[d]	80–90%	87.5%	90%
Equipment cost[e], 2-kW$_{avg}$,[f] system	NA	$1,200/kW$_{avg}$	$1,000/kW$_{avg}$
Equipment cost[e], 5-kW$_{avg}$,[f] system	$2,300–$4,000/kW[g]	$1,700/kW$_{avg}$	$1,500/kW$_{avg}$
Equipment cost[e], 10-kW$_{avg}$,[f] system	NA	$1,900/kW$_{avg}$	$1,700/kW$_{avg}$
Transient response (10–90% rated power)	5 min	3 min	2 min
Start-up time from 20°C ambient temperature	<30 min	30 min	20 min
Degradation with cycling[h]	<2%/1,000 h	0.5%/1,000 h	0.3%/1,000 h
Operating lifetime[i]	12,000 h	40,000 h	60,000 h
System availability[j]	97%	98%	99%

[a]Includes fuel processor, stack, and ancillaries.
[b]Pipeline natural gas delivered at typical residential distribution line pressures.
[c]Status varies by technology.
[d]Ratio of AC net output energy to the lower heating value (LHV) of the input fuel.
[e]Higher electrical efficiencies (e.g., 60% using SOFC) are preferred for non-CHP applications.
[f]Ratio of regulated AC net output energy plus recovered thermal energy to the LHV of the input fuel. For inclusion in CHP energy efficiency calculation, heat must be available at a temperature sufficiently high to be useful in space and water heating applications. Provision of heat at 80°C or higher is recommended.
[g]Current production volume (~30 MW per year).
[h]Includes projected cost advantage of high-volume production (totaling 100 MW per year).
[i]Time until >10% net power degradation.
[j]Percentage of time the system is available for operation under realistic operating conditions and load profile. Unavailable time includes time for scheduled maintenance.

- As a supplemental power source working in parallel with the grid, covering either the base load or the peak load
- In combined systems with intermittent renewable energy sources (such as photovoltaics or wind turbines), generating power in periods when these energy sources cannot meet the demand
- As a backup or emergency power generator providing power when the grid (or any other primary power source) is down

Accordingly, the fuel cell system, and particularly its power conditioning and interconnect module, may be designed as:

- *Grid parallel.* Allowing power from the grid to the consumer when needed but not allowing power from the fuel cell back to the grid. The fuel cell system may be sized to provide most consumer energy needs, but the grid is used to cover the short-term demand peaks. Such a system essentially does not need batteries (except for startup when the grid is down) and does not need interconnect standards.
- *Grid interconnected.* Allowing power flow in both directions, namely power from the grid to the consumer when needed and power from the fuel cell back to the grid. Such a system may be designed as load following or as constant power, because excess fuel cell power can be exported to the grid. Of course, this design option requires interconnect standards.
- *Standalone.* Providing power without a grid. The system must be capable of load following. Very often a sizable battery bank is used to enable load following.
- *Backup or emergency generator.* The system must be capable of quick startup and is also often combined with batteries or other peaking device. Batteries are typically superior for low-power/low-duration backup power, but a fuel cell system becomes competitive for higher power (several kW) and longer duration (more than 30 minutes). A backup power system may be equipped with an electrolyzer-hydrogen generator and hydrogen storage [27]. In that case the unit generates its own fuel during periods when electricity from the grid is available.

Nominal Power Output

With respect to power output, fuel cell power systems may be divided into several classes:

- 1–10 kW with applications in individual houses, trailers, recreational vehicles, and for portable power

- 10–50 kW with applications in larger homes, mansions, groups of homes, and small commercial uses such as small businesses, restaurants, warehouses, and shops
- 50–250 kW with applications in small communities, office buildings, hospitals, hotels, military bases, and so on
- For applications higher than 250 kW, PEM fuel cells may not be competitive with other high-temperature fuel cell technologies

Load Following

Depending on its application and nominal power output, the fuel cell system may be designed to operate in load following or in constant load modes. Load following requires that the fuel cell system is sized to generate the maximum required load of the user, or that it follows the load only up to its nominal power output and the load peaks are covered by either a peaking device (such as a battery or ultracapacitor) or by the grid. The latter requires that the system be designed as grid parallel. Although a fuel cell is electrically capable of load following, its functioning in this mode depends on reactant supply, both oxygen and hydrogen, which are supplied by means of mechanical devices (pumps, blowers, or compressors) and have certain inertia and time lag in responding to change of rate. It is particularly difficult to operate a fuel processor in a transient/variable load mode while keeping the quality of generated hydrogen sufficient for fuel cell operation.

A stationary fuel cell power system may be designed to operate at constant/nominal power output all the time. Such a system is either sized to cover only the base load or it is grid interconnected and thus allowed to export excess power back to the grid.

Choice of Fuel

The PEM fuel cells run on hydrogen. However, hydrogen as a fuel is not readily available, particularly not for residential applications, except if the system is to be used as a backup power system, in which case it may be equipped with an electrolytic hydrogen generator [27]. To facilitate market acceptance, fuel cell developers are forced to add a fuel-processing section to the fuel cell system. For residential and commercial applications, natural gas is a logical fuel choice because its distribution is widely developed. The majority of stationary power fuel cell systems developed to date use natural gas as fuel. Propane may be an alternative fuel for those users that are not connected to the natural gas supply line. Fuel processing of propane and natural gas is similar and usually can be accomplished with the same fuel processing catalysts and hardware. For some applications liquid fuels such as fuel oil, gasoline, diesel, methanol, or ethanol may be preferable. All of these fuels also require fuel processing. If hydrogen is available (such as in various industrial applications or in renewable energy installations equipped with an electrolytic hydrogen generator), the system may be significantly simplified.

Installation Location

The stationary fuel cell power systems may be designed for installation either outdoors or indoors. Installation indoors is more demanding with respect to codes and standards, many of which currently do not exist, therefore leaving fuel cell installers at the mercy of local authorities. On the other side, outdoor installation requires weatherproof system design, especially with respect to exposure to extreme weather conditions. Another possibility is to design a fuel cell system as a split system, where the gas processing and power generation sections of the fuel cell system are installed outdoors and the control and power conditioning sections are installed indoors.

Cogeneration

Any fuel cell system generates waste heat. The major sources of heat are the fuel cell stack, fuel processor, and tail gas burner (where the hydrogen that went through the fuel cell stack unused is catalytically combusted). The heat is rejected from the system via the heat exchangers or simply dissipated to the surroundings through radiation and convection. The heat from the fuel cell system may be captured and used for heating/preheating domestic hot water or for heating the heating medium in the space heating system combined with a natural gas boiler or a heat pump. The cogeneration systems may have more favorable economics because the overall efficiency (electrical plus thermal) may approach 90%. Figure 10-8 shows a schematic diagram of a cogeneration-capable fuel cell power system.

10.2.2 System Configuration

Unless hydrogen is available as a fuel for distributed electricity generation, fuel cells must use readily available fuels such as natural gas or propane. Natural gas is available in most densely populated areas, whereas propane may be available in remote areas. A fuel cell system for stationary power generation therefore must include a fuel processor. The efficiency of the system is even more important than in an automobile, and for that reason, system integration and optimization are necessary for achieving high

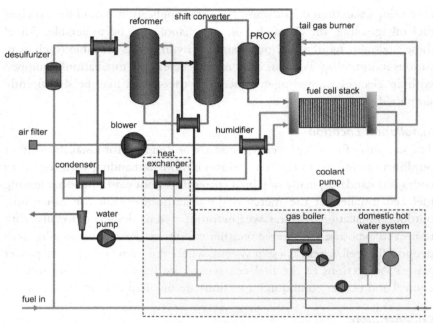

Figure 10-8 Schematic diagram of a cogeneration-capable fuel cell power system; a cogeneration subsystem, within dashed lines *(from [28])* combined with a fuel cell system from Figure 9-40.

efficiencies. Larger systems (>100 kW) can achieve efficiencies above 40%, whereas the efficiency of smaller systems (<10 kW) is typically somewhat lower (36–40%). Size and weight of stationary systems are not as critical as they are in automotive systems. Also, the cost targets for stationary systems are at least an order of magnitude higher than the cost targets of automotive systems. A stationary fuel cell can be sized at a higher cell voltage, resulting in a more efficient but larger and more expensive stack. Higher efficiencies can be achieved with systems operating at ambient pressure. Water balance, although preferable, is not as critical as it is in an automotive system, and neither is the size of the heat rejection equipment. An automotive system can utilize the air impinging on the moving vehicle, whereas the heat rejection system of a stationary system must completely rely on fans.

The biggest difference between automotive and stationary fuel cell systems is in the electric subsystem: power conditioning. The architecture of the power conditioning system greatly depends on the system operating mode, as discussed in Section 9.4, "Electrical Subsystem." An automotive system is practically a standalone system; a stationary fuel cell power system may operate as a standalone system, grid parallel, grid interactive, or as backup power.

A standalone system will require another auxiliary power source, such as batteries or supercapacitors, to provide peak power demands and to compensate for the system's inability to track rapid load changes. The fuel cell and the auxiliary power source must be sized to operate at the maximum continuous load and have the ability to handle startup load requirements, which, for motors, may be several times their rated capacity. Thus, the power conditioning system must be designed to handle the combined power outputs from fuel cell and auxiliary power sources on a continuous basis, in addition to startup load demands. The auxiliary power source is charged by the fuel cell stack during periods of low-power demands.

In a grid-tied system the power conditioning system is simplified because the grid may replace the auxiliary power source and also provide startup power. However, the system must now be capable of synchronizing with the grid and disconnecting in the event of a grid outage or if the signal quality from the utility is not within acceptable standards. A system designed to operate in both standalone and grid-tied modes is more complex because it must operate as a current source if tied to the grid and as a voltage source if operating independently. If a grid failure occurs while the fuel cell system is tied to the grid, it must disconnect quickly and maintain power to the loads without exceeding its maximum power rating. The excess loads that are dropped are the ones that are considered nonessential. The system must then monitor the grid and reconnect when the grid is operational. An efficient power management system is thus desirable for this architecture, which may be implemented by a programmable power distribution panel. An additional option for a grid-interactive system is the ability to transfer electricity in both directions. A fuel cell may be sized for peak power of an individual user, and in that case it may be preferred to export excess electricity generated during low-power periods back to the grid. Metering both incoming and outgoing electricity, or net metering, may also be a part of the electrical subsystem.

Other issues that impact the architecture and design of the power conditioning system are the amount of ripple tolerated by the fuel cell and the ability to carry unbalanced loads [29]. The ripple current on the fuel cell is the result of the switching characteristics of the electronics of the power conditioning system. Typically, two components are generated: a low-frequency component (usually 100 Hz or 120 Hz), which is twice the output AC frequency, and a high-frequency component (in the kHz range), which is due to the internal switching characteristics of the electronics. The amount of ripple can be filtered by the capacitors at the output of the fuel cell. Unbalanced loads result from the unequal load distribution on the individual lines of a multiple line system. For example, for residential homes in the United States the standard three-wire system provides two 120 V lines and a neutral. The power conditioning system should be designed so that each of the 120 V lines can handle the full current rating of the system.

10.2.3 Efficiency of Entire Fuel Cell System

The efficiency of the entire fuel cell power system is a product of efficiencies of individual components, as discussed in Section 9.5, "System Efficiency":

$$\eta_{sys} = \eta_{fp} \cdot \eta_{fc} \cdot (\eta_{pc} - \xi_p) \tag{10-7}$$

where ξ_p is defined as ratio of parasitic power and fuel cell gross power output, and η_{pc} is the power conditioning efficiency, defined as a ratio between AC power output and DC power input in the power.

If the system has parasitic loads running on DC (typically 24 V or, more recently, 42 V) and the efficiencies of DC/AC and DC/DC conversions are different, Equation (10-7) is slightly different:

$$\eta_{sys} = \eta_{fp} \cdot \eta_{fc} \cdot \eta_{AC} \left(1 - \frac{\xi_p}{\eta_{DC}} \right) \tag{10-8}$$

The efficiency at partial load may be slightly higher than at nominal power, primarily because the fuel cell efficiency, that is, voltage, is higher at lower power levels. However, this may be offset by a lower efficiency of fuel processor due to relatively higher thermal losses, and by relatively higher share of parasitic power. Figure 10-9 shows the efficiency power curve of a 3 kW residential fuel cell power unit [30]. The efficiency of more than 30% may be maintained throughout the operating range above approximately one-sixth or one-fifth of nominal power. Below that threshold the efficiency drops sharply. This means that it would not be desirable or feasible to operate such a fuel cell at a very low-power level.

In case of cogeneration, a much higher total system efficiency may be achieved (up to 90%). In this case the total efficiency is defined as:

$$\text{total efficiency} = \frac{\text{electric power output} + \text{thermal output}}{\text{fuel consumption}} \tag{10-9}$$

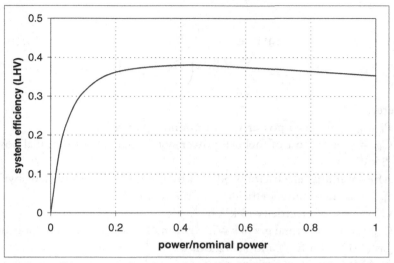

Figure 10-9 Efficiency (LHV) of small, residential, natural gas-fueled fuel cell system [30].

10.2.4 Economics of Fuel Cell Systems

The economics of automotive fuel cell power systems are fairly simple: They must compete with the internal combustion engine in terms of initial cost and efficiency. Although the efficiency is important, it has a little impact on economics. The initial cost is far more critical.

The economics of residential fuel cells are more complex. Here the purchase price must be justified with sufficient savings in expenditures for energy over the lifetime of the power system. One method that may be used to evaluate feasibility of residential fuel cell power systems is a simple payback time. This is simply a ratio between purchase price and annual savings in electricity expenditures. The result suggests to the potential user how soon her investment will pay back. This simple method does not take into account the interest lost or spent during that period, inflation, or changes in electricity and natural gas prices, but it gives a quick reference. Simple payback time is:

$$SPT = \frac{\text{purchase price of fuel cell power system}}{\text{annual savings on electricity}} \qquad (10\text{-}10)$$

or:

$$\text{SPT} = \frac{P_{fc,nom} \cdot C_{fc}}{\text{AEP} \cdot \left(C_{el} - \dfrac{C_{ng}}{\overline{\eta}_{fc}} \right)} \tag{10-11}$$

where:

$P_{fc,nom}$ = fuel cell power system nominal power (kW)

C_{fc} = specific cost of fuel cell power system per kW of nominal power ($ kW^{-1})

AEP = annual electricity produced by the fuel cell system (kW hyr^{-1})

$\overline{\eta}_{fc}$ = average annual efficiency of the fuel cell system

C_{el} = cost of electricity ($ kWh^{-1})

C_{ng} = cost of natural gas ($ kWh^{-1}), using lower heating value of natural gas (1,000 Btu ft^{-3}) and 3,412 Btu kWh^{-1} conversion factor

Both AEP, annual electricity produced, and η–fc, average annual efficiency, are functions of the load profile. Load profile varies from user to user, it varies depending on the day of the week, and it varies from season to season. The main characteristic of most residential users is high variability, that is, the load varies from almost zero to several kilowatts. The average household power varies from about 1 kW in Europe and Japan to more than 2 kW in the United States. Figure 10-10 illustrates a daily load

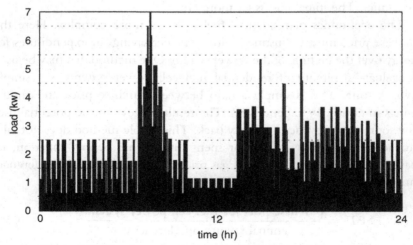

Figure 10-10 An example of a household daily electric load profile.

profile of a household, with power (minute average) ranging from 0.7 kW to 7 kW. For this particular example the total annual power consumption would be 16,000 kWh.

A fuel cell power system may be sized to generate any power between minimum and maximum load power. Obviously, if it is sized at a lower end of this range it would work with a very high capacity factor, but it would cover only a portion of the power needs. The rest would have to be supplied from the grid. If it is sized to cover all the power needs, it would provide grid independence, but it would operate with a very low-capacity factor.

The amount of electricity annually produced (in kWh) by a fuel cell system is:

$$AEP = \int_{year} P_{fc} dt \qquad (10\text{-}12)$$

where:

P_{fc} = fuel cell power (kW) at any given time (t)

$P_{fc} = P_{load}$ for $P_{load} < P_{nom}$, and

$P_{fc} = P_{nom}$ for $P_{load} \geq P_{nom}$

However, Equation (10-12) may only be used when the load profile is known. A capacity factor, CF, defined as a ratio between actually produced electricity in a time period (typically a year) and electricity that could have been produced if the system operated at nominal power for the entire period, may be more easily estimated. In that case, the amount of electricity annually produced, AEP, (in kWh), is:

$$AEP = 8760 \cdot CF \cdot P_{fc,nom} \qquad (10\text{-}13)$$

where:

8,760 is the number of hours in a year, and

CF is the capacity factor previously defined

A load itself has a certain capacity factor (i.e., a ratio between the average and peak power). For the load profile shown in Figure 10-10, the capacity factor is 25%. A fuel cell that is sized to provide all the power required by the load would therefore operate with a 25% capacity factor. If a fuel cell were sized to a lower power, it would have a considerably higher capacity factor. A fuel cell sized to cover only a base load would operate with a 100% capacity factor (providing no maintenance would be required during that period). Figure 10-11 shows the capacity factor as a function of fuel cell nominal

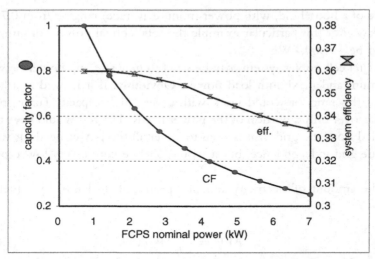

Figure 10-11 Capacity factor and efficiency of residential fuel cell power system as a function of nominal power (for a fuel cell system efficiency in Figure 10-9 and a load profile in Figure 10-10).

power for the load profile given in Figure 10-10, ranging from 100% at base load (0.7 kW) to 25% at peak (7.0 kW).

Figure 10-11 also shows the average efficiency of the fuel cell system in annual operation. The average annual efficiency of a fuel cell system is defined as:

$$\overline{\eta}_{fc} = \frac{\displaystyle\int_{year} P_{fc}\, dt}{\displaystyle\int_{year} \frac{P_{fc}}{\eta_{fc}}\, dt} \tag{10-14}$$

where P_{fc} and η_{fc} are the fuel cell system power output and system efficiency at any given time, t, respectively.

Equation (10-14) may be used only if the annual load profile is known. Instead of the exact annual load profile, which varies greatly from household to household or from user to user, an annual power distribution curve may be used.

A fuel cell that operates 100% of the time at nominal power has the highest efficiency (36% in this case, corresponding to 0.75 V/cell). The fuel cell with the lowest capacity factor also has the lowest annual efficiency

(33%) because most of the time it operates in the inefficient region (below 20% nominal power).

Substituting Equation (10-13) into Equation (10-11), the simple payback time is:

$$\text{SPT} = \frac{C_{fc}}{8760 \cdot \text{CF} \cdot \left(C_{el} - \dfrac{C_{ng}}{\eta_{fc}} \right)} \qquad (10\text{-}15)$$

In some cases, instead of fuel savings and the payback time, it is necessary to calculate the cost of electricity produced by a fuel cell. From Equation (10-15) it follows:

$$C_{el} = \frac{C_{fc}}{8760 \cdot \text{CF} \cdot \text{SPT}} + \frac{C_{ng}}{\eta_{fc}} \qquad (10\text{-}16)$$

For larger installations it is more common, instead of simple payback time, to use a capital recovery factor, CRF, which takes into account the lifetime of the fuel cell system and the interest rate:

$$C_{el} = \frac{C_{fc} \cdot \text{CRF}}{8760 \cdot \text{CF}} + \frac{C_{ng}}{\eta_{fnc}} \qquad (10\text{-}17)$$

where:

$$\text{CRF} = \frac{d(1+d)^L}{(1+d)^L - 1} \qquad (10\text{-}18)$$

where:

d = discount rate

L = fuel cell system lifetime (yr)

Capital recovery factor, CRF, has unit yr^{-1}. Note that for d $\to$ 0, CRF $\to$ 1/L, and in that case L = SPT.

A more complete economic analysis should also take into account the maintenance cost:

$$C_{el} = \frac{C_{fc} \cdot \text{CRF} + \text{AMC}}{8760 \cdot \text{CF}} + \frac{C_{ng}}{\eta_{fc}} \qquad (10\text{-}19)$$

where AMC is the annual cost of maintenance per kW of installed power ($ kW^{-1} yr^{-1}), which at this time is difficult to predict because not that many fuel cell systems are in real-life operation.

The economics of stationary fuel cells greatly depends on the prices of electricity and natural gas. From Equation (10-15) it is clear that the simple payback time has a positive value only for $C_{el} > \dfrac{C_{ng}}{\overline{\eta}_{fc}}$, which means that the ratio between the price of electricity and the price of natural gas (expressed in the same units, $\$\,kWh^{-1}$) $\dfrac{C_{el}}{C_{ng}}$, must be larger than $\dfrac{1}{\overline{\eta}_{fc}}$. For the fuel cell system efficiencies between 0.30 and 0.40, the electricity/natural gas price ratio must be higher than 3.33 and 2.50, respectively. The higher this ratio, the shorter the payback time would be.

The prices of electricity and natural gas vary over time, and they vary from region to region. Figure 10-12 shows how since the year 2000 the price of natural gas in the United States has risen sharply, causing the average residential price ratio C_{el}/C_{ng} to drop from 4 to about 2.5 [31]. The residential price ratio varies significantly from region to region, as shown in Figure 10-13 for the United States and selected countries around the world [31]. Clearly, a residential fuel cell concept may be feasible only in selected countries and regions. The price of natural gas to the utility sector is somewhat lower than the price for the residential sector, which makes the fuel cells attractive for distributed generation.

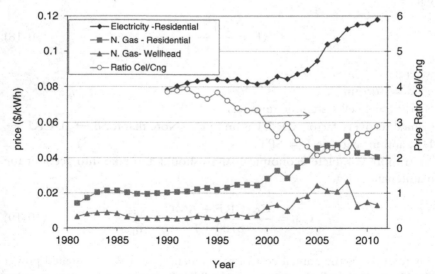

Figure 10-12 Average price of electricity and natural gas in the United States (based on data in [31]).

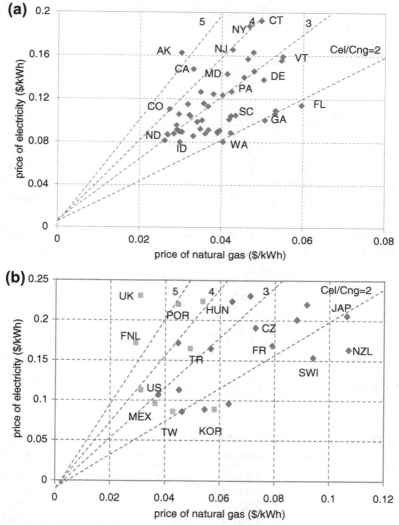

Figure 10-13 Price of electricity relative to price of natural gas for residential customers: a) United States (2010); b) selected world countries (2009-10) (based on data in [31]).

Example

A load profile of a household may be approximated by a power distribution curve, as shown in Table E1. The cost of grid electricity is $0.15 per kWh and the price of natural gas is $0.32 per m^3. Calculate the simple payback

Table E1 Example of a Household Simplified Load Profile

kW	% Time
10.0	5
5.0	10
2.0	15
1.2	40
0.733	30
Sum	100

time for a fuel cell that is sized to cover the maximum load and compare it with the payback time for a fuel cell that is sized for average power. The fuel cell cost is $1,000 per kW. Fuel cell system efficiency is shown in Figure 10-9.

The payback time is given by Equation (10-11). The missing parameters are the annual electricity produced and the average annual efficiency. The annual electricity produced is:

$$\text{AEP} = \sum P_{fc} t$$

$$\overline{\eta}_{fc} = \frac{\sum P_{fc} t}{\sum \dfrac{P_{fc}}{\eta_{fc}} t}$$

For the case where the fuel cell is sized at maximum power, $P_{fc,nom} = 10$ kW, the results are shown in Table E2.

Table E2 The Results for the Case When the Fuel Cell Is Sized at Maximum Power

kW	% Time	kWh/yr	% Power	eff	kWh/yr
10	5	4,380.0	100	0.34	12,882.4
5	10	4,380.0	50	0.35	12,514.3
2	15	2,628.0	20	0.33	7,963.6
1.2	40	4,204.8	12	0.28	15,017.1
0.7333	30	1,927.1	7.333	0.23	8,378.7
Sum	100	17,519.9			56,756.2

From the Table E2:

Annual electricity produced:

$$\text{AEP} = 17,519.9 \text{k Wh yr}^{-1}$$

Average power:

$$\overline{P} = \frac{AEP}{8760} = 2.0 \text{ kW}$$

Annual capacity factor:

$$CF = \frac{AEP}{8760 \, P_{fc,nom}} = \frac{17,519.9}{8760 \times 10} = 0.2$$

Average annual efficiency:

$$\overline{\eta}_{fc} = \frac{17,519.9}{56,756.2} = 0.308$$

Cost of natural gas:

$$C_{ng} = 0.32 \text{ \$ m}^{-3}/38,300 \text{ kJ m}^{-3} \times 3600 \text{ s h}^{-1} = 0.03 \text{ \$ kWh}^{-1}$$

Simple payback time:

$$SPT = \frac{C_{fc}}{8760 \cdot CF \cdot \left(C_{el} - \dfrac{C_{ng}}{\overline{\eta}_{fc}}\right)} = \frac{1000}{8760 \times 0.2 \left(0.15 - \dfrac{0.03}{0.308}\right)}$$

$$= \underline{10.8 \text{ years}}$$

For the case where the fuel cell is sized at average power, $P_{fc,nom} = 2.0$ kW, the results are shown in Table E3.

Table E3 The Results for the Case When the Fuel Cell Is Sized at Average Power

kW	% Time	kWh/yr	% Power	eff	kWh/yr
2	30	5,256.0	100	0.34	15,458.8
1.2	40	4,204.8	60	0.35	12,013.7
0.733	30	1,927.1	36.67	0.345	5,585.8
Sum	100	11,387.9			33,058.4

From the Table E3:

Annual electricity produced:

$$AEP = 11,387.9 \text{ k Wh yr}^{-1}$$

Annual capacity factor:

$$CF = \frac{AEP}{8760 \, P_{fc,nom}} = \frac{11,387.9}{8760 \times 2} = 0.65$$

Average annual efficiency:

$$\overline{\eta}_{fc} = \frac{11,387.9}{33,058.4} = 0.344$$

Simple payback time:

$$SPT = \frac{C_{fc}}{8760 \cdot CF \cdot \left(C_{el} - \dfrac{C_{ng}}{\overline{\eta}_{fc}} \right)} = \frac{1000}{8760 \times 0.65 \left(0.15 - \dfrac{0.03}{0.34} \right)}$$

$$= 2.8 \text{ years}$$

A 2-kW fuel cell is therefore more likely to be economical for this case. A payback time of 2.8 years will likely be appealing to an average residential customer.

The economics of a fuel cell from the previous example may be improved by either exporting excess electricity back to the grid or utilizing the heat produced by the fuel cell. In the former case the price of electricity exported back to the grid may not be the same as the price of purchased electricity. In that case, the simple payback time is:

$$SPT = \frac{P_{fc,nom} \cdot C_{fc}}{AEP_{int} \cdot \left(C_{el} - \dfrac{C_{ng}}{\overline{\eta}_{fc}} \right) + AEP_{exp} \cdot \left(C_{el,exp} - \dfrac{C_{ng}}{\overline{\eta}_{fc}} \right)} \qquad (10\text{-}20)$$

where:

AEP_{int} = amount of electricity consumed internally (kW hyr^{-1})
AEP_{exp} = amount of electricity exported back to the grid (kW hyr^{-1})
$C_{el,exp}$ = price of electricity exported back to the grid ($ kWh^{-1})

For the previous example:

$AEP_{int} = 17,519.9 kW\ hyr^{-1}$

$AEP_{exp} = P_{fc,nom} \times 8760 - AEP_{int} = 87,600 - 17,519.9 = 70,080.1$ kW hyr^{-1}

$\overline{\eta}_{fc} = 0.34$ (because the fuel cell would operate 100% of the time at nominal power)

Assuming that the price of electricity exported back to the grid will be 90% of the electricity purchase price, the simple payback time will be:

$$SPT = \frac{10 \times 1000}{17,519.9 \times \left(0.15 - \dfrac{0.03}{0.34}\right) + 70,080.1 \times \left(0.15 \times 0.9 - \dfrac{0.03}{0.34}\right)}$$

$$= 2.3 \text{ years}$$

The economics in this case greatly depend on the price of electricity exported to the grid. In some countries this price is regulated and the utilities are mandated to purchase electricity from the individual producers. The economics of grid-interactive stationary power fuel cells may be further improved if they are operated by a utility company paying a lower rate for natural gas. In fact, a natural gas utility company may become an electricity producer.

In cases where the waste heat generated by the fuel cell system is utilized, the payback time must take into account the savings in heating fuel. Heating fuel may be natural gas, heating oil, or even electricity. The efficiency of the heating systems is typically well above 80% and in the case of condensing boilers it may be 100% or more. (See Section 2.3, "Higher and Lower Heating Value of Hydrogen," for an explanation of an efficiency greater than 100%.)

The total fuel cell efficiency (both thermal and electrical) is defined with Equation (10-9). The amount of heat used per year (AHP) is then:

$$AHP = AEP\left(\frac{\eta_{tot}}{\eta_{el}} - 1\right) \times f_{hu} \qquad (10\text{-}21)$$

Note that not all the heat may be utilized over the year. Although the heat for domestic hot water is needed throughout the year, heat for space heating is needed only during the heating season, the duration of which depends on climate. A ratio between the heat actually utilized and the heat produced by a fuel cell system is included in Equation (10-21) as a heat utilization factor,

f_{hu}. In some cases, fuel cell systems for cogeneration of heat and power may be used in a heat-following mode, either using or exporting generated electricity.

Simple payback time for combined heat and power generation is:

$$SPT = \frac{C_{fc} \cdot P_{fc}}{AEP \left[C_{el} - \frac{C_{ng}}{\eta_{el}} + \left(\frac{\eta_{tot}}{\eta_{el}} - 1 \right) f_{hu} \frac{C_{fuel}}{\eta_{heat}} \right]} \qquad (10\text{-}22)$$

where:

C_{fuel} = cost of fuel ($\$ \, kW^{-1}$) used for heating (natural gas, heating oil, or electricity)

η_{heat} = efficiency of conventional heat generation

Example

If the fuel cell from the previous example also generates heat with the combined total energy efficiency of 75%, and two-thirds of that heat can be utilized, calculate the payback time. The heat was otherwise generated by burning natural gas in a 90% efficient furnace.

From Equation (10-22):

$$SPT = \frac{1000 \times 10}{17519.9 \times \left[0.15 - \frac{0.03}{0.308} + \left(\frac{0.75}{0.308} - 1 \right) \frac{2}{3} \frac{0.03}{0.9} \right]} = 6.8 \text{ years}$$

Therefore, combined heat and power generation results in further savings, and the payback time is reduced from 10.8 years for a system that generates only electricity to 6.8 years.

As shown in the previous examples, the economics of stationary fuel cells depend greatly on how those units are being used. Several different scenarios are possible:

1. *Fuel cell system for individual user (residential or commercial) sized to cover the entire load.* Such a system would have a low capacity factor and relatively low average efficiency, and for that reason the payback time would be unacceptably high. Nevertheless, this scenario may be acceptable where the user is willing to pay a premium price for being grid independent. In addition, this may pass in the new homes market, where the buyer pays for an oversized and not very efficient fuel cell, but the high cost is masked by the cost of the entire purchase.

2. *Fuel cell system sized to cover the entire load, but excess electricity sold to the grid.* This scenario may work only if the utility wants to buy excess electricity at a reasonable price or if the utility actually owns the fuel cell system and operates it as a distributed generator. The user would not see the difference in either power supply or cost, although there should be some incentives for the individual users to accept "mini power plants" in their homes or businesses. This scenario may be attractive for utilities in heavily populated urban areas where building new large power plants is out of the question. Because of their modularity, small size, ultralow emissions, and low noise, fuel cell power systems may be installed in buildings. This way utility companies would be able to gradually increase their capacity without large investments. Of course, the premise is that the fuel cell power systems are cost-effective, that is, they have a reasonable payback time.

3. *Fuel cell system sized to cover part of the load.* In this case the fuel cell system operates with high capacity factor and high efficiency and is more likely to be cost-competitive. A downside of this option is that the user depends on the grid for peak power. In case of a grid power outage the user may be able to use the fuel cell, but only for some critical loads if the wiring is arranged that way.

4. *Fuel cell system in combination with batteries (or ultracapacitors) for short-term peak power.* In this case the fuel cell may be sized to operate with a very high capacity factor, providing baseline load and charging the batteries during off-peak hours. However, analysis of such a system would have to include the economics and lifetime of the batteries.

5. *Fuel cell cogeneration system.* Economics of all of the previous scenarios may be significantly improved if waste heat generated by the fuel cell system can be utilized. In that case the payback time must take into account the amount of heat that may be recovered and fuel that would have been used to generate that amount of heat (Equation 10-22). The system may be designed and sized for either electric load following or thermal load following. Such a system would be combined with a boiler heating system.

6. *Fuel cell as standby or emergency generator.* This may be an ideal niche market for fuel cell power systems for several reasons:

- The efficiency is not important.
- The expected lifetime is only about 1,000 hours (200 hours/year for five years)—what should be achievable with today's systems.

- Quick startup is not required (batteries or ultracapacitors may be used to provide power during the fuel cell startup period).
- The fuel cell system is better than batteries in terms of size and weight (and probably cost) for applications that require more than 5 kWh of backup.
- The fuel cell system is superior to gasoline or diesel generators in terms of noise and emissions.
- Combined with an electrolyzer, such a system generates its own fuel. This application is discussed in greater detail in the following section.

10.3 BACKUP POWER

Backup power is defined as any device that provides instantaneous, uninterruptible power. The term UPS (*uninterruptible power supply*) is often used and can sometimes refer to systems that supply A/C power or systems that supply power for no more than 30 to 60 minutes. A more general definition includes all types of power outputs and all backup times. Typical applications for backup power include telecommunications systems, information technology and computer systems, manufacturing processes, security systems, utility substations, and railway applications. Backup power systems are employed in cases where the loss of power results in a significant reduction in productivity or financial loss.

Fuel cell system requirements for backup power applications significantly differ from requirements for such systems in automotive and stationary (primary) power generation markets. As Table 10-4 shows, there are only a few common characteristics between the backup power on one side and automotive and stationary (primary) power on the other.

Fuel cell backup power can use hydrogen as fuel. Hydrogen can be provided in tanks that would have to be replaced after they are emptied (which may be a logistical problem or may create new business opportunities—refilling of hydrogen bottles). A more elegant solution would be for the fuel cell to generate its own hydrogen via an electrolysis process during periods when electricity is available and the fuel cell is dormant. A combination of fuel cell and electrolyzer is often referred to as a *reversible fuel cell* (see the Section 10.5, "Regenerative Fuel Cells and Their Applications"). A backup power fuel cell thus does not depend on hydrogen infrastructure and therefore may be commercialized before the automotive and stationary fuel cells.

The operational lifetime requirement for a fuel cell in a backup power application is less than 2,000 hours. Such a fuel cell operates with continuous load requirements during the power outage, but never longer than 8 hours. This duty cycle is achievable with today's fuel cell technology. Operational lifetimes of several thousand hours have been demonstrated by almost all the major membrane electrode assembly (MEA) suppliers/developers.

The system efficiency is critical for both stationary and automotive applications. For backup power there is no imposed efficiency goal, but the fuel cell efficiency directly translates into the size and cost of both the electrolyzer and hydrogen storage. Because the fuel cell efficiency is "tradeable" with its size, and thus its cost, an optimization study is needed to determine an optimal fuel cell efficiency. However, with hydrogen as fuel the system parasitic losses could be minimized, resulting in high achievable efficiency (up to 50%).

One of the most important system requirements for backup power applications (especially for telecommunications) is the ability to start instantly upon power outage. The required response time is on the order of milliseconds. The fuel cell itself can meet this requirement as long as the supply of reactants is uninterrupted; otherwise, a bridge power (such as batteries or ultracapacitors) may be needed. System engineering solutions could significantly reduce or even eliminate the need for bridge power. For example, the fuel cell may be kept in a "ready" mode. Most of the time the backup power system is in the "idle" or "ready" mode and the operation is highly intermittent. It may only operate up to 50–200 hours per year.

Although the fuel cell generates water, the system may lose water over time. In a regenerative fuel cell system, water is lost from the system through oxygen exhaust when the system is operating in the electrolysis mode and through air exhaust when the system is operating in the fuel cell mode. However, with a proper system design the water loss is minimal and an adequately sized water tank (for at least a full year of operation) may be a more economical solution than water recovery.

Although system size and weight are absolutely critical for automotive applications, they are less important for stationary applications. However, a backup power unit must typically fit in place of the technology that it replaces (usually batteries). For some locations, the weight advantage of the fuel cell system over conventional batteries may improve its competitiveness. In those cases, the weight may be tradeable with other system elements such as efficiency and cost. Fuel cell backup power systems, including hydrogen storage for more than a few kWh, are in general much lighter than conventional batteries (lead acid) and may be lighter than most advanced batteries, too.

Clearly, the fuel cells designed for automotive or stationary power application do not meet existing backup power requirements. Indeed, in some cases they exceed the requirements, but some of the system features that drive these designs to exceed current requirements often negatively affect system cost and complexity. Therefore, a fuel cell stack and system specifically designed for backup power applications are more likely to meet the requirements at minimum cost.

For fuel cells to be offered at reasonable prices, there must be sufficient production volume with which to drive down the costs. This implies a market large enough to support the purchase of thousands or tens of thousands of units per year at a minimum. Although the automotive market holds the ultimate promise of high volume with tens of millions of units per year, the backup power market offers a significant opportunity for fuel cell commercialization in its own right.

At power levels between 1 kW and 100 kW, the backup power market worldwide is far in excess of 100,000 units annually. Large and increasing numbers of backup power systems are sold for support of computer systems, telecom systems, and other applications. These units represent both new and replacement systems. One of the positive aspects of existing technology used in backup power systems, lead acid batteries, is that the expected life is generally three to seven years. This means the replacement market is a steady and growing segment of the market. To put this in perspective, in the United States alone there are more than 100,000 cellular sites, all using backup power systems. Assuming a five-year replacement cycle, this would suggest that 20,000 replacement systems are needed each year [27].

In addition, long-term growth rates for telecommunications and computer technology well exceed average economic growth projections. Thus, the backup power market offers more than enough unit volume to make it economically viable and financially attractive for fuel cell developers.

A fuel cell backup power system must be equipped with hydrogen storage sufficient for the system operation for the required period. Empty hydrogen bottles may be replaced, or a more elegant solution is to equip the system with a hydrogen generator (an electrolyzer or a reformer). Because the system is used where electricity but not necessarily natural gas or propane is available, an electrolyzer appears to be a better option. The electrolyzer must be sized to generate the required amount of hydrogen in a given time period (typically much longer than the backup time).

The required electrolyzer power is:

$$P_{EL,nom} = \frac{P_{FC,nom} \cdot \tau_{FC} \cdot CF_{FC}}{\eta_{FC,sys} \cdot \eta_{EL,sys} \cdot \tau_{EL} \cdot CF_{EL}} \qquad (10\text{-}23)$$

where:

$P_{FC,nom}$ = fuel cell nominal power (kW)

τ_{FC} and τ_{EL} = duration of operation in fuel cell and electrolyzer modes, respectively (hours)

CF_{FC} and CF_{EL} = capacity factors of fuel cell and electrolyzer, respectively, defined as a ratio between average power and nominal power

$\eta_{FC,sys}$ and $\eta_{EL,sys}$ = system efficiency of fuel cell and electrolyzer, respectively

The electrolyzer efficiency is the reverse of fuel cell efficiency:

$$\eta_{EL} = \frac{1.482}{V_{cell}} \frac{i - i_{loss}}{i} \qquad (10\text{-}24)$$

where:

V_{cell} = electrolyzer cell voltage; similarly to fuel cell nominal voltage, electrolyzer cell nominal voltage is an arbitrary value; typically the electrolyzers operate at below 2 V per cell

i = electrolyzer current density, A cm^{-2}

i_{loss} = current and hydrogen loss, A cm^{-2}; this is typically negligible at low pressures and high operating current densities; however, it may become significant at very high pressures due to hydrogen crossover permeation

The electrolyzer system efficiency is:

$$\eta_{EL,sys} = \frac{1.482}{V_{cell}} \frac{i - i_{loss}}{i} \frac{\eta_{DC}}{1 + \zeta} \qquad (10\text{-}25)$$

where:

η_{DC} = efficiency of power conversion (either AC/DC or DC/DC)

ζ = ratio between parasitic power and net power consumed by the electrolyzer

An important feature of the fuel cell backup power system is that it separates power from energy due to recharging. In batteries, all three features are tied to the battery itself. The power output of a fuel cell backup power system depends on the size of the fuel cell. The amount of energy stored is

Figure 10-14 A prototype 1-kW backup power system. *(Courtesy of Proton Energy Systems.)*

a function of the size of the hydrogen storage. This can be changed independent of the fuel cell size, thus providing an incremental increase in energy storage at a fraction of the cost of the batteries that would be needed to provide that increase. The recharging time is a function of the electrolyzer size, which again can be tailored to fit the requirement without changing the size of the energy storage or the power output. Figure 10-14 shows a prototype 1-kW backup power system where these three features are built into three modules [27].

In applications in which the fuel cells are used as energy storage devices and compete with the batteries, specific energy is an important parameter. Specific energy or gravimetric energy density is defined as the amount of electrical energy stored (in kWh) per unit mass (kg) of the entire system. For a hydrogen/air fuel cell system, specific energy, ε, may be calculated as:

$$\varepsilon = \frac{1}{\dfrac{1}{C_f \tau \sigma_{fc}} + \dfrac{0.03}{\eta_{fc} \zeta}} \tag{10-26a}$$

For a hydrogen/oxygen fuel cell system, where both hydrogen and oxygen are stored:

$$\varepsilon = \frac{1}{\dfrac{1}{C_f \tau \sigma_{fc}} + \dfrac{1}{\eta_{fc}} \left(0.227 + \dfrac{0.045}{\zeta} \right)} \tag{10-26b}$$

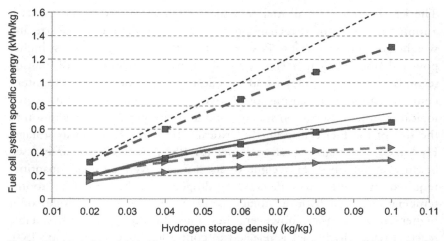

Figure 10-15 Achievable hydrogen fuel cell system energy density (solid lines are for H_2/Air and dashed lines are for H_2/O_2 systems; thin lines are for limiting cases when $\tau C_f \sigma_{fc} \rightarrow \infty$, i.e., when the fuel cell system weight is negligible in comparison to reactant gases storage).

where:

τ = duration of operation (hours); for backup power this is typically up to 8 hours, but for the systems used in conjunction with the renewable energy sources duration of operation may be up to a few days

C_f = capacity factor during operation (if the load is constant then C_f = 1)

σ_{fc} = fuel cell system specific power or gravimetric power density (kW kg-1)

η_{fc} = fuel cell system efficiency (based on lower heating value of hydrogen)

ζ = hydrogen storage efficiency (kg H_2 per kg of total storage system mass)

Figure 10-15 shows the achievable hydrogen fuel cell system energy density.

10.4 FUEL CELLS FOR SMALL PORTABLE POWER

The term *portable power system* is not sharply defined. Hoogers [32] attempted to define it as "a small grid-independent electric power unit ranging from a few watts to roughly one kilowatt, which serves mainly a purpose of convenience rather than being a primarily a result of environmental or energy-saving considerations." These devices may be divided into two main categories:

Battery replacements, typically well under 100 watts

Portable power generators, up to 1 kW

The key feature of small fuel cells to be used as battery replacements is a running time without recharging. Obviously, by definition the size and weight are also important. Power units with either significantly higher power densities or larger energy storage capacities than those of existing secondary batteries may find applications in portable computers, communication and transmission devices, power tools, remote meteorological or other observation systems, and in military gadgets. In addition to the size of the fuel cell itself, the critical issue is the fuel and its storage. Hydrogen, although a preferred fuel for PEM fuel cells, is rarely used because of bulkiness or weight of its storage, even in small quantities required by these small devices. Hydrogen may be stored in room-temperature metal hydride storage tanks. Some chemical hydrides offer higher energy density; however, they must be equipped with suitable reactors where hydrogen is released in controlled chemical reaction [33]. Most portable fuel cells use methanol as fuel, or more precisely methanol aqueous solutions, either directly (so-called direct methanol fuel cells) or via microreformers.

The military market is particularly attractive because it often can be a sympathetic early adopter of new technologies, willing to accept high prices and limited performance if other application-specific requirements can be met (such as low noise, low thermal signature, long duration both in operation and dormant, size and weight, and safety) [34]. The range of interest in small fuel cells from the military is difficult to categorize. Some examples of early military fuel cell products or prototypes include [34]:

- Battery chargers
- Soldier power
- Telecommunications
- Navigation systems
- Computers
- Various tools
- Exoskeletons
- Auxiliary power units for vehicles
- Unmanned aerial vehicles
- Small autonomous robot vehicles
- Unattended sensors and munitions
- Ocean sensors and transponders

Although a preferred fuel for military applications would be logistic fuel (JP-8), because of difficulties in reforming this fuel, particularly at such a small size, the same fuels used for nonmilitary applications, such as hydrogen, metal hydrides, chemical hydrides, and methanol, are acceptable as long as they are supplied in closed canisters or cartridges and do not have to be dispensed.

Development of small fuel cells for portable power applications has resulted in a myriad of stack configurations. Some stacks are miniaturized replicas of the larger automotive or stationary power fuel cells with the same components, MEAs, gas diffusion layers, bipolar plates, and end plates. Some use planar configurations, where the cells are connected with conductive strips. Recently, microfluidic cells manufactured on silicon-based chips have been emerging [35,36].

The fuel cell systems for these applications are extremely simplified. The simplicity of the system is more important than the cell/stack size. Power density generated is often below 0.1 W/cm^2. These cells/stacks do not need active cooling; those using hydrogen mainly operate in a dead-end mode, and air is often supplied passively.

Table 10-6 presents technical targets for various portable fuel cell systems.

Table 10-6a U.S. DOE Technical Targets[a]: Portable Power Fuel Cell Systems (<2 Watts)

Characteristic	Units	2011 Status	2013 Targets	2015 Targets
Specific power[b]	W/kg	5	5	10
Power density[b]	W/L	7	10	13
Specific energy[b,c]	Wh/kg	110	200	230
Energy density[b,c]	Wh/L	150	250	300
Cost[d]	$/system	150	130	70
Durabllity[e,f]	hours	1,500	3,000	5,000
Mean time between failures[f,g]	hours	500	1,500	5,000

Table 10-6b U.S. DOE Technical Targets[a]: Portable Power Fuel Cell Systems (10–50 Watts)

Characteristic	Units	2011 Status	2013 Targets	2015 Targets
Specific power[b]	W/kg	15	30	45
Power density[b]	W/L	20	35	55
Specific energy[b,c]	Wh/kg	150	430	650
Energy density[b,c]	Wh/L	200	500	800
Cost[d]	$/W	15	10	7
Durabllity[e,f]	hours	1,500	3,000	5,000
Mean time between failures[f,g]	hours	500	1,500	5,000

Table 10-6c U.S. DOE Technical Targets[a]: Portable Power Fuel Cell Systems (100–250 Watts)

Characteristic	Units	2011 Status	2013 Targets	2015 Targets
Specific power[b]	W/kg	25	40	50
Power density[b]	W/L	30	50	70
Specific energy[b,c]	Wh/kg	250	440	640
Energy density[b,c]	Wh/L	300	550	900
Cost[d]	$/W	15	10	5
Durabllity[e,f]	hours	2,000	3,000	5,000
Mean time between failures[f,g]	hours	500	1,500	5,000

[a]These targets are technology neutral and make no assumption about the type of fuel cell technology or type of fuel used. In addition to meeting these targets, portable power fuel cells are expected to operate safely, providing power without exposing users to hazardous or unpleasant emissions, high temperatures, or objectionable levels of noise. Portable power fuel cells are also expected to be compatible with the requirements of portable electronic devices, including operation under a range of ambient temperature, humidity, and pressure conditions, and exposure to freezing conditions, vibration, and dust. They should be capable of repeatedly turning off and on, and should have turndown capabilities required to match the dynamic power needs of each device. For widespread adoption, portable power fuel cell systems should minimize lifecycle environmental impact through the use of reusable fuel cartridges, recyclable components, and low-impact manufacturing techniques.
[b]This is based on rated net power of the total fuel cell system, including fuel tank, fuel, and any hybridization batteries. In the case of fuel cells embedded in other devices, only device components required for power generation, power conditioning, and energy storage are included. Fuel capacity is not specified, but the same quantity of fuel must be used in calculation of specific power, power density, specific energy, and energy density.
[c]Efficiency of 30% in 2013 and 35% in 2015 is recommended to enable high specific energy and energy density.
[d]Cost includes material and labor costs required to manufacture the fuel cell system and any required auxiliaries (e.g., refueling devices). Cost is defined at production rates of 50,000, 25,000 and 10,000 units per year for <2, 10–50, and 100–500 W units, respectively.
[e]Durability is defined as the time until the system rated power degrades by 20%, though for some applications higher or lower levels of power degradation may be acceptable.
[f]Testing should be performed using an operating cycle that is realistic and appropriate for the target application, including effects from transient operation, startup and shutdown, and off-line degradation.
[g]Mean Time Between Failures (MTBF) includes failures of any system components that render the system inoperable without maintenance.

10.5 REGENERATIVE FUEL CELLS AND THEIR APPLICATIONS

A *regenerative fuel cell* (RFC) is a device that can operate alternately as an electrolyzer and as a fuel cell. The electrolyzer and fuel cell functions can be integrated into a single stack of cells (unitized), or two separate (discrete) stacks may be used, one as a fuel cell and one as an electrolyzer. In an electrolysis mode, the stack generates hydrogen (and oxygen) from electricity, and in a fuel cell mode it generates electricity from stored hydrogen (and oxygen, possibly extracted from air). It is therefore an energy storage device (similar to a rechargeable battery) with hydrogen as a storage medium. The regenerative fuel cell has the highest achievable specific energy of all rechargeable energy storage options. Theoretical specific energy of H_2/O_2 fuel cell systems is 3600 Wh kg^{-1} (accounting only for the mass of reactants and assuming theoretical fuel cell efficiency), but when the masses of storage tanks and the fuel cell system are taken into account with practical fuel cell efficiencies, specific energy in excess of 400 Wh kg^{-1} may be achieved [37] (see Figure 10-15). As such, it may be used in applications where relatively large amounts of electricity need to be stored, such as:

- Energy storage for remote off-grid power sources
- In conjunction with highly intermittent renewable energy sources (solar and wind)
- Emergency or backup power generation (as discussed in the section "Backup Power")
- Unmanned underwater vehicles
- High-altitude, long-endurance solar rechargeable aircraft
- Hybrid energy storage/propulsion systems for spacecraft

Regenerative fuel cells generate their own hydrogen and therefore do not depend on hydrogen infrastructure. For that reason they may be commercialized in the niche applications listed earlier.

10.5.1 Design Tradeoffs

Several tradeoffs must be considered in the design of an RFC system, such as oxygen vs. air feed operation, unitized vs. discrete stack approach, pumped vs. static feed, and selection of operating pressure [38]. These tradeoffs are discussed next.

Oxygen vs. Air

Because oxygen is produced in the electrolysis process, it may be advantageous to store it and use it later in the fuel cell reaction. The fuel cell operation with pure oxygen results in higher voltage than its operation with air. The gain due to a higher partial pressure of oxygen is usually higher than that predicted solely by the Nernst equation, particularly because of the higher diffusion rates of pure oxygen as compared with an oxygen/nitrogen mixture. In addition, a fuel cell operating with air requires a device for pumping air through the fuel cell, resulting in additional parasitic losses. Passive air supply, based on natural convection due to temperature and concentration gradients, may be used only for very low-power applications.

The decision whether to store oxygen or vent it strongly depends on application. Systems that use oxygen may be totally closed to the environment, which is critical in applications where maintenance is difficult or costly. Obviously, for space applications oxygen storage is the only option. For many terrestrial applications the gains in fuel cell performance (or efficiency) and reduction in maintenance cost may not be sufficient to justify the capital cost and safety issues of storing high-pressure oxygen. For some applications, there is merit to accumulating a small portion of the oxygen generated during electrolysis, which is subsequently available for short-term fuel cell performance enhancement by mixing with the air stream. Often cost, efficiency, and duty cycle considerations provide significant drivers toward choice of oxidant for a given application.

Unitized vs. Discrete

In the *unitized regenerative fuel cell* (URFC), the functions of both the electrolyzer and the fuel cell are carried out by a single stack that can operate alternately in each mode. The URFC has important advantages in space applications where weight is critical. For terrestrial applications, especially those that do not include oxygen storage, separate electrolyzer and fuel cell units are generally more applicable. Inevitably there are fewer design constraints on devices designed to operate in only one mode. If, for example, a separate electrolyzer can operate at higher output pressure than a URFC, there may be sufficient savings associated with reduced tank volume to justify separate components. In some applications there is a big difference between charge and discharge power. Operation of a URFC in such a regime may not be feasible, that is, it may be outside of operational range, whereas the discrete components may be better optimized for respective charge/discharge powers. Design of an electrolyzer electrode is inherently different than design of a fuel cell electrode. Although the electrolyzer oxygen electrode (anode) is generally designed to be flooded, the fuel cell oxygen electrode (cathode) must repel produced water. Operation with pure oxygen in flooded or partially flooded electrodes may be feasible, particularly at elevated pressure, but operation with air may show severe mass transport limitations, even at very low current densities. For that reason, design of a URFC may be easier for applications in which oxygen is stored and used, and discrete units may be a better choice when air is used in the fuel cell operating mode, unless flooding issues are mitigated in the URFC design.

Pumped vs. Static Feed

Many PEM electrolyzer cell designs require high-pressure or circulating pumps, but static water-feed electrolysis can generate high-pressure gases without pumps. Static feed electrolysis transports water (by osmosis and diffusion) between the water supply and the oxygen electrode, where electrolysis occurs. Static feed systems can eliminate all moving parts (except the poppets inside valves and water expulsion containers) by suitable modifications to the electrolysis cell itself. A URFC system that employs static feed electrolysis and stores oxygen can be completely closed to the environment and has the potential to be a "maintenance-free" system. Such a system has clear advantages in locations where maintenance is expensive or prohibitive. Thermal management of static feed systems (without pumps) is a challenge, so high-performance operation can be difficult to achieve.

Although the supply of reactant gases in a static mode during fuel cell operation is feasible, particularly using pure hydrogen and oxygen, removal of water requires special cell designs.

Operating Pressure

A PEM fuel cell operates better at elevated pressure, that is, operation in higher pressure results in higher cell voltage. However, if air is used instead of pure oxygen, it must be pressurized by a compressor. Power needed to run the compressor may very well offset the voltage gain, and for that reason pressurized fuel cell systems are, in general, less efficient than the ambient pressure systems. However, if pressurized hydrogen and oxygen are available, there is no parasitic power consumption for pressurization, and that is why operation of hydrogen/oxygen fuel cells at higher pressure is preferred. PEM fuel cells operating at pressures up to 150 psig (1.0 MPa) have been demonstrated [39].

Pressurized electrolyzers result in a simpler system because there is no compressor required for storing reactant gases at high pressures. PEM electrolyzers that operate at greater than 2,000 psi (>14 Mpa) have been successfully demonstrated, and pressures up to 6,000 psi (41 MPa) appear to be achievable [40]. Operation at high pressure demands use of thicker polymer membranes (to minimize loss of hydrogen and oxygen due to diffusion), but thicker membranes result in higher ionic resistance. Although operation at higher current densities results in lower efficiency, in high-pressure systems where efficiency losses may be dominated by diffusion there is an efficiency argument to operate at higher current densities to reduce the active area. Selection of operating pressure, for both electrolyzer and fuel cell, is a complex issue and must be considered in conjunction with other tradeoff analyses.

10.5.2 Regenerative Fuel Cell Applications
Backup Power/UPS
Fuel cell use in backup power applications was discussed in the section "Backup Power." Batteries and diesel or gasoline generators currently used in these applications have severe drawbacks and limitations, such as maintenance requirements, reliability, noise, (in)convenience, and cost. The RFC may alleviate these problems. This technology has a clear advantage over batteries in applications requiring higher power outputs for longer periods. It provides greater flexibility in design because, unlike batteries, it separates charging time, storage capacity, and power output. Charging time can be varied by the size of the electrolyzer, storage capacity by the size of hydrogen storage, and power output by the fuel cell size. Therefore, various combinations are possible. The RFC also has an advantage over diesel or gasoline generators because it does not require a fuel supply—it generates its own fuel. These units generate hydrogen when grid power is available, store it, and then produce power when the grid is unavailable.

Depending on the application, hydrogen may be generated at high pressure or it may be pressurized with a compressor. Hydrogen can be stored in either high-pressure cylinders or metal hydrides. A regular bottle of hydrogen can store 0.5 kg of hydrogen at 2,000 psi (14 MPa), sufficient to operate a 1 kW fuel cell for seven to eight hours [38]. The size of hydrogen storage depends on desired system autonomy and fuel cell efficiency, although there is no explicit efficiency requirement for a backup power unit. The value of delivered electricity (when there is a power outage) by far surpasses the value of electricity used to generate hydrogen. Reliability is therefore a much more important characteristic of a backup power system than efficiency. The unit must also be capable of quick start and rapid response to load variations, and for that reason it is often combined with ultracapacitors.

The capacity of the electrolyzer is determined by allowed "recharge" time, which may be shorter, equal to, or longer than the "discharge" time, depending on the application. Consequently, the electrolyzer nominal power input could be larger or smaller than the fuel cell nominal power output.

Theoretically, the same amount of water used to generate hydrogen should be replenished by recombining that hydrogen in fuel cell reaction.

However, some water leaves the system with venting oxygen from the electrolyzer and cathode exhaust from the fuel cell.

The power conditioning subsystem consists of a regulated voltage power supply for the electrolyzer and a DC/DC or DC/AC inverter that matches the fuel cell output voltage with that of the load, which again varies with application. Depending on the number of cells in the fuel cell stack, either "boost" or "buck" voltage regulators must be a part of the power conditioning subsystem.

Backup power systems usually operate less than 200 hours per year. Both PEM fuel cells and PEM electrolyzers have demonstrated continuous and discontinuous operation well in excess of 3,000 hours. Therefore, from that aspect, PEM technology can meet the demands of this application, offering longer than a 10-year life.

Renewable Power

Renewable energy sources, such as solar and wind, are inherently intermittent. The capacity factor, defined as a ratio between average power over a period of time and maximum or nominal power, is about 20% for solar and about 30% for wind installations. An RFC may be used to store the excess energy when the renewable sources are available by converting electricity to hydrogen and then generating electricity from hydrogen when these renewable sources are not available.

The interface between the renewable source and RFC system must match their current and voltage, which differ for solar and wind systems. Photovoltaic solar systems generate direct current, which may be used directly by the electrolyzer if their polarization curves are well matched; otherwise, a DC/DC converter may be needed. Wind generators typically generate AC, so the interface must include an AC/DC inverter, similar to the electrolyzer's regulated voltage power supply.

An electrolyzer's power input vs. fuel cell power output depends on the capacity factors of both source and load and on the efficiencies of both electrolyzer and fuel cell as well as the efficiency of hydrogen storage. If the load requires constant power, the electrolyzer connected to a solar power source may have power input up to 10 times higher than the fuel cell power output [41].

The size of hydrogen storage depends on the dynamics of hydrogen production and consumption. However, if the storage must account for not only daily but also seasonal variations in renewable source availability, required hydrogen storage may be quite large. Hydrogen produced by the

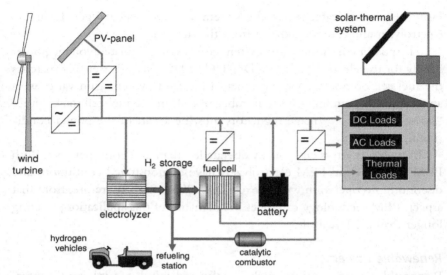

Figure 10-16 Regenerative fuel cell as a part of a renewable energy stand-alone system.

RFC may be used as a fuel for cooking, heating, or transportation. In that case, both electrolyzer and renewable power source must be sized accordingly. Such a system would provide energy independence for a device, house, farm, or a community in a remote area (Figure 10-16). This technology also has great potential in developing countries where the energy/power needs are relatively small and the power infrastructure is insufficient or nonexistent. The RFC is an enabling technology that may make the renewable energy sources more practical.

Aerospace Applications

The regenerative fuel cells, coupled with lightweight hydrogen storage, are considered as energy storage for high-altitude solar-powered aircraft. Such an airplane could fly for a long time (theoretically forever; thus it is called the *eternal airplane*), eventually replacing satellites for telecommunications, surveillance, and other applications. During the day the aircraft would be propelled by electricity produced by a photovoltaic array located on the top of the wings. The excess electricity would be used to generate hydrogen and oxygen, which are then stored in the lightweight pressure vessels that would also serve as the structural elements of the wings. During the night the aircraft would be propelled by electricity generated by recombining hydrogen and oxygen in the RFC. Energy storage systems with extremely high specific energy (>400 Wh kg^{-1}) based on RFCs have been designed to be used in high-altitude, long-endurance solar rechargeable aircraft [42,43].

Fuel cells have been used in space almost since the beginning of the space program. NASA considers highly reliable regenerative fuel cell technologies ($>10,000$ hours of continuous operation) as one of the key technologies that could enable new capabilities for future human space exploration missions.

PROBLEMS

1. The average speed and power of an automotive fuel cell system over a 400 km range are 9 m/s and 9 kW, respectively, and the average fuel cell efficiency (LHV) during that ride is 53%. Calculate the amount of hydrogen that has to be stored onboard. What is the corresponding fuel efficiency (in miles per gallon of gasoline equivalent)?
2. A power profile of a 75 kW fuel cell vehicle and corresponding drive train efficiencies are given in the following table:

Time	Power	Efficiency
3.0%	100.0%	49%
7.0%	50.0%	55%
27.5%	25.0%	57%
27.5%	12.5%	58%
35.0%	0%	

Calculate the average efficiency over such driving schedule for the following two cases:
   a. During idle periods the fuel cell is shut off and does not consume any fuel.
   b. During idle periods the fuel cell is kept running to supply the parasitic load, which is estimated at 2.5 kW. The fuel cell efficiency during that period is 50%.

3. From your home energy bills, find the price of electricity and natural gas. Also find information on the heating fuel (electricity, natural gas, or heating oil). Estimate the payback time for a 5 kW fuel cell system that operates with 50% capacity factor and average efficiency of 35% with and without cogeneration (assume total efficiency of 75%). The fuel cell system costs $4,400, including installation. How could the feasibility of this system be improved?

4. A utility company pays $6 per 1,000 cubic feet of natural gas and wants to install a fuel cell to generate electricity during peak hours. The fuel cell costs $1,200 per kW, including installation. The capacity factor will be 33% and the average efficiency also 33%. Assuming a five-year lifetime with 5% discount rate, estimate what should be the cost of generating electricity (in $/kWh). What if the same fuel cell were used as a base load operating with 95% capacity factor and 38% average efficiency?

5. A hydrogen/oxygen fuel cell system is proposed for space application where it is supposed to provide 5 kW continuous power for periods of 12 hours. Hydrogen is stored at 300 bar in lightweight tanks that provide 5% storage density (0.05 kg H_2 per kg of full tank). Assume that oxygen is stored at same pressure in similar tanks with the same weight/volume ratio. The fuel cell operates at 0.75 V and 1 A cm^{-1}. The efficiency of DC/DC power conversion is 95%. The parasitic power is 400 W. The fuel cell weighs 8 kg and its balance-of-plant weighs an additional 16 kg. Estimate the specific energy (in kWh/kg), and compare it with the best available batteries. How could the specific energy be further improved?

6. A regenerative fuel cell is installed to provide constant 2 kW power output from a PV array. The fuel cell system efficiency is 50% (LHV) and the electrolyzer system efficiency is 75% (also LHV). The fuel cell operates 12 hours during the night. During the day the PV array provides 2 kW for the load and it runs the electrolyzer. During these 12 hours the electrolyzer operates with 22% capacity factor. Calculate:
   a. The amount of hydrogen to be stored (in kg) and the size of the storage if the electrolyzer is capable of pressurizing hydrogen up to 100 bar.
   b. The power input to the electrolyzer system.
   c. The nominal power of the PV array (assuming 140 W/m^2); calculate the size of the PV array.

QUIZ

1. An automotive fuel cell system is more efficient than an internal combustion engine at:
   a. Maximum power only
   b. Partial load only
   c. Any load
2. An automobile needs to store approximately:
   a. 0.5 kg of hydrogen
   b. 5 kg of hydrogen
   c. 50 kg of hydrogen
3. One kilogram of hydrogen pressurized at 200 bar takes approximately:
   a. 6 liters
   b. 16 liters
   c. 60 liters
4. A pressurized fuel cell system with a turbine/expander:
   a. Is more efficient than a low-pressure system
   b. Has about the same efficiency as a low-pressure system
   c. Is less efficient than a low-pressure system
5. The economics of fuel cells for stationary generation of electricity depend on:
   a. A difference in price between electricity and natural gas
   b. A ratio of prices of electricity and natural gas
   c. The price of natural gas only
6. Capacity factor is:
   a. Number of hours a fuel cell system operates, divided by 8,760 (the number of hours in a year)
   b. A ratio between kilowatt hours of electricity produced in a year and kilowatt hours that would have been produced if the system operated at full power the entire year
   c. A ratio between the fuel cell average power and average power required by the load
7. Annual average efficiency of a fuel system:
   a. Is always smaller than the efficiency at nominal power
   b. Is always higher than the efficiency at nominal power
   c. May be smaller or larger, depending on how the system is operated throughout the year

8. What would be the effect of an increase in the price of natural gas on the economics of a fuel cell system that is used in cogeneration (heat and power mode)?
   a. It will make the fuel cell less economical
   b. It will make the fuel cell more economical
   c. It is hard to say; it depends on the ratio of electricity and heat utilization and respective efficiencies
9. The power needed to run an electrolyzer in a regenerative fuel cell with discrete fuel cell and electrolyzer stacks:
   a. Is always larger than the nominal fuel cell power
   b. Is always smaller than the nominal fuel cell power
   c. Depends on the required charging time and therefore can be either larger or smaller than the nominal fuel cell power
10. Specific energy of a regenerative fuel cell power system depends on:
   a. Efficiency of fuel cell
   b. Efficiency of electrolyzer
   c. Efficiency of both fuel cell and electrolyzer

REFERENCES

[1] www.netinform.net/h2/H2Mobility/Default.aspx (accessed May 2012).
[2] www.hydrogencarsnow.com (accessed May 2012).
[3] Rajashekara K. Propulsion System Strategies for Fuel Cell Vehicles. SAE Paper No. 2000-01-0369. Warrendale, PA: Society of Automotive Engineers; 2000.
[4] Tachtler, J., T. Dietsch, and G. Goetz, Fuel Cell Auxiliary Power Unit—Innovation for the Electric Supply of Passenger Cars, SAE Paper No. 2000-01-0374, in Fuel Cell Power for Transportation 2000 (SAE SP-1505) (SAE, Warrendale, PA, 2000), pp. 109–117.
[5] Konrad G, Sommer M, Loschko B, Schell A, Docter A. System Design for Vehicle Applications: DaimlerChrysler. In: Vielstich W, Lamm A, Gasteiger H, editors. Handbook of Fuel Cell Technology—Fundamentals, Technology and Applications, 4. New York: J. Wiley; 2003. p. 693–713.
[6] Masten DA, Bosco AD. System Design for Vehicle Applications: GM/Opel. In: Vielstich W, Lamm A, Gasteiger H, editors. Handbook of Fuel Cell Technology—Fundamentals, Technology and Applications, 4. New York: J. Wiley; 2003. p. 714–24.
[7] Stone R. Competing Technologies for Transportation. In: Hoogers G, editor. Fuel Cell Technology Handbook. Boca Raton, FL: CRC Press; 2003.
[8] EPA Urban Dynamometer Driving Schedule (UDDS): www.epa.gov/otaq/emisslab/methods/uddsdds.gif (accessed February 2005).
[9] Thomas CE, James BD, Lomax Jr FD, Kuhn Jr IF. Fuel Options for the Fuel Cell Vehicle: Hydrogen, Methanol or Gasoline? International Journal of Hydrogen Energy 2000;25(6):551–68.
[10] Well-to-Wheel Energy Use and Greenhouse Gas Emissions of Advanced Fuel/Vehicle Systems. North American Analysis. Report by General Motors in cooperation with Argonne National Laboratory. BP Amoco, ExxonMobil and Shell 2001.

[11] Weiss MA, Heywood JB, Drake EM, Schafer A, AuYeung FF. On the Road in 2020: A Life-Cycle Analysis of New Automobile Technologies. Boston: Massachusetts Institute of Technology; 2000.

[12] Pehnt M. Life-Cycle Analysis of Fuel Cell System Components. In: Vielstich W, Lamm A, Gasteiger H, editors. Handbook of Fuel Cell Technology—Fundamentals, Technology and Applications, 4. New York: J. Wiley; 2003. p. 1293–317.

[13] James, B. D. Fuel Cell Transportation Cost Analysis, Preliminary Results, Proc. 2012 U.S. Department of Energy (DOE) Hydrogen and Fuel Cells Program and Vehicle Technologies Program Annual Merit Review and Peer Evaluation Meeting, May 14–18 2012. Arlington, Virginia, www.hydrogen.energy.gov/annual_review12_fuelcells.html#analysis.

[14] Copeland, M. V., The Hydrogen Car Fights Back, CNNMoney.com, October 14, 2009. (http://money.cnn.com/2009/10/13/technology/hydrogen_car.fortune/index.htm)

[15] Carpetis C. Storage, Transport and Distribution of Hydrogen. In: Nitsch J, editor. Hydrogen As an Energy Carrier. Berlin Heidelberg New York: Springer-Verlag; 1988.

[16] Michel F, Fieseler H, Meyer G, Theissen F. Onboard Equipment for Liquid Hydrogen Vehicles. In: Veziroglu TN, et al., editors. Hydrogen Energy Progress XI, 2. Coral Gables, FL: International Association for Hydrogen Energy; 1996. p. 1063–77.

[17] Berlowitz PJ, Darnell CP. Fuel Choices for Fuel Cell Powered Vehicles. SAE Paper No. 2000-01-0003. In: Fuel Cell Power for Transportation 2000 (SAE SP-1505). Warrendale, PA: SAE; 2000. p. 15–25.

[18] Sadler M, Heath RPG, Thring RH. Warm-up Strategies for a Methanol Reformer Fuel Cell Vehicle. SAE Paper No. 2000-01-0371. In: Fuel Cell Power for Transportation 2000 (SAE SP-1505). Warrendale, PA: SAE; 2000. p. 95–100.

[19] Uribe FA, Gottesfeld S, Zawodzinski Jr TA. Effect of Ammonia as Potential Fuel Impurity on Proton Exchange Membrane Fuel Cell Performance. Journal of Electrochemical Society 2002;149. p. A293.

[20] Uribe FA, Zawodzinski TA. Effects of Fuel Impurities on PEM Fuel Cell Performance. In: Proc. 201st ECS Meeting, 2001-2. San Francisco; 2001. Abstract 339.

[21] Fronk MH, Wetter DL, Masten DA, Bosco A. PEM Fuel Cell System Solutions for Transportation. In: Fuel Cell Power for Transportation 2000 (SAE SP-1505). Warrendale, PA: SAE; 2000. p. 101–8. SAE Paper No. 2000-01-0373.

[22] U.S. Department of Energy. Hydrogen. Fuel Cells & Infrastructure Technologies Program, Multi-Year Research, Development and Demonstration Plan, Planned program activities for 2005–2015, Technical Plan, updated May 2012. www.eere.energy.gov/hydrogenandfuelcells/mypp/pdfs/techplan.pdf (accessed June 2012).

[23] Hoogers G. Automotive Applications. In: Hoogers G, editor. Fuel Cell Technology Handbook. Boca Raton, FL: CRC Press; 2003.

[24] Adamson, K.-A. Fuel Cell Market Survey: Buses, Fuel Cell Today, Article 916, November 2004: www.fuelcelltoday.com/FuelCellToday/FCTFiles/FCTArticleFiles/Article_916_BusSurvey%20Final%20Version.pdf (accessed February 2005).

[25] U.S. Department of Energy. Hydrogen, Fuel Cells & Infrastructure Technologies Program, Early Markets: Fuel Cells for Material Handling Equipment, www1.eere.energy.gov/hydrogenandfuelcells/education/pdfs/early_markets_forklifts.pdf (Accessed May 2012)

[26] Barbir F. System Design for Stationary Power Generation. In: Vielstich W, Lamm A, Gasteiger H, editors. Handbook of Fuel Cell Technology—Fundamentals, Technology and Applications, 4. New York: J. Wiley; 2003. p. 683–92.

[27] Barbir F, Maloney T, Molter T, Tombaugh F. Fuel Cell Stack and System Development: Matching Market to Technology Status. In: Proc. 2002 Fuel Cell Seminar. CA: Palm Springs; November 18–21, 2002. p. 948–51.

[28] Hoogers G. Stationary Power Applications. In: Hoogers G, editor. Fuel Cell Technology Handbook. Boca Raton, FL: CRC Press; 2003.

[29] Lester LE. Fuel Cell Power Electronics. Fuel Cells Bulletin 2001;(25):5–9.

[30] Barbir F, Joy GC, Weinberg DJ. Development of Residential Fuel Cell Power Systems. In: Proc. 2000 Fuel Cell Seminar. Portland, OR; 2000. p. 483–6.

[31] U.S. Energy Information Administration, www.eia.gov/electricity/data.cfmwww.eia.gov/dnav/ng/ng_pri_sum_dcu_nus_m.htm, and www.eia.gov/countries/ (accessed May 2012).

[32] Hoogers G. Portable Applications. In: Hoogers G, editor. Fuel Cell Technology Handbook. Boca Raton, FL: CRC Press; 2003.

[33] Heinzel A, Hebling C. Portable PEM Systems. In: Vielstich W, Lamm A, Gasteiger H, editors. Handbook of Fuel Cell Technology—Fundamentals, Technology and Applications, 4. New York: J. Wiley; 2003. p. 1142–51.

[34] Geiger S., and D. Jollie. Fuel Cell Market Survey: Military Applications, Fuel Cell Today, Article 756, April 2004: www.fuelcelltoday.com/FuelCellToday/FCTFiles/FCTArticleFiles/Article_756_MilitarySurvey0404.pdf (accessed February 2005).

[35] Shah K, Shin WC, Besser RS. A PDMS Micro Proton Exchange Membrane Fuel Cell by Conventional and Non-Conventional Microfabrication Techniques, Sensors and Actuators B. Chemical 2004;97(2–3):157–67.

[36] Yamazaki Y. Application of MEMS Technology to Micro Fuel Cells. Electrochimica Acta 2004;50(2–3):659–62.

[37] Barbir F, Molter T, Dalton L. Efficiency and Weight Trade-off Analysis of Regenerative Fuel Cells as Energy Storage for Aerospace Applications. International Journal of Hydrogen Energy 2005;30(4):231–8.

[38] Barbir F, Lillis M, Mitlitsky F, Molter T. Regenerative Fuel Cell Applications and Design Options. In: Proc. of (CD) 14th World Hydrogen Energy Conference; June 2002. Montreal.

[39] Murphy OJ, Hitchens GD, Manko DJ. High Power Density Proton Exchange Fuel Cells. Journal of Power Sources 1994;47:353–68.

[40] Mitlitsky, F., B. Myers, A.H. Weisberg, and A. Leonida. Applications and Development of High Pressure PEM Systems, Portable Fuel Cells International Conference (Lucerne, Switzerland, 1999) pp. 253–268.

[41] Barbir F. Integrated Renewable Hydrogen Utility System. In: Proc. 1999 DOE Hydrogen Program Annual Review Meeting; May 4–6, 1999. Lakewood, CO.

[42] Mitlitsky F, Colella NJ, Myers B. Unitized Regenerative Fuel Cells for Solar Rechargeable Aircraft and Zero Emission Vehicles, 1994 Fuel Cell Seminar. CA: San Diego; 1994.

[43] Mitlitsky F, Myers B, Weisberg AH. Lightweight Pressure Vessels and Unitized Regenerative Fuel Cells, 1996 Fuel Cell Seminar 1996. Orlando, FL.

CHAPTER ELEVEN

Durability of Polymer Electrolyte Fuel Cells

Mike L. Perry

United Technologies Research Center (UTRC), 411 Silver Lane, East Hartford, CT 06108;
perryml@utrc.utc.com

11.1 INTRODUCTION

The cost and durability of polymer–electrolyte fuel cells (PEFCs) are the major technical barriers to the widespread commercialization of this technology. PEFCs should not only be environmentally advantageous relative to the conventional power-generation devices that PEFC products seek to replace, they must also be on par with these devices on an economic basis to be commercially successful. Therefore, the cost and durability goals for PEFCs tend to be based on the capabilities of the incumbent technologies. For example, the U.S. Department of Energy (DOE) has established different targets for fuel cells based on the intended application: for automotive applications the durability goal is 5,000 hours at a cost of $30/kW, whereas stationary fuel cell systems will require about 60,000–80,000 hours of durability at a capital cost of about $1,000–1,500/kW, depending on size and application [1]. Of course, cost and lifetime are related, since customers will generally be willing to pay more for products that last longer. Therefore, one would like to develop PEFC products that not only can match the incumbent technologies with respect to durability but can, ideally, exceed them.

At first glance, it may seem that fuel cells should be inherently robust devices, since there are no moving parts to wear out and fail, at least not on a macroscopic scale within the cells. However, PEFCs are necessarily composed of components with relatively precise architectures, and small, microscopic changes can result in dramatic changes in performance. This is especially true of low-cost, high-performance PEFCs. For example, as catalyst loadings are decreased, the impact on cell performance becomes more sensitive to any changes in the limited catalyst sites that are available. (This is another aspect of how cost and durability are intimately related.) In addition, the same forces that enable a PEFC to generate electric power can also be strong driving forces for degradation reactions.

PEM Fuel Cells
ISBN 978-0-12-387710-9

435

For example, large potential differences promote adverse electrochemical reactions (e.g., corrosion reactions) as well as the desired fuel cell reactions.

Anyone who has much experience with rechargeable batteries (such as laptop batteries) also inevitably has first hand experience of how an electrochemical device can degrade with age and may have also noted that this decay strongly depends on how the device is used. PEFC durability also strongly depends on the operating conditions. However, although there are similarities in the degradation of fuel cells and batteries, there are also significant differences. Theoretically, the electrodes in fuel cells are invariant, since they are simply the sites where the reactions take place, whereas in rechargeable batteries the electrodes must undergo physiochemical changes as the state of charge (SOC) changes with charge/discharge cycles. Additionally, the operating conditions of fuel cells are inherently simpler and less variable, since they are typically designed to operate only in "discharge" mode and the concentration of the reactants is not continuously changing, as is the case in batteries with varying SOC. Therefore, fuel cells can have considerably longer lifetimes than can be obtained with rechargeable batteries. For example, United Technologies Corporation (UTC) has already developed and demonstrated stationary fuel cell systems with >60,000-hour durability and PEFC stacks with >10,000-hour durability in transit-bus applications. However, neither of these products meets the commercial cost targets, so additional improvements are required. Additional development is then required to validate that these cost-reduction improvements are also durable. Therefore, fuel cell durability is a subject that will always be of interest to developers of PEFC products and will continue to evolve as new designs or materials are introduced to enable improved performance or cost.

11.2 SCOPE AND ORGANIZATION OF THIS CHAPTER

There is a wide variety of possible degradation mechanisms in PEFCs and, therefore, PEFC durability is a large and complex field of study. A large amount of work on identifying and understanding PEFC degradation phenomena has already been done, and entire books dedicated to this topic have been published recently [2,3]. Therefore, the intent of this chapter is to provide an introduction to, not an exhaustive overview of, this broad subject. Sufficient references are provided throughout for the reader who wants to study certain aspects of this subject more extensively.

One can classify decay mechanisms in PEFCs in a number of ways—by component, by driving forces, by operating conditions that promote them, and so on. However, this chapter starts with the different types of cell *performance degradation*, which are classified according to the type of polarization that is increasing in the cells. The reason for this is because this is what can first be readily determined from relatively simple in-cell performance diagnostics, which are described in Section 11.3. Additionally, determining what types of polarization are changing can also help one in isolating what PEFC component(s) may be responsible for the degradation in performance, as shown in Section 11.4. A PEFC consists of a number of repeating components, including a membrane, catalyst layers, gas–diffusion layers, bipolar plates, and seals. Therefore, identifying which of these components is responsible for performance losses is always of paramount importance. Section 11.4 also provides a very brief summary of some of the most common degradation mechanisms for each of the key PEFC components, including decay mechanisms that can often be recoverable.

Because permanent damage is of great importance, Section 11.5 provides more details on these decay mechanisms, with an emphasis on the operating conditions that tend to promote these types of degradation. The durability of a PEFC strongly depends on operating conditions. Principal among these are potential, temperature, and relative humidity. For a given set of materials, fuel cell stacks that are subjected to less aggressive operating conditions last longer and decay less. Therefore, consideration of both materials with improved stability and design strategies that minimize adverse operating conditions provides the best path to developing PEFC systems that are optimized with respect to durability, cost, and performance.

This chapter ends with a discussion of accelerated testing. This is a related topic to operating conditions because the goal of durability testing is often to accelerate a particular decay mechanism or stress specific components within the PEFC. To do this effectively, one should understand what conditions cause PEFCs to decay, including what operating parameters (e.g., T, V) have a significant effect on the rate of each of these mechanisms. An overview of the major types of accelerated tests and the advantages and disadvantages of each is therefore the focus of Section 11.6.

11.3 TYPES OF PERFORMANCE LOSSES

As explained in previous chapters of this book, there are four major sources of performance losses, or types of polarization, in fuel cells, namely: (1) activity or kinetic polarization, (2) ohmic or resistive losses, (3) concentration polarization or mass-transport losses, and (4) internal currents or crossover losses. Therefore, a good way to categorize performance losses is according to the type of overpotential that is increasing (i.e., degrading).

One can readily assess how the different overpotentials evolve by plotting the change in voltage versus current density, as illustrated in Figure 11-1. This polarization-change plot is constructed by taking the difference between a polarization curve taken at the beginning of life (BOL) and the latest polarization curve. BOL is herein defined as the peak performance of the cell (i.e., after the break-in period where the cell performance is still increasing), since including the effects of the break-in period makes analyzing degradation more difficult. The cause of performance changes during the break-in period is not the subject of this chapter. The difference is taken this way in order to return a positive value as performance decays with age. Thus,

$$\Delta V = V_{BOL} - V \tag{11.1}$$

Operating conditions, including dwell times and procedures that recover performance, must be the same for all curves. Measuring polarization curves

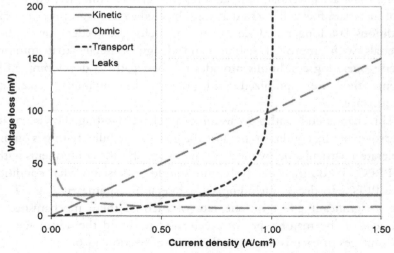

Figure 11-1 Limiting cases of polarization-change curves.

before and after applying recovery procedures should allow one to separate recoverable and irrecoverable decay.

The limiting cases depicted in Figure 11-1 were constructed by subtracting voltages calculated with a simple model of polarization; derivations of the equations are provided elsewhere [4]. However, these equations (and the resulting curves shown in Figure 11-1) are *not* expected to be *quantitatively* accurate in all cases. Instead, the different polarization-change curves depicted in Figure 11-1 should be used as simple *qualitative* indicators of the general shapes that result from different types of polarization degradation.

Polarization-change plots are constructed by taking the difference between a recent polarization curve and one taken at peak performance. Four limiting cases are shown here: (1) kinetic (solid line), (2) ohmic (dashed line), (3) transport (dotted curved line), and (4) leak (dash-and-dot curve).

11.3.1 Key Limiting Cases of Polarization-Change Curves

Four polarization-change curves are depicted in Figure 11-1; each represents the shape expected if only one type of overpotential changes. These polarization-change curves represent the difference between peak and decayed voltage at different current densities. All other independent variables are held constant.

The first limiting case depicted in Figure 11-1 is a horizontal line, resulting from an increase in kinetic losses for the oxygen-reduction reaction (ORR). In principle, decreases in either the exchange current density or the catalytic surface area could be responsible for this type of loss. Kinetic losses associated with the ORR should be independent of current density, provided the Tafel slope does not change. Practically, the Tafel slope does not appear to change significantly as platinum catalysts degrade. (For the case shown in Figure 11-1, it was assumed that the catalytically active surface area per unit volume of electrode decreased by 50%.) However, redistribution of platinum during operation may also lead to increases in ohmic and transport losses in the cathode, as well as kinetic losses.

The second limiting case is a straight line that intercepts the origin of the plot. This is the theoretical expectation for losses due solely to an increase in resistance, or ohmic losses. (For the case shown in Figure 11-1, it was assumed that the cell resistance doubled from 100 to 200 $m\Omega$-cm^2.)

If increases in mass-transport overpotential are the sole cause of decay, then one would expect a curve that intersects the origin and has a roughly exponential shape. In this case, the performance loss approaches infinity as the current approaches the new limiting current, which is necessarily lower than the BOL limiting current. The original limiting current can no longer be achieved because of the degradation in mass transport. (For the case shown in Figure 11.1, it was assumed that the limiting current had been reduced by 50% from approximately 2 to 1 A cm^{-2}.) Mass transport losses through the depth of the cathode catalyst layer may not display limiting-current behavior because reaction and transport may be in parallel, not in series. To treat this situation mathematically, an "effectiveness factor" is commonly used to describe reaction and diffusion in catalyst pellets and adoption of this approach to porous fuel-cell electrodes has been described in detail elsewhere [5].

The fourth limiting case depicted in Figure 11.1 is observed if leaks are the sole cause of decay. Gas leaks in a cell can increase due to membrane or seal degradation. Electrical shorting, which can also increase with time as electrical contact between anode and cathode components increases due to degradation of the membrane or seals, also yields a similar result on this plot. In either case, the impact of leaks on cell performance is greatest at low current densities. This behavior can be readily understood if one thinks of a leak as an additional load imposed on the cell. In the case of an electrical short, this additional load is the current being carried by the short, which means that the measured current is lower than the actual current through the active area of the cell. A gas leak across the membrane is equivalent to an electrical short in the way that it affects polarization, and the magnitude of the leak can be expressed as an equivalent current density (i.e., in A/cm^2) [6]. At low current densities this equivalent leak current will be as high or higher than the measured current density and the impact on cell performance will be significant. At higher current densities the impact will be relatively small because the leak current is small compared to the measured current. (For the case illustrated in Figure 11.1, the leak current was increased from 0 to 50 mA cm^{-2}.)

11.3.2 Analyzing Actual Polarization-Change Curves

PEFC degradation commonly results from changes to more than one type of overpotential. For example, an increase in kinetic losses often occurs early in the life of a PEFC, since the small catalyst particles employed to maximize catalyst surface area are prone to dissolve and grow larger with time. This kinetic loss may be accompanied by a transport loss since, at a given current density, the reactant may have to penetrate deeper into the degraded catalyst layer and the rate of reactant transport to each remaining catalyst site must be proportionally greater. In this case, the result would be a combination of kinetic and reactant-transport curves depicted in Figure 11.1. This would be an approximately exponential curve that intercepts the ordinate at a positive value, which is a typical result with actual cell data. In this case, the y-intercept provides a quantitative estimate of the change in polarization due to increased kinetic overpotential. With real data, which always has some degree of noise, one can use a best fit of the horizontal portion of the curve to provide a better estimate of the intercept value. Similarly, a combination of kinetic and ohmic changes can result in a sloped line with a nonzero intercept.

In reality, a polarization-change curve with an intercept of zero is uncommon because PEFCs typically experience a loss in catalytic area. In other words, one almost always measures a lower open-circuit voltage (OCV) as a cell degrades. However, if one measures a very large change in OCV and much smaller changes at higher current densities, then this is a good indication that the cell has developed a leak or short. Since OCV measurements can be noisy, it is good practice to also include at least one measurement at a low current density (e.g., 10 mA/cm^2 or less) to make this comparison.

The graphical analysis described here is simple and useful; however, it should be conducted with some caution. For example, small decreases in the limiting current may result in a roughly linear function in a polarization-change plot at currents that are a small fraction of the limiting current. One should record data out to the limiting current to obtain the best possible insight. In any case, one should always seek to verify the sources of over-potential by conducting more advanced cell diagnostics. However, this simple graphical analysis is a useful first step since it utilizes readily available data and it provides insight into what may be the most fruitful subsequent diagnostics. In summary, plotting the change in performance, instead of the absolute performance, is an efficient method to focus on how the perfor-mance is changing or degrading with age, which is the focus of this chapter.

11.4 PEFC COMPONENTS ASSOCIATED WITH DIFFERENT TYPES OF LOSSES

A PEFC consists of a number of repeating components including a membrane, anode and cathode catalyst layers (CLs), anode and cathode gas-diffusion layers (GDLs), bipolar plates, and seals. This section describes which of these components are susceptible to each kind of polarization change described in the previous section. Some common decay mechanisms for each of these components are also discussed. A summary of this section is provided in Table 11-1, where each row corresponds to one of the four limiting cases discussed in the previous section and each column corresponds to one or more major components in a PEFC. Only the most common mechanisms are mentioned here along with references that provide more in-depth discussions. Decay that is typically recoverable is also briefly discussed here, while more details on nonrecoverable decay mechanisms are summarized in Section 11.5, where the operating conditions that promote these nonrecoverable losses are emphasized.

11.4.1 Catalytic Activity Losses

As shown in Figure 11.1, voltage loss that is nearly independent of current density may indicate a loss of catalytic surface area or activity. This behavior is only true for reactions that follow simple kinetic expressions with constant Tafel slopes, which is usually true at the cathode of a PEFC (i.e., the ORR reaction). The first row in Table 11-1 corresponds to this case. Because the cell reactions are presumed to happen only within the catalyst layers, this type of loss indicates an issue within either the anode or cathode. Which electrode has the issue will depend on the particular decay mechanism.

11.4.1.1 Common Catalyst-Degradation Mechanisms

A decrease in catalytic activity can result from a loss of electrochemically active surface area (ECSA), which can result from a variety of mechanisms such as catalyst-particle growth due to dissolution and/or sintering as well as corrosion of the carbon supports; these mechanisms have been described in more detail elsewhere [7,8]. The catalytic activity of the catalyst may also change due to particle-size effects or changes in the composition of the surface, in the case of alloy catalysts. In addition, potential cycling accelerates the dissolution of platinum and alloying elements, causing these elements to be redistributed within the cell, often to areas such as the membrane, where they can no longer promote the desired reactions. This will be further discussed in Section 11.5.

A decrease in catalytic activity can also be caused by adsorption of contaminants. Common culprits include CO and H_2S on the anode and NH_3 and SO_X on the cathode. Multiple book chapters have summarized a large body of work on contamination in PEFCs [9]. The impact on performance may not be independent of current density in cases where contamination of the catalytic surface is extensive. Poisoning of platinum at

Table 11-1 Summary of Types of Performance Losses and Possible Decay Mechanisms

Type of Performance Loss	Some Possible Decay Mechanisms		
	Anode Diffusion Media or Catalyst Layer	Membrane	Cathode Diffusion Media or Catalyst Layer
Catalytic activity Performance loss nearly independent of current density	Possible causes: - Loss of catalyst area by sintering or dissolution - Contamination by adsorption* 		Possible causes: - Loss of catalyst area by sintering or dissolution - Contamination by adsorption* - Pt oxide formation*
Ohmic (ionic or electronic) Performance loss proportional to current density	Possible causes: - Dryout of ionomer* - Contamination by foreign cations* - Increased contact resistance, intra- or interlayer	Possible causes: - Dryout of membrane* - Contamination by foreign cations*	Possible causes: - Dryout of ionomer* - Contamination by foreign cations* - Increased contact resistance, intra- or interlayer
Reactant mass transfer Performance loss exponential with current density	Possible causes: - Flooding of diffusion media or catalyst layer* - Reactant channel blockage* - Carbon oxidation		Possible causes: - Flooding of diffusion media or catalyst layer* - Reactant channel blockage* - Carbon oxidation
Crossover or internal current** (electrical or reactant) Performance loss primarily at low current densities		Possible causes: - Short circuit through or around membrane (or seal materials) - Reactant leakage due to membrane (or seal) failure	

*Often substantially reversible.

**Reactant leakage, or electrical shorts, can also occur through the seal materials in the cell, which is similar to comprises in the membrane, although reactant leakage may be external (i.e., go overboard from the cell).

the anode by CO, for example, lowers the limiting current for hydrogen oxidation significantly. Oxidation of platinum in the cathode may also lead to decreased activity. However, it should be emphasized that contamination is often reversible. Raising the potential of the anode is often sufficient to remove adsorbed species. Prolonged operation of the cell in the absence of the contaminant has been shown to remove some species that adsorb on the cathode. Raising the potential of the cathode can also remove adsorbed species; this can be accomplished by stopping the fuel cell and allowing the electrodes to approach the reversible potential for oxygen reduction. Platinum oxides can be readily stripped by simply lowering the cathode potential (e.g., shorting the cells with hydrogen flowing and air flow stopped), which results in a temporary performance improvement since these oxides will inevitably appear again on these platinum sites at normal cathode operating potentials. This change in the degree of platinum oxidation on the cathode is also a large reason for the hysteresis that exists in polarization curves taken in opposite directions with respect to current density.

11.4.2 Ohmic Losses

Voltage decay that is a linear function of current density may be caused by increasing resistance. The second row in Table 11.1 corresponds to this case. One can differentiate between ohmic losses that occur in the membrane and those in the catalyst layers by conducting diagnostics that provide a more direct measurement of membrane resistance (e.g., high-frequency measurements or current interrupt). Increased ohmic losses in the membrane are obviously due to degradation in ionic resistance of this layer. However, determining the sources of ohmic losses in the catalyst layers can be challenging. The loss may be either due to changes in the electronic or ionic resistance or a result of a change in conductivity within a layer or at an interface between layers. Every layer of a PEFC is either electronically or ionically conductive, but the catalyst layers are the only layers that must be both, which can make diagnosing changes more difficult.

11.4.2.1 Common Ohmic-Degradation Mechanisms

Ionic conductivity is a strong function of ionomer hydration. Drier conditions result in higher ohmic losses within both the catalyst layers and the bulk membrane. Operating at higher relative humidity should hydrate the ionomer and reverse this type of loss. Irreversible chemical degradation of the ionomer can also take place with time, and this is often accelerated by operation under drier conditions [10]. Although this degradation will impact ionic conductivity, it is usually a highly localized phenomenon that results in loss of structural integrity and leakage through the membrane before having an appreciable impact on cell resistance.

Contamination by metal cations (such as ferrous ions) may also decrease conductivity. Similar to catalyst contamination, ionic contamination is often reversible. The ionomer in the membrane and the CLs is essentially a cation-exchange media that can be restored to the acid form by removing the source of ionic contamination and continuously operating the cell, if possible.

Increased contact resistance may also result in increased ohmic losses. This may occur if the compressive load relaxes or if gaps develop between the layers of the cell, as can happen if there is substantial carbon oxidation. Severe corrosion of carbon catalyst support will also result in other losses, such as increased ohmic and transport losses due to the change in the catalyst-layer structure.

11.4.3 Reactant Mass-Transport Losses

Performance loss that is a strongly increasing function of current density may indicate an increase in reactant mass-transport losses. The third row in Table 11.1 corresponds to this case. As with ohmic losses, the cause may be on either the anode or the cathode. However, with mass-transport losses it is a relatively straightforward process to determine which side is causing the problem (e.g., by independently varying the respective reactant concentrations and/or stoichiometry). On the other hand, it can be difficult to isolate the particular component that's responsible. Reactant transport losses may be caused by changes in transport resistance either external to the CL or internal to it, and the symptoms of some types of reactant transport losses may also be confused with ionic transport losses. Differentiating between reactant transport losses and ionic transport losses in the catalyst layers can be challenging, and the use of physics-based models and different diagnostics can be helpful to distinguish between these two sources of polarization.

11.4.3.1 Common Mass-Transport Degradation Mechanisms

Accumulation of liquid water in flow channels and GDLs may cause reactant transport losses to increase. This may result from changes in material properties, such as hydrophobicity, over time. Accumulation of liquid water within the catalyst layer related to changes in hydrophobicity may also be a problem. Loss of porosity due to the oxidation of carbon support may also diminish gas transport rates in the CL. The latter mechanism is often observed in localized areas in full-size cells due to nonuniformities in fuel or current distribution during start or normal operation, as will be shown in Section 11.5.

If reactant transport losses are due to flooding, it is sometimes possible to reverse the loss, at least temporarily, by removing the excess water. Many methods for doing so are available. Water may be removed by operating the cell at high temperature to increase the vapor pressure of the trapped water, operating the cell with drier reactants, or drying the cell out while it is not operating. Whether or not the improvement obtained by these procedures is stable, however, depends on the cause of the flooding. If the flooding is due to an irreversible mechanism such as a change in wettability or pore structure of the catalyst layer, it is likely to return shortly after the cell is brought back to normal operating conditions. If the flooding is due to an excursion to an abnormal operating condition such as a cold start, it is often substantially reversible and the recovery in performance may be persistent [11].

11.4.4 Overboard Leaks, Reactant Cross Over, and Electrical Shorts

Performance loss that primarily affects the voltage at low current density may indicate leakage, either internal or external. Internal "leaks" can be either gas crossing through the membrane or as current due to an electrical short from the anode to cathode. The fourth row in Table 11.1 corresponds to these cases. It is generally possible to distinguish between electrical shorts and reactant crossover by varying the pressure of the reactants and seeing if the leakage rate changes.

11.4.4.1 Common Mechanisms

As mentioned previously, chemical attack may compromise the integrity of the membrane, leading to pinholes through which gases can move. Seal degradation may also lead to gas leakage at the edges of the cell.

Electrical shorts may occur when chemical attack leads to membrane thinning or if fibers puncture the membrane. Degradation of perimeter seals may lead to electrical shorts forming at the edges of the cells. These types of degradation are typically not recoverable, so it is imperative that the membrane and seal materials are inherently stable under PEFC operating conditions. Additionally, one can minimize both the stress and the impact of decay by controlling the operating conditions, which is the subject of the next section. With respect to seals and membrane materials, minimizing the gas–pressure differences (between the anode and cathode as well as to ambient) as well as minimizing the number and range of temperature cycles can yield improved results with a given material set.

11.5 OPERATING CONDITIONS

The durability of a PEFC strongly depends on operating conditions. Principal among these are potential, temperature, and relative humidity. Although durability may often be improved by utilizing more stable materials, even the most stable materials cannot withstand certain adverse operating conditions. Additionally, more stable materials may have negative attributes such as higher cost and/or lower performance. Therefore, fuel cell stacks that are subjected to more benign operating conditions generally have the ability to be the most durable, cost–effective, and high–performance ones.

Successful developers of other types of fuel cells, including PAFC, MCFC, and AFC, have learned that the keys to good durability are controlling the potential and temperature as well as proper electrolyte management. Whereas a PEFC has inherent advantages relative to these other fuel cell types, including low operating temperatures and an immobilized electrolyte, PEFC stacks also have their own unique durability challenges due to the intended applications (e.g., transportation applications with dynamic load profiles and frequent start/stop cycles) as well as the stability of the membrane electrolyte in the fuel cell environment. Therefore, it is worth examining these three key operating parameters, which are essential to control in order to successfully demonstrate durable PEFC stacks.

11.5.1 Potential Requirements

The electrode potential is the main driving force for many of the degradation modes present in PEFCs, including carbon corrosion and platinum dissolution; therefore, it is important to understand what electrode potentials are likely to be encountered in operation. Table 11.2 lists the gases present and the approximate maximum electrode potentials at the two electrodes of a PEFC under different operating conditions. All reported potentials are with respect to a reversible-hydrogen electrode (RHE). The terms fuel and air electrode are used here because anodic and cathodic reactions can occur at both electrodes, depending on the local conditions. The electrode

Table 11-2 Typical Gases Present and Approximate Electrode Potentials During Various Operating Conditions of a PEFC

Operating Condition	Fuel Electrode (Anode)		Air Electrode (Cathode)		Cell Potential (V)
	Gas	Potential (V)	Gas	Potential (V)	
Normal	H_2	0.05	Air	0.85	0.8
Idle	H_2	0.0	Air	0.90	0.9
Open circuit	H_2	0.0	Air	1.0	1.0
Off	Air	1.1	Air	1.1	0
Start	H_2/Air	0.0/1.1	Air	1.0/1.5	1.0
Stop	Air/H_2	1.1/0.0	Air	1.5/1.0	1.0
Partial H_2 coverage	H_2/Inert	0.05/1.1	Air	0.85/1.5	0.8
Fuel starvation	H_2/Inert	>1.5	Air	0.85	<−0.65
Air starvation	H_2	0.05	Air/H_2	0.85/−0.05	−0.1

potentials can be controlled under some of these conditions; however, the values in the table assume no attempt to control the potential. The cell potentials shown in Table 11.2 do not rigorously account for ohmic resistance and are intended as a guide only. Boxes containing two numbers separated by a slash are used when two regions with different electrode potentials are present.

The normal operating condition prevails when excess hydrogen and air are present and power is being produced. The idle state refers to a minimal power output, corresponding to parasitic loads such as pumps and blowers, with ample hydrogen and air present. Open circuit occurs when there is no load on a cell with hydrogen and air on the fuel and air electrodes, respectively. The OCV is typically considerably lower than the thermodynamic potential difference (e.g., ~1.23 V) due to reactant crossover and mixed potentials on each of the electrodes. The off state occurs when a cell is not used for a long time, unless an inert gas is intentionally added to the electrode compartments. During the off state, the electrodes typically achieve a mixed potential that is set by oxygen reduction and corrosion or oxidation reactions.

In addition to the OCV and off states, there are at least four other power plant operating conditions that can result in even higher potentials than those described above (i.e., >1.2 V) and thereby seriously jeopardize the stability of the CLs (or even the adjacent layers, such as the GDLs, if sustained for long or repeated durations). These four conditions are:

1. Start of the power plant
2. Shutdown of the power plant (or stop)

3. Otherwise normal operation but at low fuel stoichiometry and/or with poor fuel distribution, which can result in *local* fuel starvation regions

4. Otherwise normal operation but with a fuel stoichiometry less than unity, which results in *gross* fuel starvation

The first three conditions can result in elevated potentials on the cathode (and thereby accelerate oxidation of the cathode CL), whereas the last condition leads to rapid oxidization of the anode CL (e.g., typically corrosion of the carbon to support the protons required for the current being demanded). With the exception of the start condition, all of these typically occur at normal cell operating temperature where carbon-oxidation kinetics are accelerated relative to ambient temperatures. However, even the first condition can occur at elevated temperatures in the case of a start conducted shortly after a shutdown.

The difficulty during an initial start stems from the fact that air fills the fuel system during extended shutdown periods. This air must then be replaced with fuel to operate the system, and it takes several hundred milliseconds for the fuel to traverse the fuel flow fields of individual cells of a practical size (i.e., on the order of >100 cm^2 vs. laboratory-size cells that are typically on the order of ≤ 25 cm^2). During this gas-transition time, the entering fuel causes the cell voltage to rise while air still occupies the fuel flow field at the exit locations. The voltage rise drives ionic current through the cell at the exit location in a direction opposite, or reversed, to the normal current flow; hence this CL degradation mechanism has been labeled the "reverse-current" mechanism, which is briefly described in the following. However, the net result in this case is that the air electrodes at the fuel-exit regions are locally driven (by the fuel-filled portions of the cell) to local potentials >1.4 V$_{RHE}$, which is well outside the safe operating envelope for most commonly used fuel cell catalyst supports (e.g., high surface area carbons).

A depiction of the local fuel-starvation condition is provided in Figure 11.2. The situation depicted here is during a start, with fuel (hydrogen) introduced from left to right in this simple cross-section of a PEFC. However, the same situation can result anytime there is hydrogen present in one region of the anode and is absent in an other anode region(s). (A more detailed description of this mechanism can be found in the source of this figure or in a paper that provides a more complete model of the mechanism [12].) In essence, the hydrogen/air region of the cell ("Region A" in Figure 11.2) acts as a sort of "potentiostat" for the rest of cell, since it sets a potential (typically OCV during start) that is present in the solid

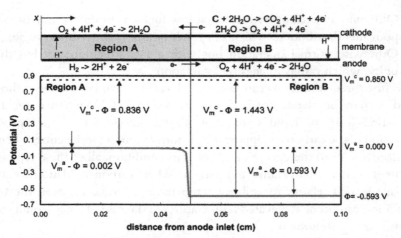

Figure 11.2 A simple depiction of the reverse-current decay mechanism. *(Reproduced with permission from [13].)*

portions of the entire cell (i.e., the "metal" potential, V_m, which is relatively uniform since the bipolar plates and electrodes have high electrical conductivity). However, since the reactants in the fuel-starved regions of the cell are totally different ("Region B" in Figure 11.2), this results in large differences between the potential in the electrolyte phases (set by the electrochemically active reactants in this region) and the potential in the solid phases of Region B. These large potential differences drive electrochemical reactions in this fuel-starved region, which are opposite of what normally occurs on these electrodes (i.e., oxidation reactions occur on the "cathode" in this region and reduction reactions on the "anode"). Hence, the local ionic current is also in the opposite direction in this region, as shown in Figure 11.2, which is why this has been labeled the "reverse-current" degradation mechanism. It should be noted that there is no electron current external to the cell; the electrons flow from the "potentiostat" region of the cell to/from the driven portions of the cell in the plane of the cell (i.e., through the bipolar plates, GDL, and other solid electronic conductors on either side of the membrane). Therefore, there is no external indication that this reverse-current mechanism is occurring, which makes it essentially impossible to detect, although the impact can be quite severe (as will be shown in the next section).

The potential distributions along the anode flow path during a condition when hydrogen is present in one part of the cell (Region A) and absent in other areas (Region B). The potentials result in reverse-current reactions in

Region B, which is often supported by carbon corrosion on the cathode in the hydrogen-starved regions.

Difficulty during shutdown is also caused by the possibility of anodes that are only partially filled with fuel during this transition, which results from the air that floods the fuel system during the shutdown period. Fuel cell developers often purge the fuel-reactant passages to safely secure the system during shutdown periods. This will also greatly shorten the transition time. However, during the purge period, the purge gas (typically either air or nitrogen) usually enters the fuel flow fields at the inlet and results in some oxygen occupying both the fuel and air flow fields at this location; consequently, a reverse-current scenario also occurs here. In the case of a nitrogen purge, there is still some oxygen, albeit a reduced amount, present on the anodes due to oxygen crossover through the membrane from the cathode. Again, this can result in driving the air electrode to >1.4 V_{RHE}. However, the reverse-current region is now located at the fuel-inlet locations instead of the fuel-exit locations. (These transitions are often shorter than a start, since the purge rates are typically faster than the fuel-introduction rates and, in the case of an N_2 purge, the local potentials are obviously more benign.)

The third condition, or *local* fuel starvation condition, occurs during normal operations and is simply another manifestation of the reverse-current degradation mechanism. Reverse current will occur if any portion of a cell becomes locally fuel starved. This occurs because the local area that receives no fuel will receive oxygen (and nitrogen) from crossover through the membrane from the air on the opposite side of the cell. Once again, the normal cell voltage (in the solid phases) will drive current in a reverse direction at the fuel-starved locations. During local fuel starvation the measured cell voltage (typically measured using the "metal," or solid, phases of the cell) will appear normal despite the presence of the reverse-current mechanism in hydrogen-depleted regions of the cell. Partial-hydrogen coverage can be caused by poor fuel distribution across the active area, which can, for example, be caused by liquid water in the fuel-flow path blocking hydrogen access to the anode catalyst layer (i.e., local "flooding" of the anode). For example, it has been shown that local fuel starvation in a region as small as $\sim$2 mm on a hydrogen-fueled anode can result in serious degradation of the cathode in a relatively short time (e.g., $\sim$100 hrs) [14].

The fourth high potential condition can also occur during "normal" operation and is caused by general, or gross, fuel starvation. This is the condition whereby any cell receiving less fuel than is commensurate with the current being drawn through the stack (i.e., total cell fuel stoichiometry <1) will result in a negative cell voltage. This condition causes the anode support (and ultimately other anode carbon components that are consumed as fuel) to be subjected to >1.2 V_{RHE}. Unlike the other conditions, the degree to which this voltage can rise is limited only by the number of cells in the stack receiving adequate fuel. Arguably, if this condition occurs, the design of either the stack and/or the system is fundamentally flawed.

It should be emphasized that, of these four cases, *gross* fuel starvation is the only condition where the measured cell voltage can be used as a good indicator of fuel starvation. In the first three cases (which are all instances of *local* fuel starvation) the cell voltage is not necessarily abnormal, even though irreversible damage is occurring on the cathode, which makes *in situ* detection very difficult.

Finally, the air starvation condition occurs when the current exceeds the limiting current for the ORR (i.e., local oxygen stoichiometry <1). In this case, hydrogen evolves from the air electrode and the potential of the air electrode approaches the reversible-hydrogen potential. The potential of the cell is slightly negative under this condition. Therefore, the primary concern in this case is not the durability of the cell components but safety, since a mixture of hydrogen and air may be present on the air electrodes (i.e., the cathodes during normal operation).

Summarizing Table 11.2, the potentials typically experienced on PEFC electrodes range from approximately 0 to ≥1.5 V. Higher potentials promote degradation modes such as carbon corrosion and platinum dissolution, which are briefly discussed in the following section. Therefore, one should strive to minimize exposure to the high potential conditions in order to minimize PEFC degradation.

11.5.1.1 Effect of Elevated Potential and Potential Cycles on Platinum

Commercial PEFC electrodes contain dispersed platinum or platinum-alloy catalysts supported on high surface area carbon. High platinum surface area is required to minimize the overpotential for the ORR. Examination of the Pourbaix diagram for platinum indicates that dissolution is expected to occur in a triangular region where pH < 0 and the electrode potential is between approximately 1 and 1.2 V at 25 °C.

Cycling the potential of a platinum electrode in acid electrolyte causes higher dissolution rates than potentiostatic experiments at similar potentials. This is an important consideration for transportation applications that require frequent, and typically rapid, load changes. When platinum surfaces oxidize, they will tend to passivate and the dissolution rate slows. This is why potential cycles accelerate dissolution, since the passive layer is repeatedly reduced and is therefore more prone to dissolution when the potential is raised again. It has been shown, as one might expect, that the rate of platinum surface area loss with repeated potential cycles depends on both the potential limits and the potential sweep rate.

Several material approaches can be taken to improve catalyst stability in PEFC operating conditions. The biggest challenge is to improve, or at least sustain, ORR activity while increasing durability. In the simplest case, use of larger platinum particles can reduce the loss of surface area; however, one must accept a penalty in surface area per gram of catalyst. As shown by Darling and Meyers [15], the potential driving force for Pt dissolution is given by:

$$U = U^o - \frac{\sigma_{Pt} M_{Pt}}{2Fr\rho_{Pt}} \tag{11.2}$$

In this equation, U is the potential for Pt dissolution, U is the standard potential for bulk Pt, r is the particle radius, σ_{Pt} is the surface tension, and ρ_{Pt} is the density of Pt. The difference between the electrode potential and U drives Pt dissolution proportionally. As seen from the this equation, U increases with an increase in r, the particle radius. This indicates that the dissolution rate decreases with increasing Pt particle size, which has been confirmed in numerous experiments. This is part of the motivation for utilizing catalyst architectures with extended surfaces, such as whiskers, since one can obtain high surface area (and good inherent activity) with less surface energy than simple spherical particles. One can also utilize a different and preferably less expensive material to reduce the curvature of the platinum surface, such as the core-shell concept with a large core, or this can be done with nonspherical cores that are not necessarily large. However, in any case, the core material should also be stable in PEFC conditions to alleviate the requirement that the shell must be pinhole-free to prevent dissolution or corrosion of the core.

Platinum alloys have been used to increase both the performance and the durability of the catalyst since the 1980s, when UTC developed ternary Pt alloys for PAFC cathodes. For example, supported PtCo and PtIrCo alloy catalysts appear to lose less surface area than Pt when cycled to high potentials. However, in PEFC, compatibility of the alloying elements with the membrane should be considered when contemplating replacement of Pt

with a *Pt* alloy, because many alloying elements are more soluble than platinum. Additionally, one should measure the propensity of any alternative catalyst to generate species that could contribute to chemical attack of the membrane (e.g., hydrogen–peroxide species).

Whenever investigating the stability of a catalyst material with respect to elevated potentials or potential cycles, one also needs to be mindful of the complete architecture of the CL, which can also dramatically impact the net result. In the case of dispersed catalysts (e.g., platinum supported on carbon, or Pt/C), a key factor is the stability of support, which can often be as problematic as the stability of the catalyst itself. This is certainly the case with conventional carbon supports, which are discussed next.

11.5.1.2 *Effect of Potential on Carbon*

Carbon is commonly used as a catalyst support, in gas–diffusion layers, and as a bipolar-plate material. Carbon is not thermodynamically stable under all conditions encountered by a PEFC. Fortunately, the kinetics of carbon corrosion are relatively sluggish, but the electrochemical oxidation of carbon is accelerated by increasing temperature and potential in an exponential manner. Hence, carbon corrosion can be lowered to acceptable levels by minimizing time at high temperature and potential in a PEFC. However, if these conditions are not controlled, carbon is not an acceptable material in a PEFC.

Figure 11.3 shows electron-microprobe images of a PEFC after repeated and uncontrolled start/stop cycling. This stack had a Pt/C cathode and a PtRu/C anode. Thinning and brightening of the cathode CL are observed near the fuel exit, whereas the inlet appears to be undamaged. Furthermore, the band of platinum in the membrane is more developed near the exit because this region has been subjected to higher potentials. A band of ruthenium is also visible outside the MEA at the exit. This occurs because the anode has been subjected to higher potentials at the exit. These patterns are generally consistent with the reverse-current mechanism that results in high cathode potentials in parts of the cell where hydrogen is absent, especially at the fuel exits, where hydrogen is absent during the initial portion of each start cycle as hydrogen is introduced into the air-filled anodes at a relatively slow rate.

Multiple methods to mitigate the reverse-current mechanism, especially during start/stop cycles, have been described elsewhere [16]. Essentially, these are system-level methods to control and limit the maximum potentials experienced by the cells during these transient conditions (e.g., applying an

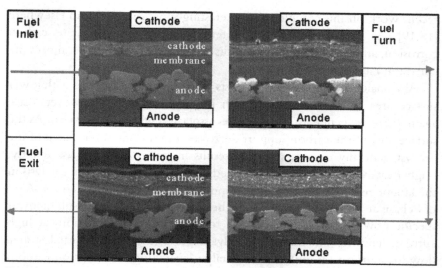

Figure 11.3 Results of a cell subjected to uncontrolled start/stop cycles. *(Reproduced with permission from [17].)*

external electrical load, or resistor, across the cells during the hydrogen transition periods). Additionally, extended periods at open circuit and idle should also be avoided to improve CL stability. In a fuel cell hybrid vehicle, these goals can be accomplished by utilizing the energy storage system (i.e., batteries) in a manner analogous to that employed in hybrid vehicles that use internal combustion engines. Additionally, due to system parasitic power requirements, PEFC power plants are less efficient at extremely low power; therefore, time spent at these conditions should be minimized in order to maximize both system energy efficiency and stack life.

Figure 11.3 shows cross-sectional electron microprobe images of four locations of a MEA from a PEFC stack that was subjected to 1,994 uncontrolled start/stop cycles. The stack utilized two fuel passes, as shown. As expected by the reverse-current mechanism, the amount of damage depends on the distance from the fuel inlet. Note the changes in the cathode catalyst layer and the presence of platinum in the membrane, especially in the second pass.

A more detailed description of carbon-corrosion mechanisms as well as the challenges still remaining in understanding these complex phenomena is beyond the scope of this chapter. Analogous to platinum, carbon also tends to form a passive layer at high potentials; however, the oxide species that are formed in this case are far more complex than platinum oxides. Reviews and recent work on this complex and interesting topic can be found elsewhere [18,19]. As one might expect, potential cycles also accelerate carbon corrosion, and both the potential limits and potential sweep rate impact the corrosion rate.

Also analogous to platinum catalysts, one can employ carbons with lower surface area to obtain supports with improved stability. However, once again, there are generally compromises associated with this approach. As the surface area of the carbon support decreases, the specific surface area of the catalyst generally decreases, which results in lower catalyst mass activity. Higher catalyst mass activity is obtained with lower mass fraction catalysts in the kinetic region (e.g., 20% Pt/C is more active than 50% Pt/C on an A g^{-1} of Pt basis at 0.9 V_{RHE}); however, higher catalyst loadings can result in lower specific power (kW g^{-1} of Pt) due to mass-transport limitations at high current densities. The competing advantages of high specific catalyst area (low loadings on carbon) vs. thin catalyst layers (high loadings on carbon) result in an optimum loading, which is typically around 40 to 50 wt % in a PEFC with conventional Pt/C catalysts. At these relatively high catalyst mass fractions, high surface area carbons are especially advantageous, since the decrease in catalyst dispersion (and metal surface area) with increased metal loadings is less severe on higher surface area carbon blacks. Therefore, to maximize kW g^{-1} of platinum one would prefer high surface area carbon blacks. As a general rule of thumb, low surface area carbons are generally more stable than high surface area carbons (there are exceptions, since carbon is a very complex material). However, a performance penalty will generally be associated with a lower surface area support, especially in a PEFC that is expected to operate at high power density (e.g., automotive applications) since these require thin catalyst layers with relatively high catalyst mass fraction. Therefore, the catalyst support selection must be optimized for a given application and overall power system design.

As shown in Table 11.2, the range of potentials to which the electrodes in a PEFC can be exposed is quite large if no attempts are made to control the potential. Therefore, the best general approach is to balance the use of materials that tolerate high potentials (perhaps at the cost of lower performance) with strategies that control the potentials experienced by the electrodes.

11.5.2 Temperature

The operating temperatures of a PEFC are typically between 65 °C and 80 °C. The low operating temperature enables quick starting and enhances power density because it reduces the need for thermal insulation. These attributes are key reasons that PEFCs are attractive for transportation applications. However, the low operating temperature makes rejection of heat to ambient surroundings more difficult than it is for internal combustion engines of comparable power. For this reason, automotive applications would prefer higher operating temperatures while maintaining or improving the performance and durability characteristics of the fuel cell system. Additionally, a higher operating temperature should improve tolerance to impurities like CO, which would be beneficial for applications operating on reformed hydrocarbons, such as many stationary power applications.

In general, the rates of thermally activated processes such as chemical reactions and transport in condensed phases increase exponentially with temperature. Thus, one might expect that the rates of proton transport in the electrolyte and oxygen reduction at the cathode should increase dramatically as the operating temperature is increased. However, the expected benefits associated with increasing the operating temperature are difficult to obtain in systems based on the current class of perfluorosulfonic acid (PFSA) membranes because system water balance deteriorates as temperature increases. Dehydration of the electrolyte membrane negatively affects proton transport and ORR kinetics and generally results in reduced cell performance. Therefore, higher operating temperatures are not generally desirable from a performance or durability perspective, since the kinetics of many decay mechanisms (e.g., carbon corrosion, Pt sintering and dissolution) obey the Arrhenius equation.

Another reason that a low operating temperature is an inherent advantage is that the impact of thermal cycles is reduced for smaller temperature ranges. Thermal cycling is an inevitable aspect of discontinuous operation, and small differences between operating and ambient temperatures help minimize thermally induced stresses. In addition, the durability of most sealing materials is a strong function of the number and magnitude of thermal cycles.

11.5.3 Electrolyte and Humidity

Water is not only the product of the PEFC reaction; it is also critical for stable operation. PFSA membranes require water to transport protons since they reside in aqueous clusters within the polymer and their mobility depends on the characteristics of the aqueous network. The more hydrated the membrane, the higher the ionic conductivity and the higher the performance. On the other hand, the presence of excessive liquid water can restrict access of reactant gases to the electrodes and result in significant performance losses. Therefore, the best operating condition for a PFSA-based PEFC is fully saturated reactants without excessive liquid water.

Maintaining a proper level of humidification can be difficult in actual power plants. The reactants are typically humidified before entering the cell in order to keep the membrane well hydrated. Condensation often occurs in the cell due to effects that include water production, gas consumption, pressure drop, electroosmotic drag, diffusion, and temperature changes. This liquid water can cause nonuniform gas distribution from cell to cell and limit gas access within cells. Additionally, pushing liquid water along the gas channels increases pressure drops, which reduces system efficiency.

Maintaining the desired humidification level throughout the cell is especially challenging in applications that require frequent changes in power. These power changes cause near-instantaneous changes in heat and water generation. If the design of the power plant incorporates external humidi-fication of the reactant gas, the capability of the control scheme and the mass of the humidification equipment will limit the time to attain steady state at the new power level. Therefore, during transients, dehydration or flooding can occur, depending on how the power is changed. This can result in reduced membrane life due to the combined effects of chemical and mechanical stresses, as discussed in the next section.

Water can condense in the gas manifolds because these manifolds are generally located at the periphery of the stack where the temperature is typically lower than the middle of the stack due to heat losses to ambient. These lower temperature regions result in the condensation of water. The condensate can block individual cell passageways, resulting in reactant gas maldistribution or, in the extreme, insufficient flow to support the current produced by the stack. As previously explained, this reactant starvation condition can cause irreversible damage to the anode CL if it occurs on the fuel side. Condensation can also occur within the active area when reactant gases flow from relatively high temperature zones into lower temperature zones. The liquid water can block flow in the channels; however, unlike manifold condensation, this condition rarely results in starvation of an entire cell. However, the blockage can cause local reactant gas starvation, resulting in a decrease in cell potential. Once again, if this occurs on the fuel side, it can result in permanent damage, in this case to the cathode CL.

Thermal cycling presents yet another water management challenge: namely, controlling water migration when the cell is stopped in order to allow the subsequent start to be successful. After the PEFC is turned off, water may be driven from the active area to the stack perimeter by temperature gradients during the cooling period. The worst condition occurs when the ambient temperature is below zero, both because the temperature gradients are largest and because ice can form within the cells. Ice formation can cause damage to cells, especially if frost-heave phenomena take place [20]. The key to avoiding permanent damage in a PEFC due to freezing is to avoid layers that completely restrict the movement of liquid water. Although hydrophobic layers in GDLs can be advantageous to manage water movement during normal operation, they can become problematic if they do not allow the flux of liquid water that needs to be accommodated during freeze/thaw cycles. For these reasons, water residing in the cells is often purged prior to a shutdown (e.g., using air as the purge gas). However, it is not possible to remove all the water residing within the small pores of GDLs and CLs. Whether this water freezes or not, it can block reactant access to the catalyst sites on the subsequent start, which may result in extending the required start time, at best, and possibly irreversible cell damage due to local fuel starvation in more severe cases. Additionally, during starts from a frozen condition, there is a "race" between the cell heating up and ice formation that can prevent reactant access to the catalyst sites and cause the cell to stop generating power (and heat). Therefore, the key to a successful start is to minimize the initial water present in the cell, design the cell to allow sufficient liquid water movement (both at low operating temperatures and during freeze cycles), and ensure that there are some "reservoirs," or void spaces, present to accommodate excess water during transients, as discussed in the next section.

11.5.3.1 Effect of Humidity and Humidity Cycles on PEFC Membranes

Generally, the durability of polymer-electrolyte membranes improves with increasing hydration. Membrane life is defined as the operating time at which the membrane loses the ability to separate the reactants due to pinholes. Continued operation beyond that point causes local reactions that lead to further increase in reactant gas mixing, which eventually results in a loss of cell efficiency as well as potentially unsafe operating conditions.

Membranes subjected to repeated changes in relative humidity (RH) undergo mechanical degradation, and it has been shown that exposing membranes to hydration cycles will result in failure. This degradation mechanism is not surprising, since PFSA membranes undergo relatively large dimensional changes when they are exposed to varying hydration levels. For example, an unconstrained, dry PFSA membrane will increase in volume by about 74% when equilibrated with liquid water. These dimensional changes are much larger than those experienced by polymeric materials subjected to thermal cycles in the temperature range of a PEFC. In a cell, where the membrane is constrained, hydration changes result in significant mechanical stress. The degree of mechanical stress depends on the change in hydration level (more specifically, the change in λ, which is the molar ratio of water molecules to ionic groups). For example, a PFSA membrane that is cycled from a supersaturated condition to 80% RH undergoes more rapid degradation than a membrane cycled between 80% and 30% RH, since $\Delta \lambda$ is greater in the former case and the swelling-induced dimensional change is roughly proportion to $\Delta \lambda$.

All membranes currently being utilized in deployed PEFCS operate best when fully hydrated. Thus, PEFC systems that maintain fully saturated (but not flooded) conditions within the cells generally demonstrate the best life, stability, and power-density metrics. Unless a membrane is developed that does not require a relatively high level of hydration, some means of storing and providing water within a PEFC should be provided to ensure stable operation. Without these water reservoirs, cell performance can suffer from changes in hydration caused by changes in operating conditions. In addition, membrane lifetime is reduced with humidity cycles, and local fuel starvation can also result. Some reservoirs for water are present in conventional PEFCs within the voids of the catalyst layers and the diffusion media. However, as developers strive to reduce the thickness of these layers to further reduce cost and/or performance (e.g., lower catalyst loadings, thinner GDLs), it can make PEFC durability more challenging. Ideally, advanced PEFC architectures should reduce cost and/or improve steady-state cell performance while also retaining water reservoirs that minimize RH variations within the cells, thereby stabilizing performance and extending life.

11.6 ACCELERATED TEST PROTOCOLS

One would like to study PEFC durability without having to conduct tests that necessarily last as long as the durability targets (e.g., 5,000 to 80,000 hours). For example, whenever a change in the design and/or materials used in PEFC products is implemented, the developer should first verify that the durability of this new design is adequate; this should ideally be done in as short a time as possible. Therefore, a key enabler to the development of advanced PEFC materials and cell designs is effective accelerated test protocols. The purpose of this section is to discuss accelerated testing of PEFCs in a general way. A detailed analysis of particular accelerated-stress tests (ASTs) like those developed and advocated by the U.S. DOE [21] is not the objective, nor is a review of the results of such tests.

Good ASTs should reduce the time and cost required to develop new products and improve existing products while also reducing technical risks to acceptable levels. The tests should provide physical insight about the nature of the failure modes likely to occur during operation and should not cause failures for reasons that would not occur in the field. The tests must be carefully crafted to achieve these goals, bearing in mind the application, because different stressors are pertinent to transportation, stationary, and other applications. Generally speaking, degradation can be accelerated by increasing stress levels above those experienced in the field or by increasing the frequency of stress application.

Table 11.3 categorizes conditions that are known to damage PEFCs (it also provides a high-level summary of the previous two sections). Detailed explanations of the degradation mechanisms, including known accelerating factors, are covered in other sources referenced in Sections 11.4 and 11.5. It is interesting to note that low potentials or high loads (i.e., high current densities) are not known to accelerate degradation in PEFCs. Therefore, most AST protocols consist of either high potential holds or highly cyclic conditions, especially cycles to low-load/high-potential conditions (e.g., OCV). The lower potential of these cycles is typically chosen to reduce either platinum or carbon oxides, depending on the decay mechanism(s) being investigated. If new decay mechanisms are discovered, then undoubtedly new AST protocols will be developed.

The extent to which a particular PEFC is exposed to any condition depends on both the application and the design of the system. For example, the magnitude and frequency of load changes imposed on a fuel cell in an automobile depends on how it is hybridized with energy-storage devices. Knowing the levels of stress present during normal operation is critical to constructing an effective testing plan. Similarly, it is important to know the bounds on stressor values. For example, using temperatures above the glass-transition point of a membrane may lead to erroneous conclusions.

One common approach to studying PEFC degradation is to use a series of test protocols to accelerate and isolate different decay modes. The ASTs given by the DOE exemplify this approach, and these ASTs cover many of the stressors listed in Table 11.3. Ideally, the damage induced by the ASTs

Table 11.3 Summary of Major PEFC Stressors and Decay Mechanisms

Stress		Decay Mode				
Type	Magnitude	Carbon Corrosion	Platinum Dissolution	Membrane Damage*	Structural Damage**	Activity Loss***
Potential	High	X	X			
	Low					
	Cycles			X		
Humidity	High	X	X			
	Low	X	X	X		
	Cycles			X		
Load	High	X				
	Low	X				
	Cycles	X			X	
Temperature	High	X	X	X		
	Low		X	X		
	(freezing)				X	
	Cycles					
Contaminants	High			X		X

*Mechanical and chemical degradation

**Physical changes, especially to the electrode layers

***Activity losses beyond those due to platinum dissolution

should be linked to damage observed in the field by known factors. If this information is not available, it should at least be possible to use the ASTs to screen new components. Two potential issues associated with applying this approach to qualify components are: (1) all sources of decay must be identified and addressed with individual tests, and (2) synergistic effects that may occur in actual operation may not be apparent. Meaningful statistics should be collected for ASTs that are highly accelerated, by running either replicate single cells or multicell stacks. The results of an AST may depend on the active area and details of the flow field and hardware designs and should be extrapolated to different designs with a high degree of caution.

One example showing the difficulty of isolating decay modes involves the degradation of platinum and carbon. Corrosion of carbon generally increases with increasing temperature, voltage, and concentration of water. Platinum catalyst also tends to increase the carbon–corrosion rate. The lower voltage limit and cycling profile and frequency are important during potential cycling experiments. Platinum dissolution and carbon corrosion tend to be accelerated by the same variables but with different sensitivities. Thus, constructing a test to accelerate only one of these corrosion rates is difficult. For conventional Pt/C electrodes, frequent cycling to potentials below 1 V tends to emphasize degradation of the metal, whereas cycling to higher potentials tends to emphasize carbon corrosion.

As is evident in Table 11.3, the stressors that are commonly employed to create ASTs are the same factors that one should strive to minimize in fielded products in order to maximize PEFC life [22]. In essence, a key goal of a PEFC developer is to minimize exposure to stressful conditions in fielded products as much as possible. If this is done, creating ASTs is also easier since one does not have to invoke extreme conditions to enable sufficient acceleration factors. Both the development of PEFCs with long life and the development of effective ASTs require an excellent understanding of all the possible PEFC decay mechanisms, including the stressors for each of these mechanisms. Since many of the conditions that are known to degrade PEFCs have now been reviewed in the open literature, it is possible to construct at least qualitative accelerated tests for individual mechanisms with an increasing degree of confidence.

Development of fuel cell products relies on information from all sources. Field trials are extremely valuable because they subject cells to intended and unintended operating conditions. Unintended conditions may arise, for example, if a strategy to mitigate a particular type of decay results in the acceleration of another type of degradation. Field trials help inform accelerated test protocols in two ways: (1) they identify the actual operating conditions, and (2) they identify why units actually fail. Therefore, regular correlation between field results and laboratory results is definitely essential in the continuous evolution of AST protocols.

11.7 CONCLUSIONS AND FUTURE OUTLOOK

Considerable advancements in the fundamental understanding of PEFC degradation phenomena have occurred in the last decade, so much so that entire books dedicated to this topic have been recently published that provide detailed reviews of the various PEFC decay mechanisms that have been discovered and studied to date. The intent of this chapter has been to simply provide a high-level overview of this still evolving field of knowledge.

Currently available and conventional PEFC materials offer a reasonably good combination of performance and stability provided that the cell, stack, and system design respect their stability limitations. In fact, the key durability targets for both transportation and stationary applications can be met with these conventional materials. However, the cost targets for the potentially largest PEFC markets have not been met (e.g., automotive and stationary power plants for commercial applications), and the design of a PEFC system based on this conventional material set remains a highly involved optimization problem. It requires at least a basic understanding of the key decay mechanisms and the use of system mitigation strategies to ensure that the cells are not subjected to aggressive operating conditions. Fortunately, this level of knowledge is now generally available in the open literature, such as the references cited herein. However, future improvements in PEFC technology are required both to meet the cost targets as well as to eliminate the need for the aforementioned system mitigation strategies, which add both cost and complexity to a PEFC system.

A major challenge in PEFC durability is that the range of conditions to which a PEFC can be exposed is surprisingly quite large. Transients, both start/stop cycles and load cycles, can result in significant changes in potential, temperature, and relative humidity. Although the exact conditions will depend on the application, all applications impose some transient conditions, and a PEFC system must be designed to either withstand or minimize the adverse conditions that can be encountered during these transients. In general, these transient conditions accelerate degradation in PEFCs; potential cycles can have an especially adverse impact on the electrode components.

It will always be desirable to develop advanced materials or cell designs that offer improved stability, especially if they do not adversely impact the performance and/or cost of the PEFC. However, it is critical for materials developers to be cognizant of the possible trade-offs that are inherent in the advanced materials being developed. For example, high-temperature membranes offer many benefits, but they also require catalysts and other cell components that are stable at the higher temperatures. Generally, such higher-stability components sacrifice some performance for stability. Success or failure of such an improved membrane is therefore coupled to the temperature sensitivity of catalyst degradation (something a membrane group may unintentionally ignore). Standardized AST protocols can help researchers develop and demonstrate advanced material sets that can potentially address the multiple PEFC decay mechanisms that have been reviewed here. However, more quantitative understanding of these decay mechanisms, as well as sophisticated physics-based models, are required before AST results can be confidently used to project PEFC life in real-world applications.

Existing membranes require hydration, and the stability of these membranes in a dynamic fuel cell environment, especially one that includes large humidity cycles, is a serious concern. In addition, the presence of liquid water within the cells can make uniform reactant delivery, which is necessary to achieve stable performance and prevent permanent cell degradation, challenging. Therefore, effective water management is imperative and can most readily be achieved by incorporating some liquid-water reservoir volume within the cells. Operation at higher temperatures (e.g., ~120 °C) offers important advantages, including elimination of the liquid phase. However, to be practical, it also requires a membrane electrolyte that functions well in dry environments as well even more stringent control of the potentials experienced by the cells, since the higher operating temperatures will tend to accelerate many of the known degradation reactions.

Advanced catalyst materials and catalyst-layer architectures are highly desirable, since improvements in this area could enable a number of desired attributes, such as higher PEFC operating temperatures, lower catalyst loadings and/or catalyst cost, higher performance, and improved cyclic stability. The stability of the catalyst support is as important as the stability of the catalyst itself. Carbon supports have their limitations, and the stability of these materials under cyclic PEFC operating conditions involves some complex phenomena that certainly deserve more fundamental investigation than has been undertaken to date. Nanostructure catalysts offer the promise of CLs without carbon supports, but they also present challenges in managing water in the resulting thin CLs without reservoir volumes to accommodate fluctuations in water saturation levels with varying operating conditions.

Improvements in PEFC durability are expected to continue to evolve at a relatively rapid pace because the field is still young. Multiple researchers around the globe are actively working on the development of a more detailed and quantitative understanding of decay mechanisms as well as new cell designs and materials that offer improved stability. In any case, consideration of both advanced materials (with inherently superior stability) and system mitigations (to minimize adverse operating conditions) will provide the best path to designing PEFC systems that continuously improve and are optimized with respect to durability, cost, and performance.

ACKNOWLEDGMENTS

The author would like to thank his many colleagues at UTC, both past and present, who have made working on many of the challenges of PEFC durability described herein an especially rewarding endeavor due to their perseverance and dedication to achieving stellar results. A special thanks to my co-authors of three previous book chapters on various aspects of this subject, which provided the foundation for this chapter.

REFERENCES

[1] Wargo EA, Dennison CR, Kumbur EC. Durability of Polymer Electrolyte Fuel Cells: Status and Targets. In: Mench M, Kumbur E, Veziroglu T, editors. Modern Topics in Polymer Electrolyte Fuel Cell Degradation. Denmark: Elsevier; 2011. p. 1–13.
[2] Buchi F, Inaba M, Schmidt T, editors. Polymer Electrolyte Fuel Cell Durability. New York: Springer; 2009.
[3] Mench M, Kumbur E, Veziroglu T, editors. Modern Topics in Polymer Electrolyte Fuel Cell Degradation. Denmark: Elsevier; 2011.
[4] Perry M, Balliet, Darling R. Experimental Diagnostics and Durability Testing Protocols. In: Mench M, Kumbur E, Veziroglu T, editors. Modern Topics in Polymer Electrolyte Fuel Cell Degradation. Denmark: Elsevier; 2011. p. 335–62.
[5] Weber AZ, Newman J. Modeling Transport in Polymer-Electrolyte Fuel Cells. Chemical Reviews 2004;104:4679.
[6] Mench M. Fuel Cell Engines. John Wiley & Sons, Inc.; 2008. p. 468.
[7] Kocha SS. Electrochemical Degradation: Electrocatalyst and Support Durability. In: Mench M, Kumbur E, Veziroglu T, editors. Modern Topics in Polymer Electrolyte Fuel Cell Degradation. Denmark: Elsevier; 2011. p. 293–329.
[8] Sasaki K, Shao M, Adzic R. Dissolution and Stabilization of Platinum in Oxygen Electrodes. In: Buchi F, Inaba M, Schmidt T, editors. Fuel Cell Durability. New York: Springer; 2009. p. 7–28.
[9] Buchi F, Inaba M, Schmidt T, editors. Polymer Electrolyte Fuel Cell Durability. New York: Springer; 2009. p. 289–366.

[10] Gittleman CS, Coms FD, Lai YH. Membrane Durability: Physical and Chemical Degradation. In: Mench M, Kumbur E, Veziroglu T, editors. Modern Topics in Polymer Electrolyte Fuel Cell Degradation. Denmark: Elsevier; 2011. p. 15–80.

[11] Srouji AK, Mench M. Freeze Damage to Polymer Electrolyte Fuel Cells. In: Mench M, Kumbur E, Veziroglu T, editors. Modern Topics in Polymer Electrolyte Fuel Cell Degradation. Denmark: Elsevier; 2011. p. 293–329.

[12] Meyers JP, Darling RD. Model of Carbon Corrosion in PEM Fuel Cells. Journal Electrochemical Society 2006;153:A1432–42.

[13] Reiser CA, Bregoli L, Patterson TW, Yi JS, Yang JD, Perry ML, et al. A Reverse-Current Decay Mechanism for Fuel Cells. Electrochemical and Solid-State Letters 2005;8:A273.

[15] Patterson T, Darling R. Damage to the Cathode Catalyst of a PEM Fuel Cell Caused by Localized Fuel Starvation. Electrochemical and Solid-State Letters 2006;9:A183.

[16] Darling RD, Meyers JP. Kinetic Model of Platinum Dissolution in PEMFCs. J. of the Electrochem. Soc. 2003;150:A1523–7.

[17] Perry ML, Patterson TW, Reiser C. System Strategies to Mitigate Carbon Corrosion in Fuel Cells. ECS Transactions 2006;3:783–95.

[18] Perry M, Darling R, Kandoi S, Patterson T, Reiser C. Operating Requirements for Durable PEFC Stacks. In: Buchi F, Inaba M, Schmidt T, editors. Polymer Electrolyte Fuel Cell Durability. New York: Springer; 2009. p. 399–418.

[20] Yu PT, Gu W, Zhang J, Makharia R, Wagner FT, Gasteiger HA. Carbon Support Requirements for Highly Durable Fuel Cell Operation. In: Buchi F, Inaba M, Schmidt T, editors. Polymer Electrolyte Fuel Cell Durability. New York: Springer; 2009. p. 29–56.

[21] Gallagher KG, Yushin G, Fuller TF. The Role of Micro-structure in the Electrochemical Oxidation of Model Carbon Catalyst Supports in Acidic Environments. JECS 2010;157:B820–30.

[22] Meyers JP. Subfreezing Phenomena in Polymr Electrolyte Fuel Cells. In: Buchi F, Inaba M, Schmidt T, editors. Polymer Electrolyte Fuel Cell Durability. New York: Springer; 2009. p. 369–82.

[23] USCAR Fuel Cell Tech Team. Cell Component Accelerated Stress Test Protocols for PEM Fuel Cells (Revised May 2010), www1.eere.energy.gov/hydrogenandfuelcells/pdfs/component_durability_may_2010.pdf.

[24] Perry M, Darling R, Kandoi S, Patterson T, Reiser C. Operating Requirements for Durable PEFC Stacks. In: Buchi F, Inaba M, Schmidt T, editors. Polymer Electrolyte Fuel Cell Durability. New York: Springer; 2009. p. 399–418.

CHAPTER TWELVE

Future of Fuel Cells and Hydrogen

12.1 INTRODUCTION

PEM fuel cells need hydrogen as fuel. Hydrogen is not a source of energy, and hydrogen is not a readily available fuel. However, in the future hydrogen could be something like electricity—an intermediary form of energy or an energy carrier. Just like electricity, hydrogen can be generated from a variety of energy sources, be delivered to end users, and at the user end it can be converted to useful energy efficiently and cleanly. Electricity infrastructure is already in place, but hydrogen infrastructure would need to be built.

It is this lack of hydrogen infrastructure that is considered one of the biggest obstacles to fuel cell commercialization. Commercialization of high-temperature fuel cells, such as solid oxide and molten carbonate fuel cells, does not depend on hydrogen, but commercialization of PEM fuel cells, particularly for transportation applications, must be accompanied by commercialization of hydrogen energy technologies, that is, technologies for hydrogen production, distribution, and storage. In other words, hydrogen must become a readily available commodity (not as a technical gas but as an energy carrier) before fuel cells can be fully commercialized. On the other hand, it may very well be that the fuel cells will become the driving force for development of hydrogen energy technologies. PEM fuel cells have many unique properties, such as high energy efficiency, no emissions, no noise, modularity, and potentially low cost, which may make them attractive in many applications, even with a limited hydrogen supply. This creates what is often referred to as a "chicken-and-egg problem"—does the development and commercialization of fuel cells come before devel-opment of hydrogen energy technologies, or must hydrogen infrastructure be in place before fuel cells can be commercialized? To answer these questions, the rest of this chapter gives a brief history of hydrogen as a fuel, presents the status of hydrogen energy technologies, and discusses the possible role of hydrogen and fuel cells in the future energy supply system.

PEM Fuel Cells
ISBN 978-0-12-387710-9
469

12.2 A BRIEF HISTORY OF HYDROGEN AS A FUEL

The first vision of an energy system based on hydrogen was provided by science fiction writer Jules Verne in his novel *The Mysterious Island* [1]:

Water decomposed into its primitive elements ... and decomposed doubtless, by electricity ... will one day be employed as fuel ... hydrogen and oxygen which constitute it, used singly or together, will furnish an inexhaustible source of heat and light, of an intensity of which coal is not capable ... Water will be the coal of the future.

After investigating the electrochemical process first described by Sir William Grove in 1839, Ostwald in 1894 predicted that the 20th century would become the Age of Electrochemical Combustion, with the replacement of steam Rankine cycle heat engines with much more efficient, pollution-free fuel cells [2].

In 1923 Haldane predicted that hydrogen—derived from wind power via electrolysis, liquefied, and stored—would be the fuel of the future [3]. This view was repeated in more technical detail some 15 years later by Sikorsky [4], who realized hydrogen's potential as aviation fuel. He predicted that the introduction of hydrogen would bring about a profound transformation in aeronautics.

Lawaczek [5] in the early 1920s outlined the concepts for hydrogen-powered cars, trains, and engines; collaborated in developing an efficient pressurized electrolyzer; and was probably the first to suggest that energy could be transported via hydrogen-carrying pipelines, similar to natural gas. In the 1920s and 1930s, Erren and his team of engineers converted more than 1,000 cars and trucks to multifuel systems using both hydrogen and gasoline as fuel [6].

The concept of a solar-originated hydrogen economy was first set down by Bockris (1962), developed and diagrammed by Justi (1965), named a Hydrogen Economy by Bockris and Triner (1970), formulated by Bockris (1971) and Bockris and Appleby (1972), and quantified by Gregory (1972) and Marchetti (1972) [7].

Bockris proposed a general plan of supplying American cities with solar-based energy via hydrogen. He suggested the use of floating platforms containing photovoltaic devices, producing hydrogen by the electrolysis of seawater, and piping hydrogen to land [8].

A similar concept for Japan, called Planned Ocean Raft System for the Hydrogen Economy (PORSHE), was later proposed and elaborated by Ohta [9].

Justi [10] proposed the thermoelectric conversion of solar energy into hydrogen in the Mediterranean area, and transportation of hydrogen to Germany through a pipeline. He envisioned hydrogen as a fuel for households and industry, large-scale and localized electricity generation, and transportation (in electric vehicles).

Bockris and Veziroglu [11] outlined the solar hydrogen energy system and discussed the real economics of potentially competitive energy systems of the future. They showed that if hydrogen utilization efficiency advantage and total fuel costs (i.e., cost of production plus the cost of environmental damage done in every step of the fuel cycle) are taken into account, the solar hydrogen energy system is the most economical energy system possible.

Scott and Hafele [12] addressed the issue of the global climatic disruption caused by excessive use of carbon fuels and concluded that the hydrogen energy system is a practical technological pathway that can mitigate and then reverse energy-sector contributions to greenhouse gas climatic disruption and at the same time bring economic growth and improvements to the quality of life. They provided a vision of the transition to the hydrogen energy system as two sequential but overlapping waves: integrated energy systems (i.e., the mix of fossil fuel and hydrogen) and neat hydrogen technologies.

In the last two decades hydrogen has received the attention of both government programs and industry. This attention has particularly been prompted by rapid development of fuel cells. Of particular interest is use of hydrogen and fuel cells in vehicles. During the last two decades all the car manufactures have demonstrated prototype fuel cell vehicles, and the most advanced ones are gearing toward commercialization as soon as 2015. This is being accompanied by development of hydrogen infrastructure, that is, hydrogen refueling stations. Currently there are more than 200 hydrogen refueling stations around the world. In 2009 Germany announced plans to build 1,000 hydrogen refueling stations to accompany its automakers' commercialization plans [13].

The most developed countries, such as the United States, Canada, the European Union, and Japan, have already come up with strategies for transitioning to a hydrogen economy and have already started its implementation. The U.S. Department of Energy (DOE) published *A National Vision of America's Transition to a Hydrogen Economy*—to *2030 and Beyond* [14]. This document summarizes the potential role for hydrogen systems in the United States' energy future, outlining the common vision of the hydrogen economy. It was followed by *The National Hydrogen Energy Roadmap,* which provides a blueprint for the coordinated, long-term, public and private efforts required for hydrogen energy development [15]. A comprehensive U.S. DOE Hydrogen Posture Plan outlines the activities, milestones, and deliverables the DOE plans to pursue to support the United States' shift to a hydrogen-based transportation energy system [16]. The Posture Plan integrates research, development, and demonstration activities from the DOE renewable, nuclear, fossil, and science offices and identifies milestones for technology development over the next decade, leading up to a technology readiness milestone in 2015. The plan also points out that the use of hydrogen as an energy carrier can enhance energy security while reducing air pollution and greenhouse gas emissions.

Reducing greenhouse gas emissions, improving the security of the energy supply, and strengthening the European economy are the main drivers for establishing a hydrogen-oriented energy economy in Europe. This is outlined in a High Level Group's report to the European Commission [17], which was the starting point for European hydrogen activities first facilitated by the European Hydrogen and Fuel Cell Technology Platform (HFP) and since 2008 by the Fuel Cell and Hydrogen Joint Undertaking, a unique public-private partnership supporting research, technological development, and demonstration activities in fuel cell and hydrogen energy technologies in Europe. A series of strategic documents have been created, such as Strategic Overview, Strategic Research Agenda [18], Deployment Strategy [19], and most recently the updated Implementation Plan [20].

12.3 HYDROGEN ENERGY TECHNOLOGIES

Technologies for hydrogen production, storage, distribution, and utilization have already been developed, though very few of them to a level at which they could compete with the existing energy technologies. The following is a brief review of technologies for hydrogen production, storage, and utilization.

12.3.1 Technologies for Hydrogen Production

Production of hydrogen requires feedstock and energy. Logical feedstocks for hydrogen production are hydrocarbon fuels, with chemical formula C_xH_y, and water, H_2O. However, extraction of hydrogen from either hydrocarbons or from water requires energy. The amount of energy required to produce hydrogen is always greater than the energy that can be released by hydrogen utilization.

Hydrogen Production from Fossil Fuels

Currently hydrogen is mostly being produced from fossil fuels (natural gas, oil, and coal). Hydrogen production technologies from fossil fuels (steam reforming, partial oxidation, and gasification) are mature and widely used, although they may still need to be optimized for large-scale production from the point of view of energy efficiency, environmental impact, and, above all, costs. Furthermore, depending on the process and primary source used, the production from fossil sources, to be sustainable in the medium–long term, should be coupled with capture and storage of the coproduced CO_2. The development of solutions for cost-effective and reliable CCS is an essential condition not only to produce hydrogen from fossil sources but in general to make possible the use of fossil fuels in a way that is compatible with a healthy environment.

The most economically advantageous and most frequently used process is the catalytic steam reforming of natural gas (but liquefied petroleum gas or naphtha can also be used in the process). Possible improvements of the process are mainly focused on recovery of thermal energy, process integration, and gas purification. The typical capacities of industrial plants range between 100,000 and 250,000 Nm^3h^{-1}, with thermal efficiencies of about 80%. Medium (200–500 Nm^3h^{-1}) and small (50–300 Nm^3h^{-1}) capacity plants are suitable for specific industrial applications, with reduced efficiencies and higher costs per product unit. Systems of lower capacity (1–50 Nm^3h^{-1}) have been developed for integration with fuel cells and are currently used in hydrogen refueling stations; in this case further improvements are needed, mainly related to the integration of different equipment, gas purification, and study of innovative catalysts [21].

At present, about 18% of the hydrogen produced worldwide is derived from coal gasification in large-scale central facilities (100,000–200,000 Nm^3h^{-1}). Substantially, three types of gasifiers are available: moving (or fixed) bed, fluidized bed, and entrained flow. The overall efficiency is about 60–65%, with a reduction of 3–6% points in plants where CO_2 capture and storage are provided. The technology is mature, even if it is more complex and less consolidated than steam reforming. The potential for further improvements is still considerable and takes into account innovative membranes for air separation, progress in gasifier configuration, hot gas purification systems, new solvents and membrane reactors for hydrogen separation, and so on. Coal gasification, integrated into combined cycles with CCS, also represents a very interesting option for centralized cogeneration of electric energy and hydrogen [22].

For heavier hydrocarbons, autothermal reforming or partial oxidation processes with thermal efficiencies higher than 75% are used. Together with the most mature processes, other technologies exist, currently less used or under development, such as:

- Thermal or catalytic cracking of methane, which converts methane in hydrogen and carbon, without CO_2 production
- Sorption enhanced reaction process, which combines in the same reactor the reactions of steam reforming and shift with CO_2 separation (through absorption in suitable materials)
- Solar steam reforming, where concentrated solar energy is utilized as a heat source for the reforming reaction

Hydrogen Production by Water Electrolysis

Electrolysis is a process of splitting water into its constituents, hydrogen and oxygen, using electric energy. In its present state, the process carries significantly higher costs than hydrogen production from fossil fuels and covers only a small share of the world production (4%). The process is mostly used to satisfy requests for high-purity hydrogen. The electrolyzers on the market are essentially of two types: alkaline electrolyzers, which use an aqueous solution of potassium hydroxide (KOH), and solid polymer electrolyzers (typically for smaller capacities), in which the electrolyte is a polymer membrane (the same as in the polymer electrolyte fuel cells). Alkaline electrolyzers represent a well-established and utilized technology, even if further R&D is needed to reduce costs (current costs are 1,000–2,000 $/kW) and increase their efficiency (currently it is 40–60% with auxiliary included) and lifetime. Significant improvements have already been obtained in advanced alkaline electrolyzers, available as prototypes, which work at higher temperatures and pressures. The solid polymer electrolyzers present some advantages (absence of corrosive liquids, higher current density and operating pressures) but have durability problems. An interesting line of development in a medium-term perspective is high-temperature electrolysis (800–1,000°C), where high-temperature heat can be used effectively to decrease the amount of electrical energy (up to 40%) needed to produce the hydrogen. The technology is similar to the one of solid oxide fuel cells, and the necessary heat can be provided from industrial, nuclear, or solar processes.

Electrolyzers, thanks to their modularity, are suitable for both hydrogen distributed production and, in perspective, centralized production. It is clear that sustainability of the process depends on the primary source used to produce electric energy. The most favorable option from this point of view is the one that foresees the coupling with renewable sources, such as photovoltaic and wind. Water electrolysis is particularly suitable for use in conjunction with photovoltaics (PVs). In general, there is a good match between the polarization curves of PVs and electrolyzers, and experience from a handful of PV/electrolysis pilot plants shows that they can be matched directly (with no power-tracking electronics) with relatively high efficiency (~93% coupling efficiency) [23].

Hydrogen Production from Biomass

Hydrogen can be produced from biomass using different thermochemical (gasification, pyrolysis) and biological processes. Among the thermochemical processes, in general more suitable for centralized production, most attention is focused on gasification. The process is similar to that used to produce hydrogen from coal (with which the biomass can also be co-gasified), even if the plant sizes are smaller. The biomass can have different origin, such as agricultural and forest residues, industrial and urban wastes, organic waste materials, or the like. Some pilot plants are currently in the demonstration phase; improvements are required to increase reliability and reduce process costs. Concerning hydrogen production from biomass, it is necessary to consider that such process is in competition with other possible biomass uses, both of conventional (combustion for production of electric energy and heat) and unconventional (biofuel production; direct use of gas in high-temperature fuel cells) type. These competing alternative uses must be taken into account in evaluating potential and perspectives of this specific chain.

Hydrogen Production from Water Through Thermochemical Cycles

Thermochemical cycles accomplish splitting of water into hydrogen and oxygen through a set of chemical reactions involving intermediate compounds that are fully recycled at the end of the process. Such cycles operate at temperatures lower than direct water thermolysis cycles (T > 2,000°C) and, in theory, without electric energy contributions. The required heat can be supplied from solar or nuclear sources and the efficiencies are estimated about 40–50%, higher than water electrolysis, if the efficiency of electricity production is taken into consideration. To date, over 200 thermochemical cycles have been investigated, mostly at theoretical level, whereas only a few have been selected to test their technical feasibility. Sulfur/iodine is one of the most promising cycles currently under evaluation; it has been demonstrated with a production of about 0.030 Nm^3/h of hydrogen. Important activities in this field are also carried out in Europe, from research structures such as CEA, Commissariat à l'énergie atomique et aux énergies alternatives (French Commission for nuclear energy and energy alternatives); DLR, Deutsche Forschungsanstalt für Luft-Und Raumfahrt (German Aerospace Research Establishment); CIEMAT, Centro de Investigaciones Energéticas, Medioambientales y Tecnológicas (Spanish Center for Research on Energy, Environment and Technology); ENEA Agenzia nationale per le nuove tecnologie, l'energia e lo sviluppo economico sostenibile (Italian National agency for new technologies, energy and sustainable economic development). Critical issues of the process are the separation of the hydrogen produced and the problems of corrosion associated with the chemicals involved in the cycle. Because of their complexity, thermochemical processes are considered for centralized hydrogen production, coupled with IV generation nuclear energy plants or with concentration solar systems. The technology development still requires considerable research effort, and the availability of commercial systems can be expected from 2030 onward.

Photoelectrochemical Hydrogen Production

A photoelectrochemical system combines in the same device both electric energy generation from solar light and its utilization to produce hydrogen from water through an electrolyzer. This device uses a semiconductor, immersed in aqueous solution, which directly converts solar light into chemical energy. It is therefore a potentially very promising process, since it can reach considerably lower costs and higher efficiencies than a photovoltaic electrolysis system. Different concepts are studied at laboratory level, with interesting results (solar/hydrogen efficiencies up to 16%), but considerable R&D efforts on materials and system engineering are needed to reach the technical and commercial maturity that can be expected only in the long term.

Photobiological Hydrogen Production

The hydrogen can be produced from water using sunlight and photosynthetic microorganisms. The process has reached, at laboratory scale, interesting conversion efficiencies (about 2% of the incident light radiation), even if it still requires important improvements, both to understand the basic mechanisms of process and for its scale-up. Photobiological water splitting is a long-term technology.

12.3.2 Technologies for Hydrogen Storage

The utilization of storage systems is necessary in different phases of the hydrogen cycle (production, service station, onboard vehicles). The availability of suitable storage systems represents one of the biggest hurdles to widespread use of hydrogen, mainly in the transport sector. In this case vehicular design, weight, size, volume, and efficiency strongly affect the amount of hydrogen that can be stored onboard.

The problems related to the hydrogen storage derive from its chemical/physical characteristics. Hydrogen is a fuel that has a high gravimetric energy density but also a low volumetric energy density, in both gaseous and liquid states. Consequently it is clear that, compared to other fuels, hydrogen requires higher-volume tanks to store the same energy content (see Table 9-2). In particular, the hydrogen storage systems for vehicular applications need to meet specific technical and economic requirements and safety standards that allow hydrogen vehicles to have similar driving range and performance as vehicles using conventional liquid fuels.

Hydrogen as an energy carrier must be stored to overcome daily and seasonal discrepancies between energy source availability and demand. Depending on storage size and application, several types of hydrogen storage may be differentiated:

1. *Stationary large storage systems.* These are typically storage devices at the production site or at the start or end of pipelines and other transportation pathways.
2. *Stationary small storage systems.* These are systems at the distribution or final user level—for example, a storage system to meet the demands of an industrial plant.
3. *Mobile storage systems for transport and distribution.* These include both large-capacity devices, such as a liquid hydrogen tanker or bulk carrier, and small systems, such as a gaseous or liquid hydrogen truck trailer.
4. *Vehicle tanks.* These store hydrogen used as fuel for road vehicles.

Because of hydrogen's low density, its storage always requires relatively large volumes and is associated with either high pressures (thus requiring heavy vessels) or extremely low temperatures and/or combination with other materials (much heavier than hydrogen itself). Table 12-1 shows achievable

Table 12-1 Hydrogen Storage Types and Densities [35]

	kg H$_2$ /kg	kg H$_2$/m^3
Large volume storage (10^2 to 10^4 m^3 geometric volume)		
Underground storage	NA	5—10
Pressurized gas storage (above ground)	0.01—0.014	2—16
Metal hydride	0.012—0.015	50—55
Liquid hydrogen	~1	65—69
Stationary small storage (1 to 100 m^3 geometric volume)		
Pressurized gas cylinder	0.012—0.035	15—30
Metal hydride	0.012—0.015	50—55
Liquid hydrogen tank	0.15—0.50	~65
Vehicle tanks (0.1 to 0.5 m^3 geometric volume)		
Pressurized gas cylinder	0.035—0.05	20—35
Metal hydride	0.012—0.020	50—55
Liquid hydrogen tank	0.09—0.13	50—60

storage densities with different types of hydrogen storage methods [24]. Some novel hydrogen storage methods may achieve even higher storage densities, but they have yet to be proven in terms of practicality, cost, and safety.

The storage system should have high density of energy (corresponding to high amounts of stored hydrogen), high density of power, good energy efficiency, low boil-off losses in liquid hydrogen storage, adequate life cycle, reduced or no environmental impact and acceptable safety features (both during operation and in the phases of manufacture and disposal at end of life), and reduced costs, and it should allow efficient, fast, and safe filling at the refueling station. Various technologies are already in use or under development for hydrogen storage. It can be stored as high-pressure gas or in liquid or chemical form or absorbed/adsorbed on special materials (metallic hydrides, chemical hydrides, carbon nanostructures). Each technology shows advantages and limitations, but all of them, even where they are already applied, still require significant R&D efforts to achieve a reliable and competitive large-scale use. The current ambitious research targets aim to develop and demonstrate hydrogen storage systems achieving 3 kWh/kg (9 wt % H2) by 2015 [25].

Future hydrogen supply systems will have a structure similar to today's natural gas supply systems. Underground storage of hydrogen in caverns, aquifers, depleted petroleum and natural gas fields, and manmade caverns resulting from mining and other activities is likely to be technologically and economically feasible [26]. Hydrogen storage systems of the same type and the same energy content will be more expensive by approximately a factor of three than natural gas storage systems due to hydrogen's lower volumetric heating value. Technical problems, specifically for the underground storage of hydrogen other than expected losses of the working gas in the amount of 1–3% per year, are not anticipated. The city of Kiel's public utility (Germany) has been storing town gas with a hydrogen content of 60–65% in a gas cavern with a geometric volume of about 32,000 m^3 and a pressure of 80 to 160 bar at a depth of 1,330 m since 1971 [27]. Gaz de France (the French national gas company) has stored hydrogen-rich refinery byproduct gases in an aquifer structure near Beynes, France. Imperial Chemical Industries of Great Britain stores hydrogen in the salt mine caverns near Teeside in the United Kingdom [28].

12.3.3 Technologies for Transport and Distribution

The use of hydrogen as N energy carrier requires its availability at large scale and at the point of utilization for a variety of applications. Infrastructures for transport and distribution, very different depending on the hydrogen production process (fossil or nonfossil sources, centralized or on-site plant), are therefore needed. To date the hydrogen is distributed as compressed gas or in liquid form.

Compressed hydrogen can be transported by pipelines (for industrial uses there are more than 700 km of pipeline in the United States and almost 1,600 km in Europe, normally operating at pressures of 10–20 bar) or on road by trucks (able to transport from 2,000 to 6,200 m^3—about 150–500 kg—of hydrogen at 200–350 bar) when quantities and distances are small. A few projects are also evaluating the possibility to use the existing gas pipelines to transport a mixture of natural gas/hydrogen.

Hydrogen in liquid form (cooled below −253°C) has a density higher than gas (about 800 times, at atmospheric pressure); therefore a cryogenic tank truck is able to transport a hydrogen liquid amount that is considerably higher than a gaseous hydrogen tube trailer (50,000 liters correspond to about 3,700 kg, i.e., almost 10 times as much).

The problem of hydrogen transport is closely linked to the distribution to the point of use and, particularly, to the possible configuration of the hydrogen filling stations. Over 200 filling stations are operational or under construction in the world in the framework of the current demonstration projects. There are different layouts: Some of them supply only gaseous hydrogen; in others liquid hydrogen is available, both as an intermediate storage for subsequent gasification and for direct use. Some so-called "total energy" stations, in which on-site production of hydrogen is integrated with an electric energy and heat generation system (usually a fuel cell), have also been realized. Some studies have indicated such solutions as the most efficient option, at least in the short to medium term.

From an economical point of view, transport by pipeline seems to be the most suitable option, even if it requires very high investments in the infrastructures. In general, it is estimated that transport and distribution imply, in the case of centralized generation, about 7–13 $/GJ of additional costs in hydrogen production.

12.3.4 Technologies for Hydrogen Utilization

Hydrogen, in addition to the traditional industrial applications, can be used both in transport (internal combustion engines and fuel cells) and in power generation.

Internal Combustion Engines

Hydrogen-powered internal combustion engines (ICEs) are, on average, about 20% more efficient than comparable gasoline engines. The thermal efficiency of an engine can be improved by increasing either the compression ratio or the specific heat ratio. In hydrogen engines both ratios are higher than in a comparable gasoline engine because of hydrogen's lower self-ignition temperature and ability to burn in lean mixtures. However, the use of hydrogen in internal combustion engines results in ~15% loss of power due to lower energy content in a stoichiometric mixture in the engine's cylinder. The power output of a hydrogen engine can be improved by using advanced fuel-injection techniques or liquid hydrogen [29].

Vehicles using ICEs can operate both with pure hydrogen or hydrogen/ natural gas blends. Significant experiments with the use of hydrogen have been carried out with modified conventional engines. To exploit the potential advantages of hydrogen at best, however, it is necessary that engines are designed to take into account hydrogen's characteristics as fuel (wide flammability range in comparison with other fuels, low ignition energy, and almost double the flame rate). Testing ICE vehicles (cars or buses) fed with hydrogen has been carried out or is in progress in the United States and Europe, even if the commitment is significantly lower than for fuel cell vehicles. Car manufactures such as BMW, Ford Motors, and Mazda are involved in the development of hydrogen-fed ICE technology.

Another solution under evaluation contemplates the use of blends of natural gas and hydrogen (HCNG) at variable content of hydrogen, but in any case not higher than 30% by volume, to avoid engine modifications. The hydrogen addition to natural gas, although at low percentages, has positive effects on engine operation, reducing exhaust emissions not only due to the substitution of the carbon with hydrogen but also because its presence allows a more complete and rapid combustion with a significant efficiency increase [30]. Both solutions (pure hydrogen and HCNG) can represent an interesting development area in the short term for the environmental benefits that they allow; hydrogen ICE technology can also contribute to promoting an early diffused hydrogen use and the construction of fueling infrastructures.

Fuel Cells

The use of fuel cells systems powered with hydrogen represents one of the most promising options in the medium to long term for the development of efficient and environmentally friendly means of transport because the use of hydrogen implies zero emissions at the local level. Besides, fuel cell vehicles offer efficiencies almost two times higher than conventional vehicles, maintaining similar performances in terms of driving range, top speed, and acceleration. Among the various fuel cell types, the polymer electrolyte membrane fuel cells (PEMFCs) are the most suitable technology for transport applications, being characterized by low operation temperature, high power density, quick startup, and rapid response to load changes. The first commercial fuel cell vehicles are expected by 2015.

Hydrogen as Fuel for Air Transportation

Hydrogen is particularly good fuel for air transportation. Actually it is the only fuel that could be produced from renewable energy sources that is suitable for air transport. Liquid hydrogen has numerous advantages as a fuel for commercial subsonic and especially for supersonic aircraft [31], particularly in terms of engine efficiency, aircraft takeoff weight, environmental impact, and safety. Weight for weight, hydrogen contains 2.8 times more energy than kerosene. On the other hand, to store the same amount of energy, liquid hydrogen needs a volume four times bigger than kerosene. A significant part of the fuel weight advantage may be eaten up by the weight of the complex fuel system, specifically by the large tanks. Nevertheless, use of hydrogen results in increased payload at a given takeoff weight.

In April 1988, the flight of a commercial airliner (a Tupolev 155) fueled with liquid hydrogen was demonstrated in the ex-Soviet Union. Many aspects of using liquid hydrogen as an aircraft fuel were studied in cooperation between EADS Airbus, the Russian design bureau ANTK Tupolev, and numerous German partners in the 1990s.

In the mid-2000s, a consortium of 35 partners from 11 European countries worked on the CRYOPLANE project—a comprehensive two-year system analysis supported by the European Commission within the 5th Framework Programme [32]. The study covered all relevant technical, environmental, and strategic aspects to provide a sound basis for initiating larger-scale research and development activities on a hydrogen-powered airliner. The CRYOPLANE study confirmed in principle the feasibility and the environmental advantages of the concept as well as identifying the need for future research and development activities. Unfortunately, work on a hydrogen-powered airliner has not continued.

Hydrogen in Electric Power Generation

Hydrogen can be used in both large, centralized power plants based on thermal cycles and in distributed energy generation with fuel cells. In the past few years some initiatives have been undertaken to develop combined-cycle gas turbine plants with precombustion CO_2 separation. (*Precombustion* is the process that allows CO_2 removal prior to combustion. Such a process is quite efficient, allowing the capture of up to 90–95% of CO_2 emissions.) These plants, particularly promising for coal gasification, allow the production of a hydrogen-rich synthesis gas and its utilization for electric energy generation in combined cycles. Such systems, also promising for the combined production of electricity and hydrogen, require significant modifications of the turbines (particularly on the burners), in addition to a considerable development and demonstration effort for the processes related to coal gasification and CO_2 separation and sequestration. It is, however, reasonable to expect that once feasibility and acceptability of CO_2 sequestration have been demonstrated, such plants will have an important role in electricity generation and, indirectly, in the hydrogen market.

Fuel cells are considered a key technology in a future economy based on hydrogen, since they are able to convert this energy carrier in a very efficient way. Each type of fuel cell has its own advantages and limitation. Stationary fuel cell systems, with a capacity ranging from a few kW to some MW, can be used for backup power, distributed power generation, and cogeneration or for portable generators.

12.3.5 Safety Aspects of Hydrogen as Fuel

Like any other fuel or energy carrier, hydrogen poses risks if not properly handled or controlled. The risks of hydrogen, therefore, must be considered relative to the common fuels such as gasoline, propane, or natural gas. The specific physical characteristics of hydrogen are quite different from those common fuels. Some of these properties make hydrogen potentially less hazardous, whereas other hydrogen characteristics could theoretically make it more dangerous in certain situations. Table 9-1 shows relevant hydrogen properties; Figure 12-1 compares hydrogen properties with other fuels and ranks their effect on safety.

Nevertheless, hydrogen's safety record has been unjustly tainted by the *Hindenburg* dirigible accident and the hydrogen bomb, although the former is essentially a proof of how safe hydrogen is, and the latter has nothing to do with either gaseous or liquid hydrogen use as fuel. Careful investigation of the *Hindenburg* disaster proved that it was the flammable aluminum powder-filled paint varnish that coated the infamous airship, not the hydrogen, that started the fire [33]. The hydrogen with which the airship was filled caught on fire considerably *after* the *Hindenburg's* surface skin started to burn and was over in less than one minute. The flames from hydrogen combustion traveled upward, far away from the crew and passengers in the cabins below.

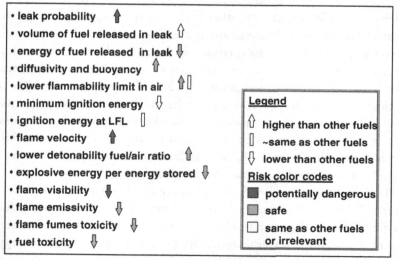

Figure 12-1 Summary of hydrogen safety-related properties compared with other fuels.

What fell to the ground with the passengers were burning shrouds from the exterior fabric, a large inventory of diesel fuel, and combustible materials that were in the cabins. Thirty-three people were killed in the *Hindenburg* fire; however, 62 people lived through the disaster by being fortunate enough to ride the *Hindenburg* down and escape the flames and wreckage that fell to the ground.

Because hydrogen has the smallest molecule, it has a greater tendency to escape through small openings than other liquid or gaseous fuels. Based on properties of hydrogen such as density, viscosity, and its diffusion coefficient in air, the propensity of hydrogen to leak through holes or joints of low pressure fuel lines may be only 1.26 to 2.8 times faster than a natural gas leak through the same hole (not 3.8 times faster, as frequently assumed based solely on diffusion coefficients). Experiments have indicated that most leaks from residential natural gas lines are laminar [34]. Because natural gas has energy density per unit volume over three times higher than hydrogen, a natural gas leak would result in more energy release than a hydrogen leak. For very large leaks from high-pressure storage tanks, the leak rate is limited by sonic velocity. Because of a higher sonic velocity (1308 m/s), hydrogen would initially escape much faster than natural gas (sonic velocity of natural gas is 449 m/s). Again, because natural gas has more than three times the energy density of hydrogen, a natural gas leak will always contain more energy [34]. If a leak should occur for whatever reason, hydrogen will disperse much faster than any other fuel, thus reducing the hazard levels. Hydrogen is both more buoyant and more diffusive than gasoline, propane, or natural gas. Hydrogen/air mixture can burn in relatively wide volume ratios, between 4% and 75% of hydrogen in air. Other fuels have much lower flammability ranges (i.e., natural gas: 5.3–15%, propane: 2.1–10%, and gasoline: 1–7.8%). However, this range has little practical value. In many actual leak situations the key parameter that determines whether a leak will ignite is the lower flammability limit, and hydrogen's lower flammability limit is 4 times higher than that of gasoline, 1.9 times higher than that of propane, and slightly lower than that of natural gas.

Hydrogen has a very low ignition energy (0.02 mJ), about one order of magnitude lower than other fuels. Ignition energy is a function of the fuel/air ratio, and for hydrogen it reaches a minimum at about 25–30% hydrogen content in air. At the lower flammability limit, hydrogen ignition energy is comparable to that of natural gas [35].

Hydrogen has a flame velocity seven times faster than that of natural gas or gasoline. A hydrogen flame would therefore be more likely to progress to a deflagration or even a detonation than other fuels. However, the likeli-hood of a detonation depends in a complex manner on the exact fuel/air ratio, the temperature, and particularly the geometry of the confined space. Hydrogen detonation in open atmosphere is highly unlikely.

The lower detonability fuel/air ratio for hydrogen is 13–18%, which is two times higher than that of natural gas and 12 times higher than that of gasoline. Because the lower flammability limit is 4%, an explosion is possible only under the most unusual scenarios—for example, hydrogen would first have to accumulate and reach 13% concentration in a closed space without ignition, and at that point an ignition source would have to be triggered.

Should an explosion occur, hydrogen has the lowest explosive energy per unit of stored energy of any fuel, and a given volume of hydrogen would have 22 times less explosive energy than the same volume filled with gasoline vapor.

Hydrogen flame is nearly invisible, which may be dangerous because people in the vicinity of a hydrogen flame may not even realize there is a fire. This danger can be remedied by adding chemicals that provide the necessary luminosity. The low emissivity of hydrogen flames means that nearby materials and people will be much less likely to ignite or be hurt by radiant heat transfer. The fumes and soot from a gasoline fire pose a risk to anyone inhaling the smoke, whereas hydrogen fires produce only water vapor (unless secondary materials begin to burn).

Liquid hydrogen presents another set of safety issues, such as risk of cold burns and the increased duration of leaked cryogenic fuel. A large spill of liquid hydrogen has some characteristics of a gasoline spill; however, it will dissipate much faster. Another potential danger is a violent explosion of a boiling liquid expanding vapor in the case of a pressure relief valve failure.

Hydrogen onboard a vehicle may pose a safety hazard. Such hazards should be considered in situations in which the vehicle is inoperable or is in normal operation and in collisions. Usually, potential hazards are due to fire, explosion, or toxicity. The latter can be ignored because neither hydrogen nor its fumes in case of fire are toxic. Hydrogen as a source of fire or explosion may come from the fuel storage, from the fuel supply lines, or from the fuel cell itself. The fuel cell poses the least hazard, although in a fuel cell hydrogen and oxygen are separated by a very thin ($\sim$20–30μm) polymer membrane. In case of membrane rupture, hydrogen and oxygen would combine and the fuel cell would immediately lose its potential, which should be easily detected by a control system. In such a case the supply lines would be immediately disconnected. The fuel cell operating temperature (60–90°C) is too low to be a thermal ignition source; however, hydrogen and oxygen may combine on the catalyst surface and create ignition conditions. Nevertheless, the potential damage would be limited because of the small amount of hydrogen present in the fuel cell and fuel supply lines.

The largest amount of hydrogen at any given time is present in the tank. Several tank failure modes may be considered in both normal operation and collision, such as:

- A catastrophic rupture due to a manufacturing defect in the tank, a defect caused by abusive handling of the tank or a stress fracture, a puncture by a sharp object, or external fire combined with failure of a pressure relief device to open
- A massive leak due to a faulty pressure relief device tripping without cause or a chemically induced fault in the tank wall, a puncture by a sharp object, or operation of a pressure relief device in case of a fire (which is exactly the purpose of the device)
- A slow leak due to stress cracks in the tank liner, a faulty pressure relief device, faulty coupling from the tank to the feed line, or impact–induced openings in the fuel-line connection

In a study conducted on behalf of Ford Motor Co., Directed Technologies, Inc., performed a detailed assessment of probabilities of the previously mentioned failure modes [35]. The study's conclusion was that a catastrophic rupture is a highly unlikely event. However, several failure modes resulting in a large hydrogen release or a slow leak have been identified in both normal operation and collisions.

Most of the failure modes discussed previously may either be avoided or have their occurrence and consequences minimized by:

- Leak prevention through proper system design, selection of adequate equipment (some further testing and investigation may be required), allowing for tolerance to shocks and vibrations, locating a pressure relief device vent, protecting the high-pressure lines, installing a normally closed solenoid valve on each tank feed line, and so on
- Leak detection by either a leak detector or adding an odorant to the hydrogen fuel (this may be a problem for fuel cells)
- Ignition prevention through automatically disconnecting battery bank, thus eliminating a source of electrical sparks which are the cause of 85% gasoline fires after a collision; by designing the fuel supply lines so that they are physically separated from all electrical devices, batteries, motors, and wires to the maximum extent possible; and by designing the system

for both active and passive ventilation (such as an opening to allow the
hydrogen to escape upward)

Risk is typically defined as a product of probability of occurrence and
consequences. The study by Directed Technologies [35] includes a detailed
risk assessment of the several most probable or most severe hydrogen accident scenarios, such as:

- Fuel tank fire or explosion in unconfined spaces
- Fuel tank fire or explosion in tunnels
- Fuel line leaks in unconfined spaces
- Fuel leak in a garage
- Refueling station accidents

The conclusion of this study is that in a collision in open spaces, a safety-
engineered hydrogen fuel cell car should have less potential hazard than
either a natural gas or gasoline vehicle. In a tunnel collision, a hydrogen fuel
cell vehicle should be nearly as safe as a natural gas vehicle, and both should
be potentially less hazardous than a gasoline or propane vehicle, based on
computer simulations comparing substantial post-collision release of gasoline
and natural gas in a tunnel. The greatest potential risk to the public appears
to be a slow leak in an enclosed home garage, where an accumulation of
hydrogen could lead to fire or explosion if no hydrogen-detection or risk-
mitigation devices or measures (such as passive or active ventilation) are
applied.

In conclusion, hydrogen appears to pose risks on the same order of
magnitude as other fuels. In spite of the public's perception, in many aspects
hydrogen is actually a safer fuel than gasoline or natural gas.

12.4 IS THE PRESENT GLOBAL ENERGY SYSTEM SUSTAINABLE?

Many scientists consider hydrogen the fuel of the future in the post-fossil
fuel era. Discussions of when and how hydrogen energy technologies will be
commercialized are probably beyond the scope of this book. Nevertheless,
some clues may be obtained by looking into the history of energy use and
transitions in energy supply.

Figure 12-2 shows the ever-growing global demand for energy and how
it has been met by a variety of energy sources. Fossil fuels make up more
than 80% of the total energy consumption in the world, which is currently
somewhat above 500 EJ/yr [36,37]. Consumption of fossil fuels increased 14

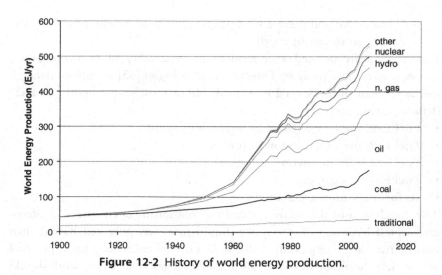

Figure 12-2 History of world energy production.

times (or about 2.6% per year, on average) and provided a basis for tremendous economic growth during the last century.

Although it was coal that started and supported the industrial revolution in the 19th century, oil became the predominant source of energy in the 20th century, particularly driven by the transportation sector. The growth of oil production slowed after the oil crises of the 1970s, but nevertheless it reached 73 million barrels per day in 2007 (or 85 million barrels per day if we include the natural gas plant liquids). Oil provides about 35% of the world's energy today.

On a larger time scale it is possible to observe transitions in world energy supply (Figure 12-3). By observing the historical patterns of energy transitions, Marchetti and Nakicenovic [38] in the late 1970s predicted that the next transition will be to natural gas and then to nuclear power. However, neither natural gas nor nuclear power ever fulfilled their promise of the 1970s and never really reached more than 25% and 6% of the total world energy supply, respectively. In the past 10 to 20 years these transitions seem to have halted, and coal seems to be gaining market share in the global energy market, primarily due to rapid expansion in India and China.

According to philosopher Ivan Illich [39], every system, process, or human activity grows or proceeds up to a threshold after which any further growth or activity becomes counterproductive, that is, the negative effects of such growth or activity become larger than the positive ones. The same idea may be applied to the global energy system. Use of fossil fuels has been

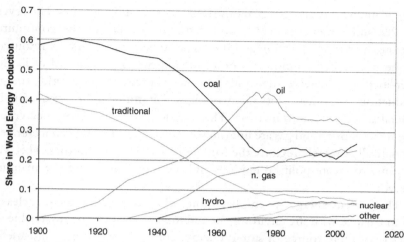

Figure 12-3 Shares of various energy sources in world energy production, showing energy transitions.

beneficial for the development of human civilization up to a certain point. However, further use of fossil fuels has become detrimental to society and even to the planet.

Exploitation of fossil fuels creates pollution on local, regional, and global scales. The quality of air in big cities, particularly in developing countries, is deteriorating due to exhaust gases from transportation and is causing serious health problems. Most of the scientific community and most of the world's countries are now convinced that the increase of carbon dioxide and other so-called greenhouse gases in the atmosphere is a direct consequence of combustion of fossil fuels and that it causes global warming, with threatening consequences such as global climate change and sea level rise [40,41].

The reserves of fossil fuels are finite. The exact remaining amount of fossil fuels is the subject of vigorous debate, but the fact that the reserves are finite is indisputable. In addition, new discoveries do not keep up with increasing projected demand. The world will eventually run out of oil and gas, quite possibly by the second half of this century [42-44]. The production of oil as its reserves diminish will also be reduced. It is possible that we have already seen the peak of oil production (see www.peakoil.org).

On the other side, demand for energy will continue to grow as described, particularly due to rapid economic development in China, India, and other developing countries. Energy consumption in developed countries is slowing due to rational use of energy, but mainly because of shifting energy-intense businesses and industries to third world countries. However, the demand for energy will continue to rise because of the continuing increase in world population and the growing demand on energy resources by the developing countries in order to improve their standard of living. Ever-increasing energy demand will keep pressure on world energy production, and prices, particularly of oil, will continue to be volatile and in general will continue to increase. This may lead to severe economic crises. The reserves of oil and gas are unevenly distributed and mainly concentrated in politically unstable regions: the Middle East and Arab countries. This will continue to create political tensions and possibly wars over the remaining reserves.

The present energy system, based on utilization of fossil fuels, is clearly not in balance with the environment and therefore cannot be sustainable. It relies on a finite source of stored energy, converts that energy into useful forms (primarily through a combustion process), and discharges the products of combustion, such as CO_2 and a myriad of pollutants, into the environment (as shown in Figure 12-4). It is obvious that such a system can only run as long as there is enough stored energy or as long as the environment is capable of absorbing pollution. The present pattern of energy use is unsustainable. It is therefore an unavoidable fact that the world energy supply will change in the future.

Figure 12-5 shows a projected gap between energy demand and fossil fuel availability. This gap represents an opportunity for nonfossil fuel energy sources, particularly renewable energy sources.

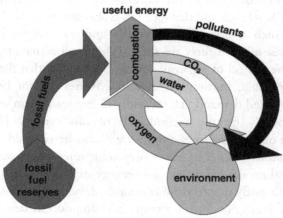

Figure 12-4 Present unsustainable energy system based on utilization of fossil fuels.

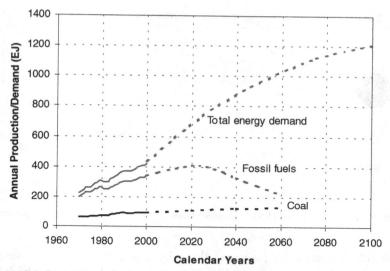

Figure 12-5 Projections of energy demand and supply. *(Sources: Historical data from [36]; world energy demand forecast from [45]; fossil fuels and coal forecasts extrapolated from [46] and [42].)*

12.5 PREDICTING THE FUTURE

Predicting the future is, of course, impossible. However, all systems in nature behave in accordance with established laws of physics, and if enough information is available, their future behavior may be predicted, at least to some extent. For example, if we throw a ball in the air, we can use a very simple model that will tell us that the ball will initially go up, reach its peak, and then come back down. If we want to predict how high the ball will fly and when and where it will fall down, a more complex model and much more information would be needed.

All systems in nature require energy. Their behavior, therefore, may be studied and predicted based on their energy use and available energy resources. The same applies to the global economy. As mentioned before, the tremendous economic growth of modern industrialized society has been based on utilization of fossil fuels, a convenient and concentrated form of energy. Although all economic models are based on growth, from a thermodynamic point of view it is clear that no system based on finite energy sources can continue to grow forever. Such a system does go through exponential growth during periods when the resources are plentiful, but

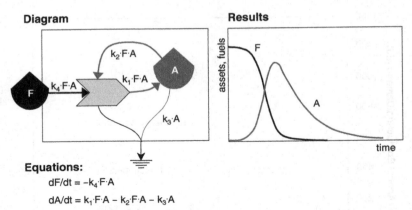

Figure 12-6 A simple model of a system based on nonrenewable energy sources.

eventually it reaches its peak and then declines as the resources become depleted, as shown in Figure 12-6. If we want to calculate how high the peak will be and when it will happen, much more information about the system and its energy use are needed. The diagram in Figure 12-6 and accompanying equations are based on Odum's energy language [47–49], originally developed to describe energy flows in ecological systems but subsequently applied to any complex system, including human economic systems on a global, national, or regional scale.

A system based on utilization of a constant flow of incoming energy (such as solar energy) behaves differently. It does not go through a peak but continues to grow and eventually reaches a steady state (Figure 12-7). The rate of growth and the steady-state level that a system can obtain depend on the rate of utilization of available solar energy and the effort required to convert solar energy into more useful forms of energy (hydrogen and electricity being only the intermediary steps, i.e., energy carriers).

Human civilization is actually a system that is based on both renewable and nonrenewable energy sources (Figure 12-8). The renewable energy sources could never have provided the growth enabled by the use of fossil fuels. The problem is that the finite reserves of fossil fuels cannot perpetuate this growth indefinitely. What cannot be predicted by such a simple model is at which level the system would reach a steady state after the nonrenewable, stored energy has been depleted. The steady-state level depends on the effort required to convert renewable energy into more useful forms of energy, that is, net energy gain from renewable energy sources.

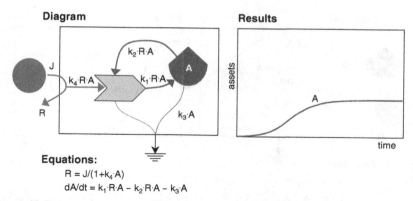

Figure 12-7 A simple model of a system based on a constant-flow (renewable) energy source.

Various modeling studies, such as those conducted by a Massachusetts Institute of Technology (MIT) team and published in a popular book, *Limits to Growth* in 1972 [50], and its updates, *Beyond the Limits* in 1991 [51] and *Limits to Growth: The 30-Year Update* in 2004 [52], indicate that the global economy will peak sometime in the first half of the 21st century and after that will continue to decline. The main reasons for that decline would be environmental stress and depletion of natural resources (including fluid fossil fuels). A modeling study conducted at the University of Miami [45] came up with similar results, although using a different method. However, both studies concluded that it would be possible to reverse the negative trends by timely introduction of clean, new technologies that will ease and actually reverse the burden on the environment and that will not depend on exhaustible natural resources. Although the MIT study was not specific about those technologies, the University of Miami researchers identified a hydrogen economy based on renewable energy sources as a solution to global economic and environmental problems [45]. This study indicated that the timing and the rate of introduction of a new energy system might be critical. As shown in Figures 12-9 and 12-10, early introduction of the solar hydrogen energy system will have long-term beneficial effects on both the global economy and the environment. If this transition starts only when the economy begins to decline, it may be too late to reverse the trend, because the economy would no longer be able to afford to invest in a long-term project such as establishing a new energy system.

Historically, in the context of a longer time span, the fossil fuels era may well be considered just a short interlude between the solar past and the solar

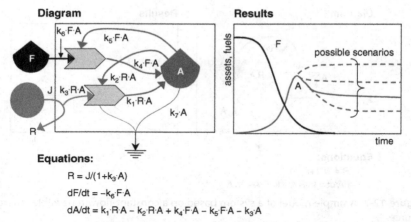

Equations:

$$R = J/(1+k_3 \cdot A)$$
$$dF/dt = -k_6 \cdot F \cdot A$$
$$dA/dt = k_1 \cdot R \cdot A - k_2 \cdot R \cdot A + k_4 \cdot F \cdot A - k_5 \cdot F \cdot A - k_3 \cdot A$$

Figure 12-8 A simple model of a system based on renewable and nonrenewable energy sources.

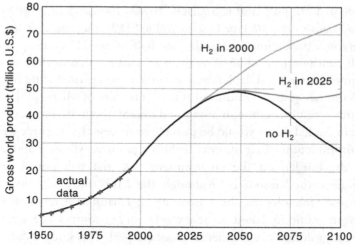

Figure 12-9 Effect of timeliness of a transition to a hydrogen energy system on the global economy (1990 US $) [45].

future. In that short period (about 300 years), fossil fuels made possible a tremendous development of human civilization. If fossil fuels are used to support the establishment of a permanent energy system such as one based on renewable energy sources, primarily solar, they could be considered a spark that provided a transition from the low-level solar energy past to the higher-level solar energy future. Solar energy is steadily available in

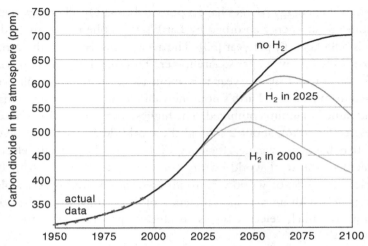

Figure 12-10 Effect of timeliness of a transition to a hydrogen energy system on CO_2 concentration in the atmosphere [45].

quantities that exceed human needs by several orders of magnitude. However, solar energy (both direct and indirect) is so dispersed that it requires a lot of effort (high-quality energy) to convert it to usable energy forms. The rate at which renewable energy could support the growth of the economy and the eventual steady state depends on the magnitude of efforts required to harvest it and present it to end users. In economic terms this is equivalent to the cost of energy at the user end. More studies are required to determine the net energy ratio of solar electricity and solar hydrogen technologies.

12.6 SUSTAINABLE ENERGY SYSTEM OF THE FUTURE

It is clear that our civilization is facing an unavoidable transition from convenient but environmentally not so friendly and, ultimately, scarce energy sources to less convenient but preferably clean and inexhaustible ones. There are several known energy sources that satisfy cleanliness and abundance requirements, such as direct solar radiation, various forms of indirect solar energy (wind, waves, ocean currents, ocean thermal, biomass), and geothermal, tidal, and nuclear energy. Technologies for utilization of these sources are at various stages of development, and their (direct economic) competitiveness with existing energy technologies varies from case to case. Human ingenuity may add new sources to this list in the future.

Renewable energy sources have a potential to satisfy all energy needs in the future. Solar energy absorbed by Earth's atmosphere, oceans, and land masses is 3,850,000 EJ per year [53]. The average insolation at the surface is approximately 240 watts per square meter. This may look small, but this is actually so vast that in one year it is about twice as much as will ever be obtained from all of the Earth's nonrenewable resources of coal, oil, natural gas, and mined uranium combined. The highest insolation at the surface is in the deserts. Covering about 2% of all the world deserts with photovoltaic panels that have an average efficiency of 10% would generate electricity equivalent to the total world energy consumption (96,000 TWh). If, for example, only 10% of worldwide rooftops were covered by solar photovoltaic panels with an average efficiency of 10%, they would be able to generate as much electricity as is today produced in the world (18,800 TWh).

Wind is a derivative of solar energy and has much smaller global potential. Researchers at Stanford University's Global Climate and Energy Project recently estimated that the global wind power potential at locations with mean annual wind speeds ≥ 6.9 m/s at 80 m above ground is ~72 TW (630,000 TWh/yr) [54]. Of course, not all of this potential is extractable. A study by the German Advisory Council on Global Change (WBGU) calculated that the global technical potential for energy production from both onshore and offshore wind installations was 278,000 TWh per year [55]. The report then assumed that only 10–15% of this potential would be realizable in a sustainable fashion and arrived at a figure of approximately 39,000 TWh supply per year as the contribution from wind energy in the long term (which is about twice the present world electricity generation from all sources).

Hydropower is also a derivative of solar energy. Total global hydropower potential is estimated at about 40,000 TWh/yr (144 EJ), but the technically exploitable amount is about 15,000 TWh/yr (54 EJ) [56], which is still several times the present electric power generation from hydropower (which is 3,000 TWh/yr).

Biomass is also considered a form of renewable energy with huge potential. Indeed, world biomass production is some 400 billion tons per year (equivalent of 3,000 EJ/yr). Technically it could be possible to produce about 1,000 EJ/yr (278,000 TWh/yr) if all newly produced biomass is transformed to useful forms (electricity, fuel, heat) [53,56]. Clearly this is not possible, because biomass is used for food and lumber, but more importantly, it plays a crucial role in ecosystems. In addition, growing and processing biomass for energy purposes require a lot of inputs (in the form of labor, fertilizer, and energy), resulting in very low net energy ratio. (This is why biomass for energy production is only viable when it's heavily subsidized.) For this reason, use of biomass for energy generation is a very controversial issue. What is not controversial is the fact that the efficiency of photosynthesis is very low (>1%), even for the best "energy crops," which would result in vast agricultural areas being used for energy production purposes. This area would be much more than 10 times larger than in the case of other renewable energy sources, namely solar and wind.

Other forms of renewable energy (ocean currents, tides, waves, geothermal) have much smaller potential and could not make bigger impacts on a global scale, but locally they could certainly contribute to the energy mix.

From Table 12-2, where the available and technical potential of the major renewable energy sources is summarized, it is clear that the renewable energy sources have the potential to satisfy all the energy needs of today's society. However, some problems with utilization of renewable energy sources impede their widespread use. One of the obvious problems is the high cost of equipment needed for utilization of renewable energy sources. It is reasonable to expect that with time, as technology develops and matures and as the manufacturing volumes increase by several orders of magnitude, the cost of equipment will decrease. At the same time the cost of fossil fuels will continue to increase, and it is likely that at some point the renewable energy sources will become cost competitive. In some cases—for example, wind energy in windy locations—they are cost competitive even today.

The other set of problems with renewable energy sources is that they are available intermittently and with variable and often unpredictable intensity,

Table 12-2 Availability and Technical Potential of Renewable Energy Sources

	Potential (EJ)	Technical Potential (TWh/yr)
Solar	3,850,000	400,000[*]
		94,000[**]
Wind	1,000	39,000
Hydro	144	15,000
World present electricity generation		18,800
World energy final consumption		96,000[***]

[*] $\frac{1}{10}$ world deserts covered by PV panels

[**] $\frac{1}{2}$ world rooftops covered by PV panels

[***] Not counting losses in energy transformations

and the best resources are usually far from human settlements, where energy is needed the most. In other words, there is a temporal and spatial mismatch between availability and demand. On top of that, renewable energy is not suitable for transportation (which uses about one third of total energy consumption).

These problems with renewable energy sources may be solved by converting renewable energy to energy carriers or forms of energy that can be stored, transported, and delivered to end users, where they can be converted to useful forms of energy. One such carrier is electricity, which is already being used worldwide, although it does not satisfy all the requirements. Electricity is a convenient form of energy—it can be produced from various sources, it can be transported over large distances, it can be distributed to end users, it is clean (although its production from fossil fuels is not), and at the user end it may be very efficiently used in a variety of applications. However, electricity cannot be stored in large quantities and as such cannot solve the problem of the intermittent nature of renewable energy sources. Hydrogen is another clean, efficient, and versatile energy carrier that supplements electricity very well. It can be produced from all forms of energy, it can be stored and transported, and it can be converted to useful forms of energy quite efficiently while producing no harmful emissions in the entire fuel chain.

Figure 12-11 shows a hypothetical energy system of the future completely based on utilization of renewable energy. Various forms of renewable energy, primarily solar, wind, and hydro, can be converted to electricity and heat. Although this transformation may be accomplished in centralized power plants (probably a better term would be *energy plants*), a significant portion may be accomplished locally or individually. In some cases economy of scale favors larger units (for example, large wind turbines cost less per kW then small ones), but in some cases mass-produced small units may also result in lower cost.

In such a system there are obvious needs: (i) to store energy for solving fluctuations in availability of renewable energy sources; (ii) to transport energy over mid- and long distances; and (iii) to fuel the transportation sector (land, sea, and air). Hydrogen could satisfy those needs much better than electricity. Figure 12-12 shows a vision of a future energy system primarily based on the renewable energy sources in which hydrogen plays a significant role [14].

Together electricity and hydrogen may be able to satisfy all future energy needs and form a permanent energy system that would be independent of

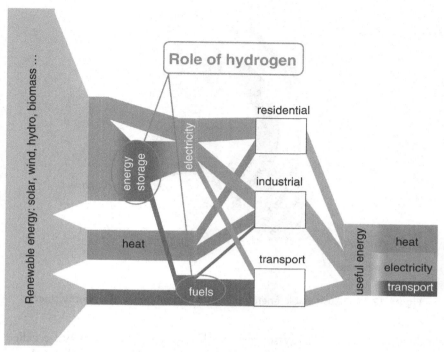

Figure 12-11 Future energy system based on renewable energy sources.

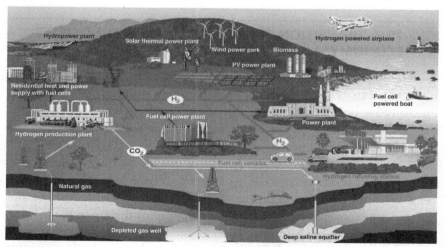

Figure 12-12 Vision of an energy system of the future with hydrogen playing a significant role (from [14]).

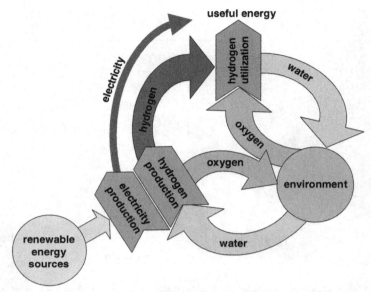

Figure 12-13 A sustainable energy system based on renewable energy sources.

individual primary energy sources. Such an energy system may be called a *hydricity economy* [57,58]. This type of system would be in complete balance with the environment, as shown in Figure 12-13, and could run for as long as the energy sources are available.

12.7 TRANSITION TO HYDROGEN OR A "HYDRICITY ECONOMY"

A transition from convenient but environmentally not so friendly and ultimately scarce energy sources (fossil fuels) to less convenient but clean and inexhaustible ones (renewable energy sources) seems imminent. Hydrogen may play a significant role in this transition by allowing renewable energy sources to be used in virtually any application.

Whereas the future energy system in which hydrogen could play a significant role can be depicted, a more difficult task is to define a path to get there from here. Replacement of a global energy system will not and cannot happen overnight. There is enormous capital tied up in the existing energy system. Building another energy system that would compete with the existing one is out of the question. The new energy system must gradually replace the existing one. Because businesses are too often concerned with short-term profits, governments and international organizations must realize the long-term benefits of the renewable energy system and support the transition both legislatively and financially. Introduction of "real economics" and elimination of subsidies for the existing energy system would help in that transition. The term *real economics* refers to economics that takes into account past, present, and future environmental damage associated with the use of a particular energy source or fuel, depletion of environmental resources, military expenses for keeping energy resources accessible, and other hidden external costs.

One problem, which we've already discussed, is the high cost of renewable energy. Hydrogen makes the renewable energy even more expensive. An energy system based primarily on renewable energy sources would be impossible without the storage, transport, and fuel features that hydrogen offers. However, renewable energy sources can start penetrating the energy markets without those features, and so at this stage they do not need hydrogen.

On the other side, hydrogen without renewable energy does not make much sense, although it can be produced from fossil fuels. Hydrogen produced from fossil fuels, particularly natural gas, as a fuel cannot compete in today's market with the very fuels from which it is produced. Hydrogen production from fossil fuels makes sense only in a transition period, to help establish hydrogen supply infrastructure and to help commercialize hydrogen utilization technologies. Because coal has by far the largest reserves of fossil fuels, hydrogen production from coal would make sense in this transition period, but only providing that carbon capture and sequestration is applied.

The nonexistence of a hydrogen infrastructure is a major obstacle for market penetration of hydrogen as a fuel and hydrogen-utilizing technologies such as PEM fuel cells. This infrastructure can be built, but it would take decades to do so. During this period there would actually need to be a dual infrastructure. Yet another major difficulty is interrelation and interdependence between hydrogen technologies. For example, it is extremely difficult, if not impossible, to introduce hydrogen-powered automobiles or hydrogen-powered airplanes into the market without reliable and economically feasible technologies for hydrogen production, distribution, storage, and refueling. On the other hand, significant development of hydrogen production, distribution, and storage technologies will never happen without a large demand for hydrogen.

As with any new technology, hydrogen energy technologies such as fuel cells are in most cases initially more expensive than the existing mature technologies, even when real economics is applied. Hydrogen energy technologies are expensive because the equipment for hydrogen production and utilization is not mass produced. It is not mass produced because there is no demand for it, and there is no demand for it because it is too expensive. This is a closed circle, another chicken–and–egg problem.

The only way for hydrogen energy technologies to penetrate the major energy markets is to start with those technologies that have niche markets, where the competition with the existing technologies is not as fierce or where they offer clear advantage over the existing technologies, regardless of the price. Another push for commercialization may be gained through governmental or international subsidies for technologies that offer some clear advantages. Once developed, these technologies may help reduce the cost of other related hydrogen technologies and initiate and accelerate their widespread market penetrations. One example is fuel cell buses in major third-world cities, where hydrogen could be produced from clean and renewable energy sources. These buses could replace heavy-polluting diesel buses in regular service, providing extreme environmental and health benefits and thus justifying the international subsidies.

Hydrogen being a long-term option, and actually being a burden in the short term, is not universally accepted as a part of an energy vision. Therefore, there is a constant need to inform the general public, policy-makers, politicians, governments, and intergovernmental organizations of the benefits, potential, and technological developments of hydrogen energy technologies. However, it is important that the right message is always conveyed and the right picture of the role of hydrogen in the energy future depicted to avoid creating false expectations.

Because of their low net energy ratio, the renewable energy sources most likely will not be able to provide the economic growth comparable to the growth that was fueled by relatively cheap fossil fuels. This fact alone makes the transition toward a new energy system even more difficult, particularly if it is left to the market forces. It will be difficult to convince an industrial subject or a government to invest in something that initially will be less lucrative than investing in "business-as-usual" energy technologies, although business as usual is not sustainable in the longer term. The financial hardship due to the higher cost of renewable energy sources may be partially offset by the fact that renewable energy utilization does not require big power plants that must be planned and built long in advance. The modular concept of renewable technology may allow adding these technologies to the local or nation's energy system in small increments.

However, the fact is that endless economic growth will be impossible without limitless sources of energy. Unfortunately, our sources of energy are not limitless; on the contrary, they are limited either in quantity (such as physical reserves of fossil fuels and uranium) or in flux (such as solar energy and its derivatives). What one can hope for in the longer term is some kind of a steady-state economy based on a constant available flux of renewable energy sources converted to useful energy forms via energy carriers such as electricity and hydrogen. Improvements in energy efficiency in every step of energy transformation, from the sources to the end form, will therefore result in more useful energy and thus a steady state at a higher level. In other words, economic growth may still be possible, not based on more use of free natural resources but rather on technological developments.

The key question is how to make a stagnating or descending economy prosperous. First, our definition of prosperity will have to be readjusted. Prosperity does not mean more assets but higher quality of life. In an economy that is growing, the key drivers are *more, faster,* and *competition.* In such an economy progress is measured by growth, and therefore "growth is (was) good." Today we talk more and more about sustainable development. Development does not have to mean growth. It is possible to develop society, technology, cooperation, knowledge, culture, health care, and so on but without physical growth. In such a stagnating or shrinking economy, the key aspects are *less, more efficient,* and *cooperation.* This economy will require a major shift in our mindsets, culture, and policies—a shift from the goals of continuous growth to the goals of sustainable development, promotion of energy and resource conservation, and prioritizing protection of the environment.

12.8 THE COMING ENERGY REVOLUTION?

The recent history of human civilization is characterized by technological revolutions (Figure 12-14). The industrial revolution started with the invention of the steam engine, which allowed utilization of coal and revolutionized manufacturing and had a profound effect on economic and social systems. The electricity revolution brought convenient energy to almost every home and allowed the development of many electrical devices and gadgets, which in turn caused remarkable changes in lifestyle. The automobile revolution started with Ford's mass-manufactured, affordable Model T automobile. Automobiles changed city layouts and city dwellers'

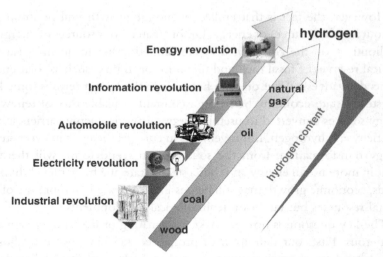

Figure 12-14 History of technological revolutions and history of fuels.

way of living. Most recently, the information revolution, which is still ongoing, started with the invention of computers. Although originally invented for computing only, computers are now being used in everyday life for storing and disseminating information, communication, entertainment, art, and so forth. In a short time, life without computers has become unimaginable.

What is common among these revolutions is that each started with a technological invention so powerful that it made changes in everyday life and allowed development of new products and services unimaginable before its implementation. The fuel cell is likely one of these powerful technologies that could create the next revolution—the energy revolution. It will change the way energy is converted to useful power. It will allow greater utilization of renewable energy sources, it will promote rational use of energy, it will decentralize power generation, and it will allow power generation at various scales while not harming the environment.

On the same timescale, it should be noted that different fuels have been used alongside the technological revolutions. Before the industrial revolution, the main fuels were wood and other traditional forms of energy. With the coming of the industrial revolution, coal became the main fuel for more than a century until it was replaced by oil, the main energy source that fueled the automobile revolution. In recent years, natural gas has become more and more significant. The trend is clear: Each fuel has been substituted with a fuel that has a higher hydrogen-to-carbon ratio. Whereas wood is primarily carbon, coal has a ratio of hydrogen to carbon of ~1:1, oil or gasoline ~2:1, and natural gas 4:1. Therefore, it can be extrapolated that the logical choice for an ultimate fuel is pure hydrogen—the fuel of the coming energy revolution.

12.9 CONCLUSIONS

- Transition from the fossil fuel-based energy system seems inevitable due to depletion of liquid fossil fuels (oil and natural gas) and due to environmental concerns.
- Hydrogen and electricity are the most probable energy carriers of the future because they can be produced from any energy source and they are clean and versatile. Produced from renewable energy sources, they result in a permanent energy system.
- Hydrogen technologies, that is, technologies for hydrogen production, storage, and utilization, have already been developed. Although these technologies are not mature, there are no major technological obstacles to widespread utilization of hydrogen.
- The transition to a hydrogen energy system will be difficult for a variety of reasons, such as competition from an established infrastructure, lack of policies favoring "real" economics, interdependence of hydrogen technologies, dependency on the renewable energy technologies, and so forth.
- More analyses of solar hydrogen and solar electricity net energy gain are required.
- Fuel cells may be the first hydrogen technologies commercialized on a large scale, with applications ranging from power generation to transportation. This technology has a potential to revolutionize the energy business.

REFERENCES

[1] Verne J. The Mysterious Island. New York: Airmont Publishing Company; 1965.
[2] Ostwald WZ. Elektrochemie 1894;1, p. 122. Cited by A. J. Appleby, From Sir William Grove to Today: Fuel Cells and the Future, lecture at the Grove Anniversary Fuel Cell Symposium (The Royal Institution, London, 1989).
[3] Haldane JBS. Daedalus or Science and the Future. London: Kega, Paul, Trench, Truber and Company, Ltd.; 1923.

[4] Sikorsky II. Science and the Future of Aviation. Steinmitz Memorial Lectures; 1938. Schenectady Section, AIEE.

[5] Lawaczek F. Technik und Wirtschaft im Dritten Reich. Munich: Eher Verlag; 1932.

[6] Erren RA, Compell WH. Hydrogen: A Commercial Fuel for Internal Combustion Engines and Other Purposes,. Journal of the Institute of Fuels 1933;6(No. 29).

[7] Bockris JO'M. Energy: The Solar-Hydrogen Alternative. Sydney: Australia and New Zealand Book Co; 1975.

[8] Bockris JO'M. Memorandum to Westinghouse Company. C. Zenner; 1962.

[9] Escher WJD, Ohta T. Direct Solar Energy Conversion at Sea. In: Ohta T, editor. Solar-Hydrogen Energy Systems. Oxford: Pergamon Press; 1979. p. 225–48.

[10] Justi E. Conduction Mechanisms and Energy Transformation in Solids. Verlag Vandenhoeck and Ruprecht, Göttingen; 1965.

[11] Bockris JO'M, Veziroglu TN. A Solar-Hydrogen Energy System for Environmental Compatibility. Environmental Conservation 1985;12(No. 2):105–18.

[12] Scott DS, Hafele W. The Coming Hydrogen Age: Preventing World Climatic Disruption. International Journal of Hydrogen Energy 1990;15(10): 727–38.

[13] Copeland MV. The Hydrogen Car Fights Back, CNNMoney.com; October 14, 2009, (http://money.cnn.com/2009/10/13/technology/hydrogen_car.fortune/index.htm); October 14, 2009.

[14] A National Vision of America's Transition to a Hydrogen Economy—to 2030 and Beyond. U.S. Department of Energy. Washington, DC: Office of Energy Efficiency and Renewable Energy; February 2002.

[15] The National Hydrogen Energy Roadmap. U.S. Department of Energy. Washington, DC: Office of Energy Efficiency and Renewable Energy; November 2002.

[16] Hydrogen Posture Plan. U.S. Department of Energy. Washington, DC: Office of Energy Efficiency and Renewable Energy; February 2004.

[17] Hydrogen Energy and Fuel Cells, A Vision of Our Future. Final Report of the High Level Group. Brussels: European Commission (EUR 20719 EN). Available online, www.fch-ju.eu/page/documents; 2003.

[18] European Hydrogen and Fuel Cell Technology Platform. EC, Brussels: Strategic Research Agenda. Available online, www.fch-ju.eu/page/documents; July 2005.

[19] European Hydrogen and Fuel Cell Technology Platform. EC, Brussels: Deployment Strategy. Available online, www.fch-ju.eu/page/documents; August 2005.

[20] Fuel Cells and Hydrogen Joint Undertaking, Multi-Year Implementation Plan, 2008–2013, Document FCH JU 2011 D708, Brussels, 2011. Available online: www.fch-ju.eu/page/documents.

[21] Conte M, Di Mario F, Iacobazzi A, Mattucci A, Moreno A, Ronchetti M. Hydrogen as Future Energy Carrier: The ENEA on Technology and Application Prospects. Energies 2009;2:150–79.

[22] HYPOGEN Pre-feasibility Study. EUR 21512 EN. Final Report prepared by ENEA, Fraunhofer ISI and Risoe National Laboratory. Available online, http://isi.fraunhofer.de/isi-media/docs/x/publikationen/HYPOGEN.pdf; January 2005.

[23] Steeb H, Brinner A, Bubmann H, Seeger W. Operation Experience of a 10kW PV-Electrolysis System in Different Power Matching Modes. In: Veziroglu TN, Takahashi PK, editors. Hydrogen Energy Progress VIII, vol. 2. New York: Pergamon Press; 1990. p. 691–700.

[24] Sherif SA, Barbir F, Veziroglu TN, Mahishi MM, Srinivasan SS. Hydrogen Energy Technologies. In: Kreith F, Goswami DY, editors. Handbook of Energy Efficiency and Renewable Energy. Boca Raton, Florida: CRC Press/Taylor & Francis Group; 2007. pp. 27.1–27.16.

[25] Hydrogen, Fuel Cells and Infrastructure Technologies Program Multi-Year RD&D Plan, U.S. Department of Energy, Oct. 2007. Available online: www.eere.energy.gov/hydrogenandfuelcells/mypp/

[26] Taylor JB, Alderson JEA, Kalyanam KM, Lyle AB, Phillips LA. Technical and Economic Assessment of Methods for the Storage of Large Quantities of Hydrogen. Int. J. Hydrogen Energy 1986;11(1):5–22.

[27] Carpetis C. Storage, Transport and Distribution of Hydrogen. In: Winter C-J, Nitsch J, editors. Hydrogen as an Energy Carrier. Berlin Heidelberg: Springer-Verlag; 1988. p. 249–89.

[28] Pottier JD, Blondin E. Mass Storage of Hydrogen. In: Yurum Y, editor. Hydrogen Energy System, Utilization of Hydrogen and Future Aspects, NATO ASI Series E-295. Dordrecht, The Netherlands: Kluwer Academic Publishers; 1995. p. 167–80.

[29] Nornbeck JM, Heffel JW, Durbin TD, Tabbara B, Bowden JM, Montano MC. Hydrogen Fuel for Surface Transportation. Warrendale, PA: SAE; 1996.

[30] Ortenzi F, Chiesa M, Scarcelli R, Pede G. Experimental Tests of Blends of Hydrogen and Natural Gas in Light-Duty Vehicles. Int. J. Hydrogen Energ. 2008;33:3225–9.

[31] Brewer GD. Hydrogen Aircraft Technology. Boca Raton, FL: CRC Press; 1991.

[32] Klug HG, Faass R. CRYOPLANE: Hydrogen Fuelled Aircraft: Status and Challenges. Air & Space Europe 2001;3(3/4).

[33] Bain A. The Freedom Element: Living with Hydrogen. Cocoa Beach, FL: Blue Note Books; 2004.

[34] Swain MR, Swain MN. A Comparison of H_2, CH_4, and C_3H_8 Fuel Leakage in Residential Settings. International Journal of Hydrogen Energy 1992;17(10): 807–15.

[35] Thomas CE. Hydrogen Vehicle Safety Report, prepared by Directed Technologies, Inc., for Ford Motor Company. Contract No. DE-AC02–94CE50389. Washington, DC: U.S. Department of Energy; 1996.

[36] Annual Energy Review 2010, U.S. Department of Energy. Washington, DC: Energy Information Administration; July 2011.

[37] World Energy Outlook 2009. OECD, Paris: International Energy Agency; 2009.

[38] Marchetti C, Nakicenovic N. The Dynamics of Energy Systems and the Logistic Substitution Model. Research Report RR-79–13. Laxenburg, Austria: International Institute for Applied Systems Analysis; 1979.

[39] Illich I. Energy and Equity. London: Marion Boyars; 1974.

[40] Pearce F. Special Report Climate Change. NewScientist.com news service. see also, http://environment.newscientist.com/channel/earth/climate-change/dn9903; 2006.

[41] McCarthy JJ, Canziani OF, Leary NA, Dokken DJ, White KS, editors. Climate Change 2001: Impacts, Adaptation & Vulnerability, Contribution of Working Group II to the Third Assessment Report of the Intergovernmental Panel on Climate Change (IPCC). UK: Cambridge University Press. See also, www.ipcc.ch; 2001.

[42] Campbell CJ, Laherrère JH. The End of Cheap Oil. Scientific American March 1998:78–83.

[43] Roberts P. The End of Oil: On the Edge of a Perilous New World. Boston: Houghton Mifflin; 2004.

[44] Goodstein D. The End of the Age of Oil. New York: W.W. Norton & Company; 2005.

[45] Barbir F, Plass Jr HJ, Veziroglu TN. Modeling of Hydrogen Penetration in the Energy Market. International Journal of Hydrogen Energy 1993;18(3):187–95.

[46] International Energy Outlook 2010. U.S. Department of Energy. Washington, DC: Energy Information Administration; July 2010.

[47] Odum HT. Systems Ecology. New York: John Wiley & Sons; 1983.

[48] Odum HT. Simulation Models of Ecological Economics Developed with Energy Language Methods. Simulation August 1989:69–75.

[49] Odum HT, Odum EC. Modeling for All Scales: An Introduction to System Simulation. San Diego, CA: Academic Press; 2000.

[50] Meadows DH, Meadows DL, Randers J, Behrens WW. Limits to Growth. New York: Universe Books, Publishers; 1972.

[51] Meadows DH, Meadows DL, Randers J. Beyond the Limits. Post Mills, VT: Chelsea Green; 1991.

[52] Meadows DH, Randers J, Meadows DL. Limits to Growth: The 30-Year Update. 3rd edition. White River Junction, VT: Chelsea Green; 2004.

[53] Smil V. General Energetics: Energy in the Biosphere and Civilization. New York: John Wiley & Sons; 1991.

[54] Archer CL, Jacobson MZ. Evaluation of Global Wind Power. J. Geophys. Res. 2005;110:D12110.

[55] World in Transition: Towards Sustainable Energy Systems. German Advisory Council on Global Change (WBGU), 2003.

[56] Heinberg R. Searching for Miracle: Net Energy's Limits & the Fate of Industrial Society, A Joint Project of the International Forum on Globalization and the Post-Carbon Institute (False Solution Series #4); September 2009.

[57] Scott DS. Smelling Land: The Hydrogen Defense Against Climate Catastrophe. Enhanced Edition. Victoria, BC: Queen's Printer; 2008.

[58] Nakicenovic N. Carbon-Free Hydricity Age, CAN Europe, The Policy and Environmental Implications of CO Capture and CO2 Storage. Brussels: Hydrogen and Fuel Cell Technologies; May 2004. 27–28.

INDEX

Note: Page numbers with "*f*" denote figures; "*t*" tables.

Printed in the United States
By Bookmasters